Knowledge Management

Kai Mertins
Peter Heisig · Jens Vorbeck
Editors

Knowledge Management

Concepts and Best Practices

Second Edition
with 130 Figures

Springer

Professor Dr.-Ing.Kai Mertins
Head of Division Corporate Management

Peter Heisig
Head of Competence Centre Knowledge Management
Chairman Global Benchmarking Network (GBN)

Fraunhofer Institute IPK
Pascalstraße 8-9
10587 Berlin, Germany

Jens Vorbeck
Consultant IBM Unternehmensberatung GmbH
Kaiserswerther Staße 117
404774 Düsseldorf, Germany

ISBN: 3-540-00490-4 2nd Edition
Springer-Verlag Berlin Heidelberg New York

ISBN 3-540-67484-5 1st Edition Springer-Verlag Berlin Heidelberg New York

Cataloging-in-Publication Data applied for.

A catalog record for this book is available from the Library of Congress.

Bibilographic information published by Die Deutsche Bibliothek:
Die Deutsche Bibliothek lists this publication in the Deutsche Nationalbibliografie; detailed bibliographic data is available in the Internet at *http://dnb.ddb.de*

Cover design: Eric Kirchner, Heidelberg

Mertins & Vorbeck: Knowledge Management: Concepts and Practices

First Indian Reprint 2009

ISBN 978-81-8489-188-1

This edition is manufactured in India for sale in India, Pakistan, Bangladesh, Nepal and Sri Lanka and any other country as authorized by the publisher.

This edition is published by Springer (India) Private Limited,
A part of Springer Science+Business Media, Registered Office: 906-907,
Akash Deep Building, Barakhamba Road, New Delhi – 110 001, India.

Printed in India by Rashtriya Printers, Delhi.

Foreword

By Leif Edvinsson

Professor of Intellectual Capital

University of Lund

Lund, Sweden

In a marketplace, characterized by the growing importance of intangibles and knowledge, as we have not only in Europe today, it has become more and more evident how important knowledge management and Intellectual Capital management has become. Some of us who started this knowledge exploration journeys some decades ago could perhaps realize how quickly this theme should become so important. Today it is on almost every executive's agenda. The timing, the context and the comprehensive content of this second edition could not have been better.

In this second edition the authors have extended their overview with case studies and facts that confirm what we intuitively knew. Knowledge and intangibles are a top strategic issue. It is vital for an organization if it wants to be competitive, sustainable and profitable. The solid overview and the holistic review and testing of the theories with real cases, gives the reader a lot of facts and models of how management could invest in this area with very good return, in one case with a return of investment of 200 % in one year!

It is a very extensive work on the growing field of knowledge management. In its extension guidelines and emerging standards are already visible, as e.g. the guidelines from the Danish government on IC reporting or of the Norwegian Association of Financial Analysts. This book is also a platform for emerging strategic insights on the cracking of the old theory of the firm. The valuable and strategic assets are more and more in the context and external perspective of the legal constituency of the firm. Something I started to view and call the corporate longitude, i.e. the external networks, alliances as well as the internal tacit dimensions of culture and knowledge recipes. www.corporatelongitude.com highlights that the new theory of the firm is in lateral value creating dimensions such as innovations and renewals instead of management of the past and in focus on time flow instead of cash flow. Therefore it becomes more and more essential to include a new type of rating of intellectual capital components, a kind of second generation of KM, like IC rating (cf. www.intellectualcapital.se) that focuses on bench learning of efficiency, renewal and risks. A method already applied in more than 200 cases around the world. This rating is something that also highlights the importance of structural capital as a leverage and springboard for the releasing of the human capital and talent potential. The IC multiplier effect is the strategic indicator for how the enterprise creates the platform for sustainable future earnings for its stakeholders. Furthermore the field of application of KM is now growing outside the corporations into applications of IC of Nations and Regions.

Foreword

By Laurence Prusak
Former Executive Director Institute for Knowledge Management
IBM Global Services,
Cambridge, MA, USA

Knowledge management, both as a practice and as a discipline, is entering its second decade, happily, without all the false expectations of any beginning movement, but with a growing and solid base of practice and theory on which to build.

Much of this valuable base consists of cases, models, histories and practice descriptions - in other words: stories. All innovative management movements progress through the development of narratives. These are usually developed by consultants, professors, and practitioners (in some cases all three functions embodied in the same person) who seek to have a message conveyed (KM works!) through a supporting, rule-based, narrative structure i.e. when firm X did Y, A occurred, which boosted productivity or efficiency or effectiveness or innovation. While theorists may argue about the veracity, reliability or appropriateness of these stories, there is little doubt that they are the major "meme" by which knowledge about knowledge management spreads around the world. Therefore, we should try to produce the most truthful and useful stories that we can in order to inspire others to further the cause of KM in which we believe in.

Best practices, of course, are a form of narrative that have a significant pedigree within the management literature. It differs, but slightly, from a case in that its intent is perhaps more analytic than purely pedagogic. However, it is a powerful and much used tool for two specific reasons: It lends credentialism and legitimization to management discourse.

Credentialism simply means "best practice" guarantees that serious thinkers working with serious executives have produced some substantive, documented outcome. By "naming names" this process produces verifiable credentials which surely adds much impetus as to whether the practice will spread and be adopted by others.

Legitimization is a parallel activity to credentialism. It offers "legitimacy" to an argument by showing how the argument was produced by well-known thinkers from academia or prominent consulting firms. How the argument is being published by an established press, reviewed and discussed in established KM trade publications, and perhaps even presented at one of the ubiquitous KM conferences.

Credentialism and legitimization are counters to the skeptical argument: Why should we try this new approach? How can we judge if it really works? Who did it and what happened when they did it?

The volume that follows is a particularly fine example of how solid research, good thinking, and a narrative approach can produce a very substantial argument for the continuing saga of KM.

Acknowledgment

We would like to thank the following participants for their assistance in the realization of this book.

We highly appreciate the fruitful cooperation with Larry Prusak (formerly IBM, USA), who wrote an additional foreword for the second edition. Moreover we would like to thank Robert C. Camp, PhD, PE (Best Practice Institute, USA), Dr. Christoph Haxel (Henkel KGaA, Germany) and Dr. Andreas Spielvogel (Continental AG, Germany) for their contribution of the forewords to the first edition.

The success of this book relies heavily on its case studies. We took very much pleasure in the inspiring interviews with knowledge management practitioners from industry. We appreciate the dedicated support and the co-authorship of the following persons: Frank Spellerberg and Patricia Spallek (Arthur D. Little), Dr. Jürgen Oldigs-Kerber, Alla Shpilsky and Stephen Sorensen (Aventis), Cornelia Baumbach (Hilti), Dr. Peter Schütt and Andrea Martin (IBM), Anja Schulze (Institute for Technology Management, University of St. Gallen), Christian Berg and Peter Drtina (Phonak) as well as Martina Müller (formerly Siemens).

In addition to the authors we wish to place on record our thanks to a considerable number of contributors, who took significantly part in the success of the compilation of the second edition.

Grateful thanks go to Gerd Harrie for editorial support. For the bibliography we thank Katja Kuldszun and Karin Boeck.

We would also like to thank our team members Sebastian Heide and Johannes Niebuhr and also our graduates Miriam Lahkämper, Ago Herrmann, Andreas Baumüller, Andrea Maack and Thorsten Roser for their support.

Regarding the extensive translation we thank Annelie Geier for her committed assistance and dedication.

This first "German Benchmarking Study Knowledge Management" was supported by the following companies: Continental AG, Hannover; Daimler-Benz Aerospace – MTU Motoren- und Turbinen Union München GmbH; Eternit AG, Berlin; Henkel KGaA, Düsseldorf; Merck KGaA, Darmstadt; PSI AG, Berlin. We would like to cordially thank our sponsors.

Last but not least we would like to thank our publishing partner, "Springer Verlag" and in particular Dr. Martina Bihn and her team, for the great support and outstanding cooperation.

The Editors

Preface to the Second Edition

Kai Mertins, Peter Heisig, Jens Vorbeck

Knowledge management did not become a hype and out of fashion, but is still an emerging discipline. Since we have published our research results and case studies in the first edition of this book we have had the opportunity to discuss them with practitioners and KM experts in China, Denmark, Finland, France, Hungary, Ireland, Mexico, The Netherlands, Norway, South Africa, Spain, Sweden, Switzerland, USA, United Kingdom, United Arab Emirates and Germany. The interest and the feedback have encouraged us to work on this second edition.

Furthermore we have applied our methods and the identified best practices in several projects within public and private organizations, among them a German car manufacturer, the German organization for Air Traffic Control, a medium-sized software company, the Ministry of Internal Affairs of North Rhine-Westphalia (Germany), a Police Department and, last not least, our own organization, the Fraunhofer Society. These projects ranged from coaching the KM project team, designing a knowledge management concept for implementation to conducting knowledge management audits and short-term workshops. In the course of these practical experiences we acquired new insight in potentials and barriers for knowledge management and were able to develop new solutions as well as to improve our methods.

The Fraunhofer Framework for Knowledge Management has proved a very helpful basis for analyzing, finding and developing holistic KM solutions. It has been considered as one of the most valuable holistic KM Framework in Europe.[1] In autumn 2002 the Fraunhofer Competence Center Knowledge Management has been nominated with seven other European KM expert teams to work on "The European Guide to Good Practice in Knowledge Management"[2]. Furthermore, we contributed to an EU roadmap project on "Next Generation Knowledge Management" with other partners from Europe to define future research topics. The business process orientation has become the guiding approach of several KM methods as well.[3]

The German Ministry of Education and Research has funded a research project "Wachstum mit Wissen"[4] (Growth with Knowledge) to improve KM methods. Based on our business process orientation we expanded our methods towards KM change management, a process and content based intellectual capital approach and a network analysis to set up communities of practice. The sum of these experi-

---

1 Weber et al. (forthcoming), Kemp et al. (2002).

2 For further information about KM Standards and Guidelines cf. http://www.cenorm.be/isss/Workshop/km/Default.htm, cited 1-20-2003.

3 Cf. chapter 2 and in detail Abecker et al. (2002).

4 Support code 01 HW 0119.

ences encouraged us to rework the content of the first edition and complement it with the latest developments in KM research.

How it all started

In 1990, just after the unification of Germany, Fraunhofer IPK started two research projects, partly funded by the German Federal Ministry of Research and Technology with companies, universities and research institutes from both parts of Germany.[5] The aim of the projects was to discover and to describe the tacit knowledge of skilled mechanical workers, to develop processes and the task environments and technical prototypes to support the creation, sharing and use of tacit knowledge and to implement and test these solutions with the workers on the shop floor. With the introduction of CNC-Machines in mechanical workshops, experienced and highly skilled workers often felt insecure about their ability to control the process. They missed the 'right sound' of the metal and the 'good vibrations' of the machine. These signals were absorbed by the new CNC-Machines and hence workers were not able to activate their tacit knowledge in order to produce high quality products. Within a second project[6] we observed similar problems with the introduction of other CIM-Technologies, such as CAD/CAM in the design and the process-planning department and with MRP systems for order management. The information supply chain could not fully substitute the informal knowledge transfer chain between the different departments.

Some years later, in spring 1997, the book by Nonaka and Takeuchi "The Knowledge-Creating Company"[7] helped us to look at our results from a different angle. We discovered that we had to bundle our research capabilities coming from engineering, computer science, psychology, business administration and social science in order to create a powerful team addressing all issues of knowledge management.

Two events set the ball rolling: The newly-founded Competence Center Knowledge Management, a profit center at the Fraunhofer IPK in Berlin, and the concept for the first German benchmarking study on the state of the art of knowledge management in Europe conducted by the Information Center Benchmarking Berlin in cooperation with the Global Benchmarking Network and sponsored by six German companies.

5 Martin (1995); Carbon, Heisig (1993).

6 Mertins et al. (1993); Fleig, Schneider (1995).

7 Nonaka, Takeuchi (1995).

Applied research results (Part I: Design Fields)

Part I *"Design Fields"* integrates both our results from the benchmarking studies, our experiences from knowledge management projects in industry and administration as well as the discussion within the scientific KM community. The diversity of the individual sections reflects once more our integrated approach to knowledge management.

The chapter *"Business Process Oriented Knowledge Management"* describes the method GPO-WM© for analysis and design of business processes from a knowledge management perspective. The GPO-WM© implementation model is described and exemplified for an implementation project.

The chapter *"The Fraunhofer Knowledge Management Audit"* reviews the current approaches to assess the handling of knowledge and describes the practical application of the Fraunhofer Audit approach on the examples of a service industry enterprise and a research institute.

The chapter *"Motivation for Knowledge Management"* addresses the challenges of successfully implementing knowledge management into daily routines from the psychological perspective. An overview of the latest research results in the field of motivation and incentive systems is followed by the development of a change management approach focusing on sustainable change of behaviour to overcome knowledge management barriers. Improved communication, supportive leadership and the development of competencies for knowledge management are combined with a motivational task design and individual participation in this "pro-active change management" approach. The chapter closes with the summary of experiences from a recent change management project for the implementation of knowledge management in a large company.

A fundamental consideration when designing a KM system is the development of human resources. Based on the Business Process Oriented Knowledge Management approach the article on *"Role Models, Human Resources and Strategy"* describes the conditions for synchronizing processes and personnel. The definition of knowledge management roles depends to a significant extent on the strategic disposition of the firm. Enterprises, which seek to create new knowledge, will tend to follow personalized approaches whereas firms with a focus on using existing knowledge will tend to rely IT-tools to manage knowledge. Business processes and strategic positioning provide therefore elements for the definition of KM roles.

The article *"Knowledge Management Tools"* seeks to support the selection and implementation of Knowledge Management Tools. A reference to the non-technical means for knowledge exchange introduces to the Fraunhofer IPK approach towards KM-Tools. In the following section methods for structuring knowledge and information are presented. In the third section basic KM-Technologies are described briefly and more than hundred KM-Tools are classified. The chapter concludes with a checklist that supports the choice of an appropriate KM-Tool.

The chapter *"Intellectual Capital: Measuring Knowledge Management"* successes the chapter on 'Intellectual Capital' in the first edition. Promising new approaches as well as traditional concepts to measuring Intellectual Capital are introduced and evaluated according to their strengths and weaknesses as well as their scope of application. In addition this chapter gives an introduction to measuring the success of a knowledge management project and point out how practitioners deal with the challenge.

Empirical research results (Part II: Surveys)

The results of the first German benchmarking study on Knowledge Management in Europe from the first edition are summarized in the introductory chapter followed by some new empirical data.

In autumn 2001 we initiated the First Global Delphi Study on "The Future of Knowledge Management" with the Institute of Psychology at the Humboldt University (Berlin) co-funded by the Donors' Association for the Promotion of Science in Germany ("Stifterverband für die Deutsche Wissenschaft"). The results from this Delphi-Study have been discussed during an international conference in March 8 – 10, 2002 in Berlin (Germany) with the participation of KM experts like Larry Prusak and David Snowden. An overview of the key findings is presented in the chapter *"Delphi Study on the Future of Knowledge Management – Overview of the Results"*.

EFQM[8] and CIBIT[9] conducted in winter 2001/2002 a follow-up screening survey in order to identify the current state of knowledge management practice across Europe and a short list of good practice organizations. The chapter *"A survey on good practices in Knowledge Management in European companies"* gives an overview of their results.

From an ongoing international project[10] of the OECD[11] we publishing a first overview of the results from a survey carried out by our colleagues from the Fraunhofer Institute for Systems and Innovation Research (ISI) on *"How German Companies Employ Knowledge Management. An OECD Survey on Usage, Motivations and Effects"*.

---

8 European Foundation for Quality Management – www.efqm.org, cited 1-20-2003.

9 CIBIT Consultants | Educators – www.cibit.nl, cited 1-20-2003.

10 Cf. www.oecd.org -> Education -> Economics and Management of Knowledge, cited 1-20-2003.

11 Organization for Economic Cooperation Development – www.oecd.org, cited 1-20-2003.

Qualitative research results (Part III: Case Studies)

From November 2000 to March 2001 the Fraunhofer Competence Center Knowledge Management has been invited by the Benchmarking center TECTEM[12] at the University of St. Gallen (Switzerland) to support their benchmarking exercise as subject matter expert. The chapter *"Knowledge Management – Results of a benchmarking study"* describes the procedure and the good and best practices identified by the research team in six leading European companies.

Beside these examples we have asked two price-winning companies, Siemens Medical Solutions and Aventis, to describe their knowledge management approach. This includes the implemented knowledge management solution as well as selected results and lessons learned.

The contributors of the best practice companies identified in our benchmarking study were asked to update their case studies. The updated versions have been included in Part III; the others were withdrawn from this edition.[13]

European KM Landscape (Part IV: "KM – Made in Europe")

In the first edition seven leading research institutes in Europe gave a two-page description of their knowledge management approach and a brief outlook on future developments in KM (cf. www.km-in-europe.org, cited 1-20-2003).

In this edition we take a look on the research projects carried out all over Europe. The article *"Building Communities. Organizational Knowledge Management within the European Commission's Information Society Technologies Programme"* gives an overview on the research projects funded between 1998 and 2002 and an outlook at future research areas in knowledge management in the 6th Framework Programme.

Recommended Further Readings

The book closes with an updated overview of scientific journals and business reviews in the wider field of knowledge management as well as some suggestions of recommended further literature and further readings in knowledge management.

---

12 Transfer Center for Technology Management; cf. www.tectem.ch, cited 1-20-2003.

13 This includes the case studies of Booz, Allen & Hamilton; Celemi; Skandia and Thomas Miller & Co. Ltd..

Foreword First Edition

By Robert C. Camp, PhD, PE

Chairman Global Benchmarking Network (GBN), Best Practice Institute™,

Rochester, NY, USA

The perception, sharing, and adoption of best practices are mostly attributed to the activity called benchmarking. Obtaining maximum value from best practices is usually attributed to knowledge management. One is an extension of the other. Knowledge management can be looked upon as the management of knowledge about best practices whether in the mind as human capital or as intellectual assets or property.

Most organizations now recognize the absolute imperative for the identification and collection of best practices through benchmarking. It can be a strategic strength when practiced and a fatal weakness if not pursued. But there is a serious disconnection in the exchange and adoption process.

Despite significant advances in the approaches and technology that pursue improvement (six sigma, process redesign, customer relationship management, etc.), organizations continue to experience great difficulty in successfully transferring leading practices. Some would say these are exemplary, proven, observed, or promising, but, in the final analysis, they are best practices – with the objective of becoming world class.

More insight is needed into how leading, or best practices are transferred and adopted – said differently, best practices for knowledge transfer or knowledge management.

The subject of knowledge management has gained considerable prominence in the United States through consortia studies, articles, and conferences. But there have been few substantive studies about the experience of knowledge management in Europe, until now. This study is commendable for its findings based on fact, for its use of definitive questionnaires, and its verifications through site visitations. It documents the recent (3-5 years), most prominent experience of organizations across all sectors in Europe.

The framework of the individual, group, and organizational sharing of best practices – from the explicit or documented to the tacit or experience-based developed through research – is very helpful in understanding the process. The many case studies in this study bring that research to life.

This is a substantive research effort. It should provide both qualitative and quantitative insight for those embarking on or pursuing knowledge management best practices for years to come. I recommend it highly for the study of its implications and the use of its findings to get results.

Foreword First Edition

By Dr. Christoph Haxel

Henkel KGaA, Düsseldorf

Henkel InfoCenter

Using knowledge in a structured and organized way is one of the key factors that determine corporate success. The goal is to share and apply knowledge faster and more efficiently than your competitors.

In Henkel's research departments, "knowledge management" has a tradition going back more than 100 years. Sharing knowledge also means publishing and protecting knowledge. Henkel applied for its first patent in 1896. Since then, thousands of inventions have been patented and many research findings have been published. In Research and Technology, we recognized early that ever the increasing speed of industrial reorganization and regrouping would require new ways of managing knowledge. In order to identify the best ways, we decided in 1997 to take part in a benchmarking study.

Within Henkel, the decision to enter a new era in information and knowledge policy within the group was made at a meeting of top managers in May 1998. Our concept for the global collection, distribution, and application of knowledge was called "The power of shared knowledge." The intention was to make the knowledge and experience of Henkel's 56,000 employees worldwide more accessible. It was not just a question of gathering facts and figures, but of making the views and experience of experts available.

We agreed that despite Henkel's modern and open information culture, a lot could still be learned from others and that our existing methods could be refined. In the benchmarking study, business consultants were of special interest, since their entire "production" capital is in the minds of their staff. We therefore asked how consultants handle knowledge management within their companies.

One business consulting firm involved in the benchmarking study particularly impressed us. We visited them to see first hand how they handle knowledge management and to discuss their experience in this area. This firm now advises our company worldwide. Henkel's Management Board also called them in, independently of the benchmarking study, to assist our Düsseldorf headquarters in knowledge management. The benchmarking study was not the only influencing factor in choosing them, but it may well have helped the decision-making process.

Knowledge management at Henkel means exchanging experience and information that is orientated to the business process. The new management tool is now being introduced and built up group-wide.

The core of the tool consists of knowledge “pieces”. These are items of knowledge relating to a particular business and task. To “mine” knowledge pieces, the experience of one of our experts is skillfully tapped by an interviewer. During the debriefing process, nuggets of wisdom are entered into databases. The existing infrastructure at Henkel, where more than 25,000 employees communicate via an international information network running on the Lotus Notes platform, eliminated the need for any major investments in hard- or software. Push and pull information technologies now help to transport these valuable pieces of knowledge around the world to the right place at the right time.

Five pilot projects – 'Technology', 'Product Launches', 'Brands', 'Systems' and 'Suppliers' – are now almost complete. The task ahead is to distribute this new management tool throughout the group and to develop it further. Knowledge management has become an established feature of management practice at Henkel.

Topics such as 'e-commerce', 'innovation' and 'time to market' are the new focal points. In meeting these challenges, our experience with knowledge management and its tools is enormously valuable.

With the results already achieved in knowledge management, Henkel is well equipped for the future and on the road to becoming a model of best practice.

Foreword First Edition

By Dr. Andreas Spielvogel
Director Development Processes & Tools
Business Unit Original Equipment
Continental AG

Tire engineering is not a classical engineering discipline and lies on the borderline between mechanical engineering and chemistry. Few books have been published on tires. Even publications are not seen very often compared to other disciplines, and I have never seen documentation describing how to design a modern tire.

It is therefore crucial that a tire engineering company does its own research and knowledge documentation. Nevertheless, the documentation of know-how had at one time low priority when compared to other R&D activities. When we started with knowledge management, most know-how was located in the minds of people who have been with the company for many years.

But what happens when R&D is relocated and undergoes significant restructuring? Older engineers and chemists retire. Younger ones quit or move into other functions and hopefully the majority stays.

In our case, the threat was so great that preventative action against loosing know-how seemed necessary.

So we started in 1995 with knowledge management without knowing that it would be so popular in the near future. Even before 1995, a lot of activities existed that nowadays have the attribute "knowledge management". Most of our approaches have been derived from in-house research and use "simple" technology. The basis was our culture, processes, and procedures. We had very little advice from outside the company.

It was therefore a challenge for us to participate in the benchmark with Fraunhofer Gesellschaft. We had no idea what our performance would be. Nevertheless we found ourselves in the upper third.

Our greatest expectation was to get into contact with other companies and to learn from best practices. The major benefits for us were our improved sensitivity to the need for knowledge management, faster implementation of new ideas, the ability to avoid of mistakes, and better overall performance. We also got a lot of attention for our approach from the public. This was more than we had expected.

Now, in the year 2000, knowledge management is one of the key elements in corporate basics (http://www.conti-online.com) and we are proud to be mentioned as a case study in the recent literature.[14] We are continuously striving for improvement.

I recommend this book for practitioners, specialists, and managers who are looking for ideas and examples of how to implement or improve knowledge management. The book gives a broad overview and best practices from well known European companies in a variety of branches. It emphasizes the important role of culture, motivation, and skills. A review of tools that support knowledge management has also been included.

---

14 Edvinsson, Brünig (2000).

Brief Contents

Part IV: KM – Made in Europe

Contents

Part III: Case Studies

1 Introduction

Kai Mertins, Peter Heisig, Jens Vorbeck

1.1 Knowledge

"Knowledge is nothing natural" as philosophers claim[15], while from the historian's perspective[16] the accumulation of knowledge is the basis of the human evolution since its very beginning thousands of years ago. The discussion on knowledge has a very long tradition. As early as the 5th century B.C. the philosopher Socrates dealt with the question of the limits of knowledge.[17]

The importance of knowledge for the competitiveness of companies, organizations and even economies is widely accepted nowadays. Nevertheless any publication on knowledge management will need a clarification of the term.

The most common association with the term knowledge is scientific knowledge. This includes knowledge, which stems from academic research facilities such as universities and research institutes. This knowledge is developed by using scientific methodologies and standards. It is tested and validated by the scientific community. It is explicitly described in research papers, reports and books. Nearly the same association is linked with the knowledge produced by the research and development departments of companies. Their knowledge is however embedded into products and services.

The other association with the term knowledge is the knowledge an experienced person possesses. For example, when experienced lathe operators hear the 'right sound' and feel the 'good vibrations' of the machine, they know that the process is going right[18]. A similar observation is quoted from a worker at a paper mill: "We know the paper is right when it smells right."[19] However, you do not only find this kind of knowledge in craft work settings, you find it also in high tech chip production[20] environments as well as in social settings, like schools[21].

---

15 McGinn (2001), pp. 46.

16 Fried, Süßmann (2001), Bennet, Bennet (2003).

17 Platon (1981).

18 Martin (1995); Carbon, Heisig (1993).

19 Victor, Boynton (1998), p. 43.

20 Luhn (1999).

21 Bromme (1992).

Still, the term knowledge is not easy to define. Today there are numerous descriptions and definitions of knowledge. Romhardt[22] found 40 dichotomies of knowledge like explicit versus implicit or tacit, individual versus collective. Von Krogh/Venzin[23] created seven categories of knowledge to be used in management and organization theory: tacit, embodied, encoded, embrained, embedded, event and procedural.

A very popular classification of knowledge derived from computer science distinguishes between signals, data, information and knowledge. Some even add wisdom to this hierarchy. This classification, if not properly understood and used, could lead to an unproductive discussion of the 'right' distinction between the categories. The next difficulty arises from the often conflicting definitions of information and knowledge.

Instead of a clear hierarchy, we found that the understanding of a continuum ranging from data via information to knowledge proved to be the most practical scheme for knowledge management.[24] The distinction between data and information on the one hand and knowledge on the other is easy to understand and communicate using the following typical questions. For data and information the questions are: who – what – where – when? etc. while typical questions for knowledge are: how? and why?[25]

From the practical perspective important aspects that have to be considered are the criteria of validity and reliability. In academia we have quality standards which apply also i.e. to pharmaceutical, chemical and similar research-intensive industries. But in several businesses these standards are not useful or even too expensive and too slow. Therefore specific validation criteria have to be defined. In a business context this criteria is practicability, which is based on personal experiences like "a practice or a process that has worked"[26] or on calculated improvements like "it's proven and it's providing value"[27].

In our business process-oriented approach knowledge is understood as a resource to fulfill a process task and as a product generated by this task or process. The process task itself is seen as a knowledge processing task and analyzed by a number of empirically validated and practically proven criteria. A holistic business process-oriented knowledge management solution has to approach the various elements – like data, information and knowledge – in an integrative perspective. In daily business processes the lack of only one element can hamper the

---

22 Romhardt (1998), pp. 28.

23 Von Krogh, Venzin (1995); for a comprehensive overview cf. Maier (2002), pp. 50.

24 Probst et al. (1998), pp. 36; Heisig (2000).

25 Eck (1997).

26 Cf. chapter 14.

27 Wolford, Kwiecien (2003), pp. 503.

achievement of the overall performance goals. Therefore, a distinction between data, information and knowledge is crucial.

1.2 Empirical Foundation: The Benchmarking Survey

Our approach towards knowledge management should not only be based on extensive research and practical experience from project work but also on empirical quantitative and qualitative data. Therefore, we decided to base our approach on a comprehensive survey of the German TOP 1000 and European TOP 200 companies and in-depth case studies of European best practice companies. In autumn 2000 the results of this survey and ten of thirteen[28] case studies were initially published. Many of our survey results are still valid today and serve as the empirical basis of our Knowledge Management Model. Therefore, we would like to summarize in the following paragraphs the most important results.

Our survey questions were organized in three sections: (a) general understanding and status of knowledge management activities in the whole company; (b) status of knowledge management activities in the business process the company considers as their best practice; and (c) demographic data about the respondent and the company. 148 questionnaires (12.3%) were returned. The following chart illustrates the top five branches the participating companies came from.

Line of Business	1. Chemistry and pharmaceuticals (16,3 %)
Our company belongs to the following industry	2. Consulting (15,4 %)
(n=146)	3. Automotive and aircraft (15,4 %)
	4. Computers and telecommunications (15,4 %)
	5. Machine engineering and metal processing (14,4 %)

Fig. 1.1: TOP 5 Industry Branches Participating in the Benchmarking Survey

1.2.1 Understanding of Knowledge Management: People and Processes

Knowledge management was understood as a *"part of corporate culture, which 'supports the active exchange of information, knowledge and experiences between employees and departments'"* (55,5 %)[29] and as *"a company-oriented method: 'the sum of procedures that determine the generation, distribution and application of knowledge to achieve organizational goals'"* (47,3 %). 11,0 % contributed their own definition, like one from a metal processing company: "The sum of procedures that serves to create, evaluate, distribute, and apply knowledge in order to

---

28 Three companies only wanted to share their experiences within the group of benchmarking partners and did not give us the permission to publish their case study.

29 Multiple references possible, n = 146.

attain predefined goals". It is understood neither as a "*technological term*" (14,4 %) nor as an "*immaterial asset*" (5,5 %).

According to the survey results knowledge management was viewed to be most important in processes such as "*understanding markets and customers*", "*development of products and services*", and "*management of improvements and change*" (Fig. 1.2). Clearly the companies began with knowledge management in those fields that they considered as their core competencies.[30] Large deviations were apparent between service and manufacturing firms. About 40% of the service companies started knowledge management with "*information management*", while this was the case with only a quarter of the industrial companies. In the industrial sector, the greatest focus of knowledge management was initially the "*development of products and services*". This was the case with only a third of the service companies.

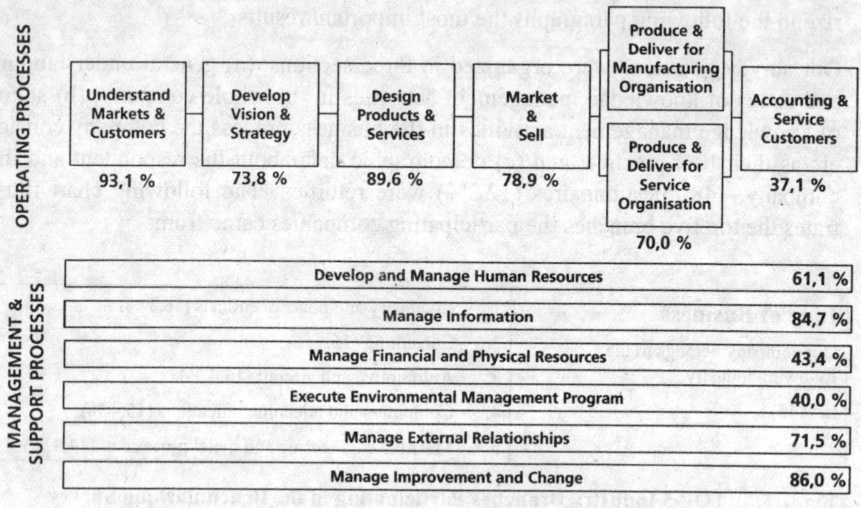

Fig. 1.2: Importance of Knowledge Management in Business Processes[31]

Knowledge management was viewed to be a means to achieve company goals such as an "*increase of customer satisfaction*" (74,5 %)[32] and "*improvement of innovative capability*" (72,5 %), "*increase product quality*" (70,9 %) and "*reduction of costs*" (64,0 %).

The central knowledge objects were "*methods*" (52,7 %)[33], like "How to do it (best)?" and "*products*" (50,4 %) followed by knowledge external to the organiza-

---

30 Cf. Fig. 2.4 in chapter 2.

31 The percentage in Fig. 1.2 summarizes the two answering options "essential" and "important".

32 The percentage summarizes the two answering options "very strong" and "strong".

33 Multiple references possible, n = 131.

tion like "*customers*" (40,5 %) and "*markets*" (29,8 %). Clear differences were again evident between service and manufacturing companies. While service firms saw their customers as the most important source of knowledge (55,7 %), for the majority of manufacturing firms, this source was their products (62,9 %).

1.2.2 The Knowledge Management Core Process

All approaches towards knowledge management emphasize different knowledge management tasks. Besides the wording the approaches differ mainly by the number of tasks. The range is between three and eight. Davenport and Prusak[34] use knowledge generation, knowledge codification and coordination and knowledge transfer. The building blocks proposed by Probst et al.[35], which have received most attention in the German language area, count eight: identification, acquisition, development, sharing, utilization, retention and assessment of knowledge and knowledge goals.

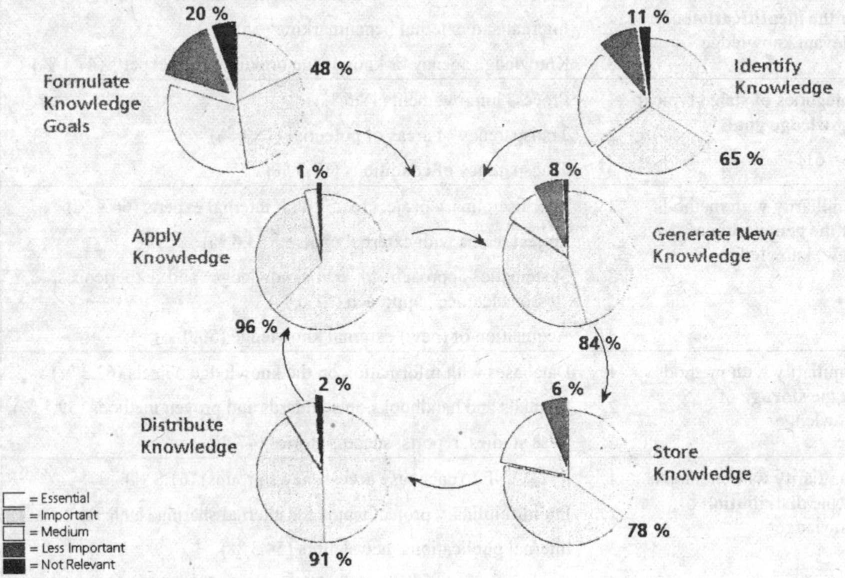

Fig. 1.3: Relevance of Core Process Activities

Therefore we wanted to know, *How important are the following core activities of Knowledge Management for your company?*[36]: generate (new) knowledge, store

---

[34] Cf. Davenport, Prusak (1998).

[35] Probst et al. (1998).

[36] Likert-scale: 5 = essential, 4 = important, 3 = medium, 2 = less important, 1 = irrelevant.

and preserve knowledge, distribute knowledge, apply knowledge, identify knowledge, formulate knowledge goals, other core activities (Fig. 1.3).

The survey result shows that four core activities: "*apply*", "*distribute*", "*generate*" and "*store*"; received the highest confirmation from the respondents. This result led us to use these four core activities as the main analytical lenses and design criteria within the Fraunhofer Knowledge Management Reference Model. In order to be more operational for our benchmarking clients regarding these abstract core activities, we classified more than forty knowledge management methods gathered from secondary research and asked *how good is your command of which methods*[37]. In Fig. 1.4 the knowledge management methods, which were applied, are shown.

The core process of knowledge management		
Familiarity with methods for the **identification** of relevant knowledge	1.	Production of manuals and handbooks (56,0 %)
	2.	Internal and external benchmarking (55,6 %)
	3.	Knowledge agency or knowledge broker, senior experts (47,1 %)
Categories of stated typical **knowledge goals** (n = 61)	1.	Process improvement (18,5 %)
	2.	Transparency of areas of potential (15,1 %)
	3.	Transparency of conditions (12,3 %)
Familiarity with methods for the **generation** of (new) knowledge	1.	Interdisciplinary project teams with internal experts (66,9 %)
	2.	Project teams with external experts (53,6 %)
	3.	Systematic approach to edit knowledge and experiences, e.g. "lessons learned" approach (51,6 %)
	4.	Acquisition of (new) external knowledge (50,0 %)
Familiarity with methods for the **storage** of knowledge	1.	Databases with information on the knowledge objects (62,3 %)
	2.	Manuals and handbooks on standards and proven methods (59,1 %)
	3.	Case studies, reports, success stories (47,2 %)
Familiarity with methods for the **distribution** of knowledge	1.	INTRANET (data base access, news groups) (61,8 %)
	2.	Interdisciplinary project teams for internal sharing (60,8 %)
	3.	Internal publications, newsletters (54,3 %)
	4.	Coaching and mentoring by senior experts (51,2 %)
Familiarity with methods to promote the **application** of knowledge	1.	Autonomous interdisciplinary teams (56,6 %)
	2.	Coaching of teams by senior experts (48,0 %)
	3.	Obtain internal expert opinion on results (30,9 %)

Fig. 1.4: Survey Results on the Core Process of Knowledge Management

37 Likert-scale: 5 = excellent, 4 = good, 3 = fair, 2 = poor, 1 = very poor, percentage = Σ (5+4).

The results prove that there is nearly always a combination of human-based and technical-based methods to effectively deal with knowledge management tasks. For the core activity *distribute* the *Intranet* (61,8 %) is the best-commanded method, followed by *interdisciplinary project teams for internal sharing* (60,8 %), *Internal publications, newsletters* (54,3 %), and *coaching and mentoring by senior experts* (51,2 %). This relatively even distribution of excellent and good command of the mentioned methods could be interpreted in a way that the crucial task regarding the design of a knowledge management method is to find the right and effective combination and balance between human-based and technical-based methods and tools.

1.2.3 The Knowledge Management Design Fields

In order to implement a successful knowledge management solution our secondary research indicated that the corporate culture and leadership are two main success factors, despite the fact that most knowledge management projects started with the implementation of a technological solution. Therefore, we asked an open question about the "*essential success factors (enablers) for efficient Knowledge Management*". The answers where categorized by external staff and frequencies were calculated. Now, we added the category "*Clear definition of goals*" as a management task to the category "*Promotion by TOP Management*". The categories "*Staff motivation and qualification*", "*Training and further education*" and "*Rewards*" have been aggregated into one single human resource category named "*motivation and skills*". After this new categorization the frequencies have been recalculated (Fig. 1.5).

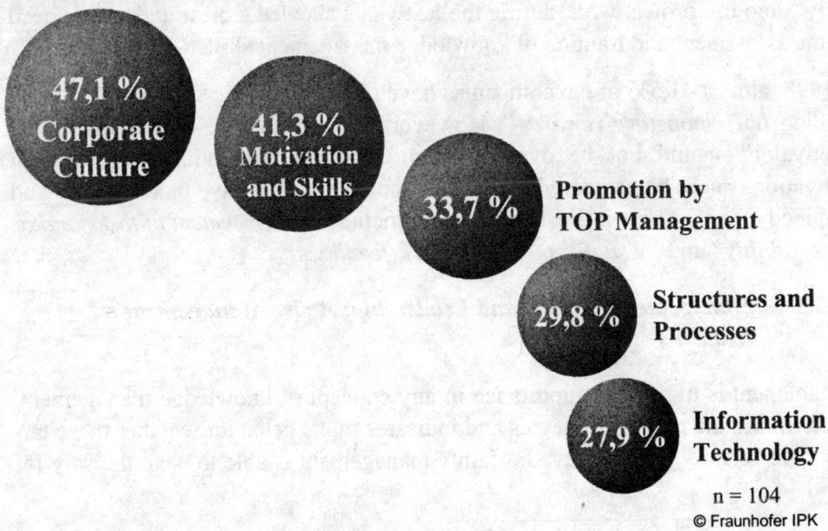

Fig. 1.5: Five Critical Success Factors for Knowledge Management

The result confirms the importance of a "*corporate culture*". Cultural aspects have been named by 49 companies out of 104 as an essential enabler for efficient knowledge management, followed by "*motivation and skills*" (43) and "*promotion by top management*" (35). "*Information technology*" was only named by 29 companies as an essential enabler. The importance of "*structures and processes*" was emphasized by 31 companies. Despite the empirical findings and the evidence experienced by the pioneers, companies still start with a technology-driven approach and do not address the skill aspects at all. Some additional results regarding the design fields of knowledge management are briefly described in the following paragraphs (Fig. 1.6).

1.2.3.1 Corporate Culture: Structure – Openness – Autonomy

The organizational culture of a company is reflected in its philosophy and vision, management style, and in its physical organizational structures, such as the architecture of buildings and the layout and design of rooms. Corporate culture is essentially characterized by three aspects: 60% of the companies mentioned that *"errors – up to a certain extent – are tolerated"*, *"the exchange of knowledge is encouraged"*, and *"autonomous actions and learning are encouraged"*. *"Development of a common language"* and *"openness and trust"* are mentioned by every third company. Knowledge management is facilitated by aspects of the *"organizational structure"*, the *"environment"* and *"open communication"*.

1.2.3.2 Human Resource Management: Skills and Feedback

In our initial approach we did not put much emphasis on the skill aspect. One case study[38] and our project work during the last years showed a clear and strong need for the assessment and training of knowledge management skills[39].

In 1998, almost 41,5% of the companies have tried to improve staff motivation by handing out *"monetary rewards"*. However, the significance of this traditional "motivator" should not be overestimated. A financially induced increase in motivation cannot be compared to intrinsic motivators that have much greater and sustained effects. These intrinsic motivators include *"enhancement of tasks and/or responsibility"* and *"visualization of success (feedback)"*.

1.2.3.3 Management Systems and Leadership: Active Management is Required

Management is of pivotal importance in any concept of knowledge management. It determines the path of strategies and indicates by its behavior whether these targets have actually been achieved. Mainly management is able to pave the way for

---

[38] Unfortunately, the company merged after having participated in our benchmarking study and we have been denied the permission to publish their case study.

[39] Cf. Heisig, Finke, Jaitner (2003).

a knowledge-oriented corporate culture. According to the respondents, management could facilitate a more effective knowledge handling by *"promoting autonomous activities of employees"*, *"open communication"* and *"coaching"*.

Management has to communicate tasks and create a motivating environment based on trust and credibility. By setting goals and offering support, managers enable individual staff members to coordinate their own work and to develop a learning organization through their teamwork.

1.2.3.4 Information Technology Connect and Access

Information technologies support communication, cooperation and coordination, and the timely access of information and the sources of knowledge. The standards for knowledge management in our survey have been *"databases with information on the knowledge objects"* and *"Intranet"*. Companies only commanded applications of *"expert systems"* and *"intelligent agent technology"* to limited extend.

1.2.3.5 Control and Measurement with 'Hard' and 'Soft' Indicators

Controlling knowledge management activities in organizations is still one of the biggest challenges. Nevertheless every third company used *"'hard' and 'soft' indicators"* to evaluate the outcomes of their knowledge management initiative.

The design fields of knowledge management		
Corporate culture It is characteristic of the corporate culture that	1.	Errors are tolerated up to a certain extent (60,3 %)
	2.	Our employees act and learn autonomously (58,9 %)
	3.	Our employees are encouraged to exchange knowledge (57,5 %)
The following aspects of **corporate culture** facilitate efficient knowledge management	1.	Organizational structure (41,5 %)
	2.	Environment (39,0 %)
	3.	Open communication (32,2 %)
Staff motivation Staff has been motivated through the following measures	1.	Monetary rewards (41,5 %)
	2.	Enhancement of task and/or responsibilities (26,4 %)
	3.	Visualization of Success (feedback) (15,1 %)
Management The following managerial aspects facilitate knowledge management	1.	Promote autonomous activities of employees (76,7 %)
	2.	Open communication (50,9 %)
	3.	Coaching (17,5 %)
Methods of **evaluating knowledge management**	1.	Success is not evaluated (35,6 %)
	2.	Through "hard" and "soft" indicators (35,6 %)
	3.	Through "soft", qualitative, non-monetary indicators (16,4 %)

Fig. 1.6: Survey Results on the Design Fields of Knowledge Management

1.3 The Reference Model for Knowledge Management

A thorough analysis of the State-of-the-Art led us to design a working model for knowledge management that guided us when we conducted the survey. Our industrial partners on the other hand were very attentive regarding the applicability of our results. One major characteristic of our working model is thus based on input and their need for added value.

The reference model for knowledge management as it exists today is a result of our clients' requirements for a practical framework and the permanent consideration of academic and applied research on knowledge management from different disciplines. It is based on our benchmarking and Delphi survey results that validated our first working model in 1999 and confirmed our approach in 2002. And least but not last it has proven its practicability within implementing knowledge management in manufacturing and service companies and public administration and improved with our lessons learned. These considerations and practical experiences lead us to develop an approach…

- …that is easy to understand, communicate, and apply to the entire company, in order to enhance acceptance, effectiveness and efficiency,
- …in which all actions are directed towards the value-adding core processes of the organization.

The Fraunhofer reference model for knowledge management has three layers. The first and permanent focus of all knowledge management activities is the *value-adding business processes*. The business processes are the application area of knowledge and within these processes knowledge is generated. This is nothing new, but we have and can improve the knowledge handling with well-known and new methods and tools as well. Furthermore the business process represents the field of knowledge or knowledge domain or the content that the stakeholders require to perform their tasks.

The second level is the *knowledge management core process*, which can be further broken down into the four empirically validated core activities, i.e. activities that specify the handling of knowledge: "create (new) knowledge", "store knowledge", "distribute knowledge", and "apply knowledge". These core activities have to form an integrated process. Not the excellent performance of one single core activity is required, but to implement an integrated process using all four activities. We consider all four activities as equally important. This helps us to see knowledge as a resource to be applied, an asset to be stored, a product to be generated and something that flows from one process to another process step. Furthermore these activities have to be integrated into the business processes. A knowledge management activity should be nothing apart!

The third level forms the six so-called *design fields of knowledge management*: corporate culture, leadership, human resources, information technology, organization and roles and control. The first five are derived from the critical success factors of our benchmarking survey. The sixth design field control results from the requirements of clients and managers to have some measures and indicators to assess the outcome of knowledge management activities.

Our understanding of knowledge management is as follows:

"Knowledge management includes all methods, instruments and tools that contribute to the promotion of an integrated core knowledge process – with the following four core activities as a minimum, to generate knowledge, to store knowledge, to distribute knowledge and to apply knowledge – in all areas and levels of the organization in order to enhance organizational performance by focusing on the value creating business processes."

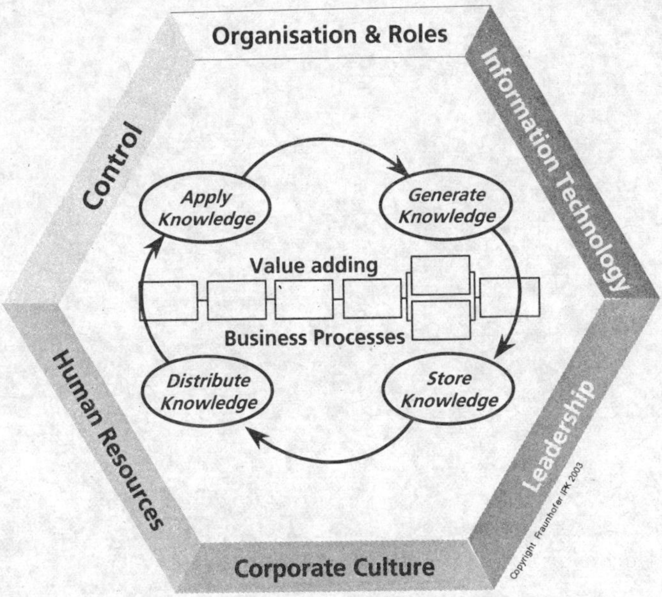

Fig. 1.7: Core Process and Design Fields of Knowledge Management

The reference model for knowledge management has been recognized as one of the few holistic KM frameworks for standardization in Europe.[40] As research continues on the issue of knowledge management we will attentively adjust the model where it proves to be necessary.

---

40 Cf. Kemp et al. (2002); Weber et al. (forthcoming).

upon the mass and quantity of the water [illegible] the dense [illegible] of the [illegible] [illegible]
[illegible] the [illegible] [illegible]. The [illegible] [illegible] [illegible] [illegible]
[illegible] of [illegible] [illegible] [illegible] [illegible] and [illegible]
as the [illegible] [illegible] [illegible] [illegible].

[illegible] [illegible]

[illegible]

Fig. 1.? [illegible] [illegible] Knowledge Management

[illegible] [illegible] [illegible] [illegible] [illegible] [illegible] [illegible] [illegible] [illegible] [illegible] [illegible] [illegible] [illegible] [illegible] [illegible] knowledge management [illegible] [illegible] [illegible] [illegible]

1.? [illegible] [illegible] Knowledge [illegible]

Part I

Design Fields

2 Business Process Oriented Knowledge Management

Peter Heisig

In today's organizational practice, knowledge has been recognized as one of the most important resources for success. Therefore, the lack of theoretical understanding of knowledge and practically proven methods for efficient knowledge management is surprising. Only recently have academics and practitioners begun to publish their approaches and experiences under the concept of knowledge management.[41]

The Fraunhofer CCKM approach to business process oriented knowledge management (GPO-WM®)[42] is based on the following assumptions:

1. We use our knowledge and the know-how of our colleagues, suppliers, clients, competitors and other resources on a daily basis to solve problems and get our work done.
2. "I have no time" or "my team has no time" is one of the most frequently mentioned barriers[43] for knowledge management in organizations. Therefore, knowledge management tasks have to be combined with daily work tasks and integrated into the daily business processes. Knowledge management should not be treated apart from the daily routine.
3. The ways in which knowledge is generated, stored, distributed and applied differ according to the particular business processes. These specific requirements have to be integrated into the knowledge management approach.
4. The business process provides the context for knowledge, leads to a focused and user-oriented structure of the knowledge used and produced in business tasks.
5. Aspects of corporate culture are the most frequently mentioned success factor for knowledge management. However, corporate culture is not a homogenous entity. It is an interwoven network of different professional cultures (e.g. en-

41 Skyrme, Amidon (1997); Probst, Raub, Romhardt (1998); Willke (1998); Davenport, Prusak (1998); North (1998); Krogh et al. (2000); Collison, Parcell (2001); Denning (2001); Abecker et al. (2002); a good overview is given by Remus (2002); Holsapple (2003).

42 GPO-WM in the German original: Methode des **G**eschäfts**p**rozess**o**rientierten **W**issens**m**anagements.

43 Bullinger et al. (1997).

gineer, economist, lawyer), functional cultures (e.g. sales, R&D, production) as well as underlying corporate traditions and values. The business process has the potential to provide a neutral and linking element for the different perceptions. The definition of knowledge related behaviour leads to a more rational discussion about relevant changes than a discussion on the right values.

6. The business process oriented knowledge management approach enables greater involvement of the employees, helps to achieve their commitment and facilitates their intrinsic motivation generated from improvements of the daily knowledge handling.

Our approach uses the well-developed business process engineering (BPE) method but does not aim to implement a workflow solution as it has been misunderstood by some colleagues in several discussions. Other KM approaches often require a new additional method to implement KM. In contrast, we only add another perspective to the basic BPE method in order to enable practitioners to reuse their existing know how on process analysis and design and to use proven and practice-oriented KM-specific criteria to build KM solutions.

2.1 Knowledge Management is Business and Process Oriented

The concept of knowledge management has been used in different disciplines, mostly in knowledge engineering[44] and artificial intelligence (AI)[45]. AI research often reduced the concept of knowledge management to the description of the development and use of expert systems.[46] The current KM practice shows that IT based approaches towards knowledge management are still dominant. IT based KM approaches focus mainly on the storage (data bases, DMS) and distribution (intranet and internet applications, Push-Pull-Point) of explicit, electronically documented knowledge, thus overlooking the tacit dimension of knowledge. Recently, we observed a trend towards more process- and human-based KM approaches focusing more on networking and the exchange of tacit knowledge.[47]

In contrast to the technology-driven conceptions, the results of the Fraunhofer company benchmarking survey show that knowledge management is mainly understood by practitioners from manufacturing and the service industry as part of corporate culture and a business-oriented method, as one answered: "The sum of procedures to generate, store, distribute and apply knowledge to achieve organizational goals."

---

44 De Hoog (1997); Schreiber et al. (2000).

45 Göbler (1992); Forkel (1994).

46 Gödicke (1992), p. 68; however Davenport et al. (1996) found only one expert system application within 30 knowledge work improvement projects.

47 Cohen, Prusak (2001), Denning (2001).

The Cranfield survey[48] supports this interpretation with 72 % of selected European managers claiming that knowledge management is 'The collection of processes that govern the creation, dissemination, and utilization of knowledge to fulfill organization objectives'.

The business process focus was also confirmed by the recently concluded First Global Delphi-Study on "The Future of Knowledge Management" carried out by the Humboldt University Berlin and Fraunhofer Competence Center Knowledge Management at IPK Berlin: *Integrating knowledge management into business processes is the dominant recommendation and the most pressing research issue.*[49]

Nearly all approaches to knowledge management emphasize the process character with inter-linked knowledge processing tasks or core activities. The approaches show mainly three differences: firstly, the wording of the single KM core activities differs from one to another. Synonyms used for knowledge *sharing* are: distribute, communicate, diffuse, transfer and get. The knowledge activity *generation* is also named creation, development, production, and acquisition. The second difference relates to the number of knowledge activities each approach uses[50]. It ranges from two or three[51] up to eleven[52] process steps or activities. The eight building blocks proposed by Probst et al. [53] have received most attention in the German language area: identification, acquisition, development, sharing, utilization, retention and assessment of knowledge and knowledge goals. In the Fraunhofer company survey , four activities have been assessed as "essential" and "important": *apply* knowledge (Σ 96 %), *distribute* (Σ 91 %), *generate* (new) knowledge (Σ 84 %) and *store* (Σ 78 %).[54] The third difference relates to the emphasis given to single knowledge management tasks. The approach by Nonaka and Takeuchi[55] focuses mainly on the creation of knowledge while Bach et al.[56] emphasize the development and distribution of explicit, electronically documented objects of knowledge.

The close relationship between processes and knowledge management is underscored by the critical success factors named by the companies in our survey. Nearly one out of three companies (29,8 %) mentioned aspects of the design of

---

48 Cranfield School of Management (1998).

49 Cf. chapter 8.

50 For an overview of English language approaches cf. Beckman (1999) and Despres, Chauvel (2000); for German language approaches cf. Remus (2002).

51 Two: Bach et al. (1999), Firestone, McElroy (2002); three: Amelingmeyer (2000), Davenport, Prusak (1998).

52 Like the ADL-process: cf. chapter 12.

53 Probst et al. (1998).

54 Cf. introduction, Fig. 1.3.

55 Nonaka, Takeuchi (1995).

56 Bach et al. (1999).

structures and processes as a critical success factor for knowledge management (cf. chapter 1).

Experiences and study results show a clear need to have a closer link of KM with processes and the corporate strategy in the future than today. Nevertheless, more research has to be carried out on the strategic alignment of KM.[57] In short as one KM-Pioneer at BP-AMOCO expressed it: "In the past, KM was started where the ‚energy' was. In the future KM will be aligned with the strategy and integrated into the business processes."

2.2 Approaches to the Design of Business Process and Knowledge Management

One primary design object in private and public organizations are the business processes which structure work for internal and external clients. Known as Business Process Engineering (BPE) or Reengineering[58] (BPR) or Business Process Optimization (BPO)[59], the design of business processes became the focus of management attention in the 90's. Various methods and tools for BPR have been developed by research institutes, universities and consulting companies.

One reason of the failure of BPR projects in the 90's was that they overlooked knowledge as an important factor. "BPR`s focus is typically on studying and changing a variety of factors, including work flows and processes, information flows and uses, management and business practices, and staffing and other resources. However, most BPR efforts have not focused much on knowledge, if at all. This is indeed amazing considering that knowledge is a principal success factor – or in many judgement, the major driving force behind success. Knowledge-related perspectives need to be part of BPR."[60]

On the contrary, only very few approaches to knowledge management have explicitly acknowledged this relation. And even fewer approaches have tried to develop a systematic method to integrate knowledge management activities into the business processes.

Since our first edition, considerable improvements have been made to integrate the knowledge-related perspectives into methods for process design and process improvement[61] while other approaches have not been enhanced. In the following paragraphs, some selected approaches are presented and discussed briefly.

---

57 Krogh et al. (2000), p. 71.
58 Hammer, Champy (1993).
59 Diebold GmbH (1993).
60 Wiig (1995), p. 257.
61 Cf. the articles in Abecker et al. (2002); Rabrenovic (2001); Schnurr et al. (2001), El Sawy, Josefek (2003).

CommonKADS Methodology (Schreiber et al. 2000), originating from knowledge engineering, claims to support the development of knowledge management solutions. It integrates an organizational model, describing critical success factors such as "general context (mission, strategy, environment)", "Structures", "Processes", "People", "Culture & Power", "Resources" and "Knowledge Assets" and present a knowledge management cycle with seven activities covering the complete life cycle of knowledge within the organization: identify, plan, acquire and/or develop, distribute, foster the application, control and maintain, dispose.

The approach emphasizes the value and process view of knowledge management: "As outlined previously, knowledge is a prime enabler to successfully carry out the business processes within the organization, which in turn create value for the recipients of its products and services."[62] However, the method could not show how to integrate these knowledge management activities into the business processes. The knowledge assets analysis addresses typical questions of knowledge logistics ("right place" and "right time") and quality ("right form" and "right quality"), although it does not cover the cultural aspects influencing these items.[63] Therefore, the main advantages provided by the CommonKADS methodology for knowledge management are the techniques for knowledge- oriented analysis of work tasks and the methods to enhance knowledge sharing and reuse.[64] The method does not support the design of processes to deal with tacit knowledge. The elicitation methods help to make explicit the knowledge concepts and reasoning requirements of a prospective system. Knowledge management as a meta-level activity addresses mainly process improvement aspects of the life cycle of knowledge. The link to business processes has to be made more explicit to cover all knowledge management activities and not only the sharing and reuse activities.

The **Business Knowledge Management Approach** (Bach et al. 1999, 2000) tries to relate knowledge management activities to business objects and business processes that are based on multimedia document processing. The approach distinguishes between business processes, the knowledge structure, which represents the knowledge domains (content), and the knowledge base, which includes knowledge management processes, roles and responsibilities as well as IT systems and documents. Some indicators are proposed to help the management control knowledge management processes. In their second book (Bach et al. (2000)), they present a few instruments to improve the use of tacit knowledge for business processes and an evaluation of these instruments with recommendations for their integration in business processes. With this enhancement, they partly overcome our criticism made in the first edition, that one limitation of their approach is that it does not take the tacit dimension of knowledge into account.

---

62 Schreiber et al. (2000), p. 72.
63 Lullies et al. (1993).
64 Schreiber et al. (2000), p. 72.

Their methodology PROMET®I-NET[65] for intranet-based knowledge management focuses on explicit, electronically documented data and information (similar approaches, which also advocate intranet-based knowledge management, are commonly found under the term "Content Management"). The knowledge management process named "development"[66] is limited to the identification of unknown knowledge and the improvement of existing objects of information. Furthermore, the process does not involve the aspect of the generation of new knowledge, even though the term "development" might lead to this understanding.

The **Model-Based Knowledge Management Approach** (Allweyer 1998) adds a new perspective to the modeling of existing business processes, especially of knowledge-intensive processes. Allweyer affirms that knowledge-intensive processes are less structured, not exactly foreseeable and, in most cases, not repeatable. Knowledge management activities are considered as an integral part of existing business processes. The existing four-level architecture of business process management[67] is adopted for knowledge management and the method is renamed knowledge process redesign.

The approach is limited to the description of required and used knowledge as well as generated and documented knowledge. Knowledge is understood as information in connection with the value for the owner of this information, which allows him to act. Moreover, the approach claims to support the structuring of knowledge into categories and the construction of a knowledge map to locate who knows what inside the organization. Easy-to-understand pictograms are proposed to help users describe the use of documented and tacit knowledge within their business processes. The approach does not make explicit how to integrate the knowledge management activities into business processes and the criteria to analyze and improve the knowledge processing within the business process.

The **PROMOTE Methodology** (Hinkelmann et al. 2002), developed by an EU funded project, aims to provide a method and software tool to model business processes and knowledge processes. The methodology is based on the Business Process Management Systems approach with the five phases 'strategic decision', 're-engineering', 'resource allocation', and workflow 'and' performance evaluation.[68] PROMOTE adapts these five phases for the introduction of knowledge management: aware enterprise knowledge, discover knowledge processes, modeling knowledge processes and organizational memory, making knowledge processes and organizational memory operational, evaluate enterprise knowledge. This approach models the selected business process and identifies the knowledge intensive tasks (KIT) in order to analyze the kind of knowledge and to identify the knowledge flow between these processes. In the next phase, the knowledge proc-

---

65 Kaiser (1999).
66 Bach et al. (1999), Bach et al. (2000).
67 Scheer (1997).
68 Cf. Karagiannis et al. 1996.

esses, such as "how to find an expert", are modeled and assigned to the business process. The knowledge structures are defined, too.

The PROMOTE methodology has adopted some ideas from the Fraunhofer approach[69], but is lacking criteria for the knowledge analysis and the design of processes. There is no explication what the elements of the business and knowledge processes that have to be modeled are and the level of detail is not clearly defined: work task or business process. Furthermore, the approach does not clearly address the dimension of knowledge itself and their repercussion to the process and required methods to improve the handling of knowledge. Nevertheless, the PROMOTE methodology is one of the most engineered approaches towards an integration of knowledge managements tasks into business processes.

The **Building Block Approach** (Probst et al. 1998), considered here as the most widespread KM approach in German speaking areas, specifies eight building blocks to manage knowledge – knowledge goals, knowledge identification, acquisition, development, sharing, utilization, retention and assessment. Knowledge as a resource is considered to be the only integrative pattern of their approach, which follows no other external logic than the inherent logic of knowledge. Furthermore, they do not systematically include categories such as leadership, culture and technology within their concept, which has already been criticized by practitioners as a deficit.[70]

The idea of building blocks for knowledge management has been proposed by Wiig with examples of building blocks for knowledge creation and dissemination.[71] While Wiig emphasizes the connection of these building blocks with the (re-)design of business processes, the approach of Probst et al. does not provide any hints of how to integrate the proposed building blocks into the business processes and work processes.

The improvement of existing methods and development of new methods to integrate knowledge management and business processes has increased and some first useful results are available to practitioners. Their main origins stem from the Knowledge Based System and the development of information systems as well as from the workflow design of intranet development and business process (re-) engineering. Depending on their original focus, the mentioned approaches still show their current strengths within their particular originating areas. Knowledge and knowledge activities have become part of nearly all approaches to business process design. The dominance of explicit and documented knowledge is still valid, but instruments to improve the usage of tacit knowledge are considered in some approaches. Aspects of process design and knowledge structuring, like in ontology's, are gaining more and more attention. However, detailed criteria for the

---

69 Cf. Hinkelmann, Karagiannis, Telesko (2002) p.70, 81.

70 Vogel (1999), p. 124.

71 Wiig (1995), p. 291.

analysis and design of business processes from the knowledge perspective and the perspective of knowledge processes are still missing.

2.3 Integrated Enterprise Modeling for Knowledge Management

Since the late 80's, the division Corporate Management at the Fraunhofer Institute for Production Systems and Design Technology (Fraunhofer IPK) has developed the method of **I**ntegrated **E**nterprise **M**odeling (IEM) to describe, analyze and design processes in organizations.[72] Besides projects of traditional business process design, this method has been used and customized for other planning tasks such as quality management (web and process-based quality manuals for ISO certification), simulation, the design and introduction of process-based controlling in hospitals, benchmarking and Performance Management Systems (Fig. 2.1).[73] The IEM-method is supported by the software tool MO^{2}GO (**M**ethode zur **O**bjektorientierten **G**eschäftsprozess**o**ptimierung = Method for Object-Oriented Business Process Optimization).

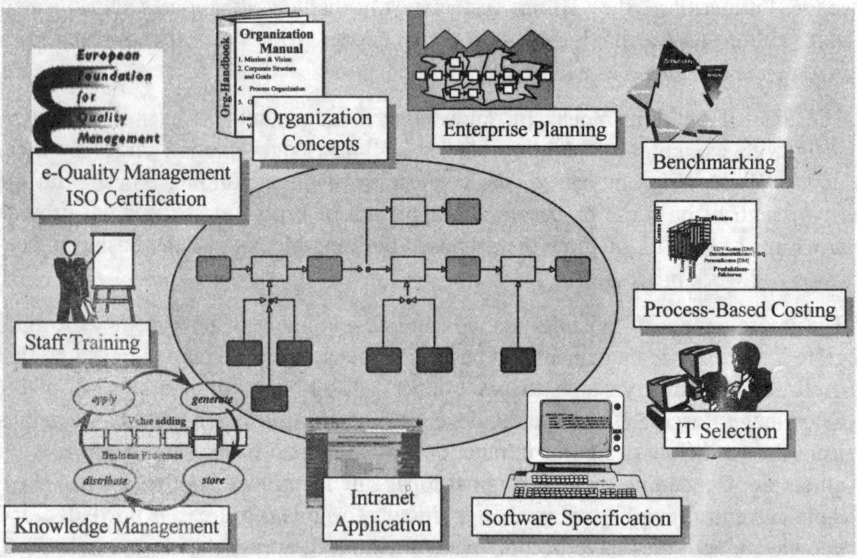

Fig. 2.1: Application Fields of the IEM Business Process Models

The method of integrated enterprise modeling (IEM) distinguishes between the three object classes "Product", "Order" and "Resource." These object classes are combined by the construct "Action" within a generic activity model (Fig. 2.2).

---

72 Süssenguth (1991); Spur et al. (1993); Schwermer (1998).

73 Mertins, Jochem (1999); Siebert (1998); Jäkel et al. (2000); Krause (2000).

Five elements are provided to link the objects to the actions: sequential, parallel, case distinction, junction and loop.

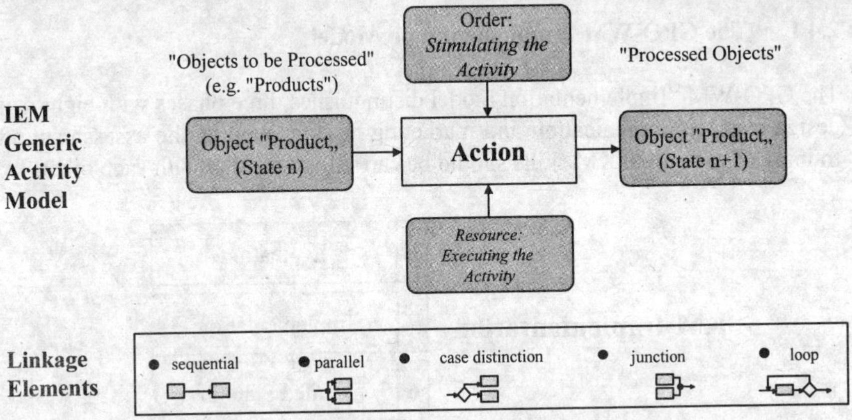

Fig. 2.2: The IEM Generic Activity Model and IEM Generic Activity Model for Knowledge Management

Since process modeling has been used as a method for analysis and design in several projects during the past years and companies are thus familiar with the methodology, the integration of a knowledge management perspective into business process modeling should not cause any practical problems. The ISO 9000:2000 certification requires the description of processes as well. These process models could be reused and enhanced in order to save effort and costs.

2.4 Method for Business Process Oriented Knowledge Management

Based on our empirical research and KM projects carried out for our clients in the public administration and in the service and manufacturing industry, we developed the Method for Business Process Oriented Knowledge Management (GPO-WM®). The Method consists of the GPO-WM® Implementation model, the Fraunhofer KM Audit to perform a quick scan of the current state at the beginning and identify the strengths and weaknesses of the current utilization of knowledge, the GPO-WM® Analysis of the selected business process as well as some KM Best Practice Building Blocks for the design and development of an integrated KM solution.

Our aim is to synchronize a more systematical application and generation of knowledge in business processes with the activities of the involved people and the supporting information technology with tools giving global access to the explicit knowledge base and facilitating just-in-time connection between the owners of the tacit knowledge.[74]

2.4.1 The GPO-WM® Implementation Model

The GPO-WM® Implementation model distinguishes three phases with eight main design steps. Communication and marketing tasks as well as the assessment and training of individual KM skills should be carried out parallel with each other.

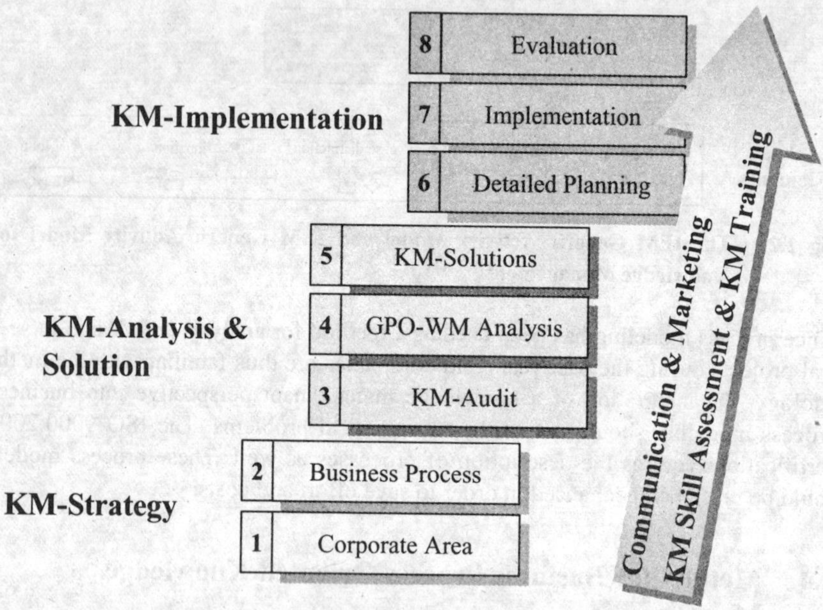

Fig. 2.3: The GPO-WM® Implementation Model

The first phase focuses on the definition of the KM-Strategy by choosing the corporate area where KM activities will contribute to the improvement of the company performance as well as the identification of the specific business process, which should be improved with operational KM methods and instruments. The second phase aims at developing the KM solution. Therefore, the KM audit provides relevant data about the current state situation at the start. In the GPO-WM Analysis step, the business process will be described and analyzed in detail according to specific KM criteria. Alternative KM solutions should be specified and presented for management decisions in the fifth step. The third phase, the KM implementation, starts with the detailed planning, followed by the implementation of

[74] Heisig (2001b).

the selected KM solution and ends with the evaluation of the outcomes of the KM project.

Every change management process has to be supported by communication and marketing activities in order to clearly inform and involve the stakeholders about the vision and the detailed objectives of the KM initiative, the advantages for the company, the middle management and the involved employees.[75] Despite our assumption that we use knowledge to perform our daily business tasks, our research results suggest that there is a great need for the development of specific KM skills for nearly every manager and employee. Therefore, we suggest to accompany the introduction of KM with the assessment of KM skills and training measures.[76]

2.4.2 KM Strategy

While the corporate strategy defines long term goals, policies and business strategies to accomplish these goals, the KM Strategy describes the KM vision, specifies the KM goals and focuses on the area where KM activities will be started. Unfortunately, there is still a lack of theory and strategic KM frameworks, with few exceptions[77], and in practice the operational short-term view is dominant.

Regarding the required KM Strategy, one first management task is to reframe the way knowledge is viewed mainly by senior and middle management in the company. Instead of regarding knowledge as something vague and difficult to measure, it should be considered as an "asset tied to specific actions and business results". Top management has to transmit a clear vision about the importance of knowledge assets for the companies' goals, sustainability and success: "the increase in the stock of useful knowledge and the extension of its application are the essence of modern economic growth".[78] Secondly, management has to name the relevant knowledge and decide about priorities because "not all knowledge has strategic value"[79] and "no firm can manage all forms of knowledge equally well"[80]. Despite the lack of evidence about which types of knowledge led to which forms of strategic advantage[81], the knowledge domains named most often are few in number, such as[82]: customer knowledge, knowledge in people, knowledge in products, knowledge in processes, organizational memory, knowledge in relation-

---

75 Cf. chapter 4.

76 Cf. the Training Program *"Knowledge Master"* at Siemens, in: Erlach et al. (2000); Reinmann-Rothmeier, Mandl, (2000); Heisig, Finke, Jaitner (2003).

77 Hansen et al. (1999); Davenport (1999); Krogh et al. (2000); Metternich (2001); Spek et al. (2003).

78 Kuznets, S. (1966), cit. in Teece (1998) p. 55.

79 Krogh et al. (2000), pp. 71.

80 Davenport (1999), pp. 2-7.

81 Davenport (1999).

82 Skyrme (2000), Spek et al. (2003), p. 450 "consider 8-12 knowledge areas as the optimal number to be used during one analysis".

ships and knowledge assets. According to the Fraunhofer survey, there are five main knowledge domains: methods (52,7 %), products (50,4 %), customers (40,5 %), markets (29,8 %) and competitors (15,3 %).[83] The resource-based view of the firm postulates four aspects, strategic resources like knowledge must have to lead to competitive advantage: (a) valuable, (b) rare, (c) imperfectly imitable and (d) cannot be strategically equivalent substitutes.[84] The third important management task regarding the KM Strategy is to keep track of the development of the relevant knowledge inside and outside the organization. Therefore, the knowledge environment has to be scanned continuously. Tools and instruments still have to be developed to monitor the knowledge development.

Based on the business strategy – whether the company produces standardized products and services or delivers customized products – Hansen et al. (1999) proposed two different KM approaches: Codification or Personalization. Standardized products and services are mainly based on the reuse of explicit, codified knowledge while customized products rather tend to incorporate tacit knowledge. The codification of knowledge leads to a higher emphasis on processes and IT whereas customization requires more emphasis on people and face-to-face sharing methods.

In 1996, the APQC Benchmarking Study identified six KM strategies in US-based organizations: transfer of knowledge and best practice (11 out of 11 best practice companies), personal responsibility for knowledge (8/11), innovation and knowledge creation (7/11), KM as a business strategy (5/11), customer-focused knowledge (4/11) and intellectual asset management (2/11).[85]

In the Fraunhofer survey , a comparison between the business processes in which knowledge management is most commonly started and the processes in which the questioned companies locate their core competency shows that knowledge management follows core competencies (Fig. 2.4).

It might not be necessary to embark in a great effort to create an explicit KM Strategy. What seems more important is that managers explicitly include in their strategic thinking, their dialogue and their strategy documents, the category of knowledge, as Davenport (1999) suggests. One consultant at a software company expressed this need during our KM audit interview as follows: "It is clear to me that our business needs knowledge and experiences, but what types of knowledge are important according to our business strategy? On what should I focus, what should I document and what should we share?"

---

83 Heisig, Vorbeck (2001), p. 105.
84 Probst, Büchel, Raub (1998).
85 APQC (1996).

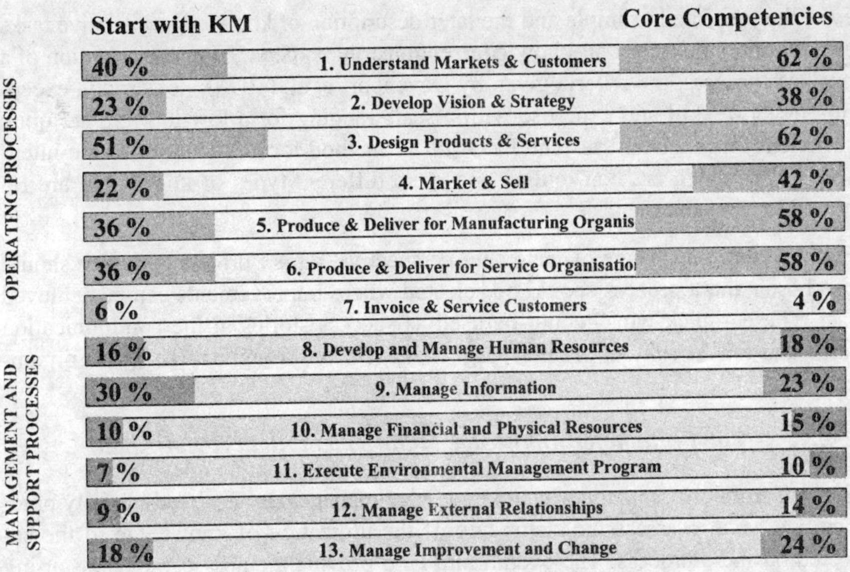

Fig. 2.4: Where Companies Start with Knowledge Management and where They Locate Their Core Competency

2.4.2.1 Selection of the Business Process

From the business process perspective, the selection of a focused company area, in which the KM initiative has to start, and the definition of the relevant knowledge are not enough, because there are still several business processes involved. Therefore, the management has to decide about the concrete business process or processes that should be improved by KM methods and instruments.

Eppler et al. (1999) proposed the two categories *knowledge intensity* and *process complexity* with ten attributes for the selection.[86] The approach distinguishes four classes, and only processes with high process complexity and stronger knowledge intensity should be improved by knowledge management methods. This understanding is similar to that of knowledge work processes, which are "characterized by variety and exception rather than routine" and "performed by professional or technical workers with a high level of skill and expertise. Knowledge work processes include such activities as research and product development, advertising, education, and professional services like law, accounting, and consulting. We also include management processes such as strategy and planning."[87] The CommonKADS method suggests in the first process breakdown worksheet the "Boolean indicating whether the task is considered knowledge-intensive." How-

---

86 Eppler et al. (1999).

87 Davenport et al. (1996), p. 54; similar also Allweyer (1998), p.39.

ever, the presented example and the later description of knowledge-intensive tasks are of limited help for non-knowledge engineering experts.[88] For the selection of a business process, the criteria given by Davenport et al. (1996), i.e. variety, exception, level of skill and expertise with a scale ranging form low to high, are quite sufficient for practical use. The PROMOTE method focuses on knowledge-intensive tasks) which are non-routine and quite different types of knowledge are required to carry out the task.[89]

Finally, from the change management perspective, a pilot project approach should be chosen and a process should be selected where improvements can be achieved fast. These "quick wins" could be used as success stories in the communication tasks and for the roll out of knowledge management throughout the whole organization.

2.4.2.2 The Fraunhofer Knowledge Management Audit (FKM-Audit)[90]

Starting from our assumption that knowledge management is already a daily practice, we have to assess the status quo of the utilization of knowledge in the selected business process. The overall aim is to provide a quick scan of the starting situation with its strengths and weaknesses as well as a specific roadmap for the implementation of KM.

One important aspect of the KM audit is that some individual good practice methods will be identified by interviewing the process-owners. Integrating and building the KM solution on these good practices will increase the commitment and acceptance of KM by the involved staff and reduce the project costs.

Data about the general set-up, like corporate culture, leadership, incentive systems and IT infrastructure will be gathered by additional questionnaires. The general set-up has to be known in order to adjust the proposed KM methods and tools to the starting situation. This is because the implementation of KM tools normally takes less time than a cultural change program and companies could not wait to implement KM until the 'right' culture is in place. The audit will also provide information about existing barriers and enablers for knowledge management in the company, which are relevant for change management efforts.

The third important aim of the KM audit is quantitative and qualitative data about the status quo. Current performance indicators do not consider knowledge related data. Even basic data is often not available – like the time spent for searching the

---

88 Schreiber et al. (2000), p.33: process breakdown, p.56: example with scale: yes – medium – high – very high, p. 125: 11 types based on the kind of problem being solved.

89 Hinkelmann et al. (2002), p. 77.

90 For a detailed description cf. chapter 3 and Mertins, Heisig, Finke (2001); similar approaches are proposed by Bukowitz, Williams (1999); Weggemann (1999); Langen, Ehms (2000); Liebowitz, Rubenstein-Montano, McCaw (2000).

best knowledge or the time spent to document the best solutions. Without these data, the evaluation of the KM solution will be quite difficult.

The results of the KM audit are: (a) assessment of current state of used methods and tools in the business process for KM, (b) identification of barriers and enablers, (c) identification of individual or team-based good practice methods for KM, (d) a roadmap for the company and business process specific implementation of KM.

2.4.3 The GPO-WM® Analysis of Business Processes

Based on the information about the general set-up, a detailed description and analysis of the business process will be performed within the fifth step of the KM implementation. The following two tasks are undertaken:

- Knowledge-oriented modeling and
- Knowledge-oriented analysis of the business process

2.4.3.1 IEM Generic Activity Model for Knowledge Management

The IEM approach offers the possibility to describe knowledge as an object *and* as an activity within the process model. Based on the generic activity model, the IEM generic activity model for knowledge management (Fig. 2.5) has been defined.

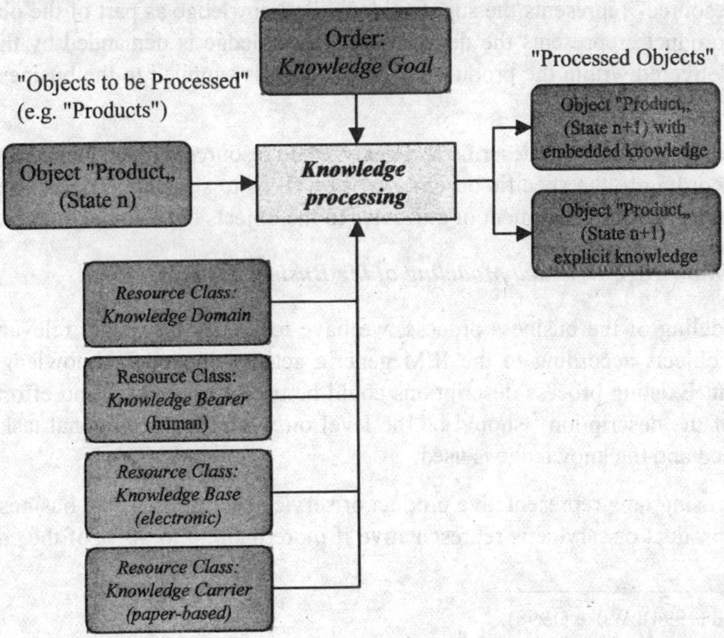

Fig. 2.5: IEM Generic Activity Model for Knowledge Management

From the resource-based view, knowledge is modeled as a sub-class of the superordinated class "resource". The sub-class "knowledge domain" represents the knowledge content required by the action. "Knowledge bearer" describes the human expert who carries the knowledge or the responsible organizational unit in which the expert is located. This sub-class supports the representation of the tacit dimension of the knowledge domain. Under the term "knowledge base" the electronically documented knowledge is subordinated, which is stored in EDP-Systems or other databases. The importance of paper-based elements is recognized with the sub-class "knowledge carrier". These two sub-classes represent the explicit dimension of knowledge. From the knowledge management perspective, the action is considered as a knowledge-processing task.

Knowledge is required to produce or/and deliver a service or/and product and thus becomes implemented in the object "product".[91] As knowledge is the only resource in business, which increases during usage, the actions in the process model are at the same time regarded as potential knowledge generating activities. In order to support awareness of this knowledge generating function, the frequently hidden knowledge results are modeled as a sub-class "explicit knowledge products" within the process model. The object "order" which triggers the actions could be used to describe the knowledge goals of a process or/and task.

This approach enables us to combine the two different strategic views of the firm, the resource-based view with the market-based view. Knowledge as an object of the class "resource" represents the supply side while knowledge as part of the object class "product" represents the demand side. Knowledge is demanded by the client and delivered within the product and knowledge is supplied to the business process from the resources.

Furthermore, it is possible to describe and analyze the resources, orders, tasks and products according to the specific design goal as well as to simulate different design alternatives by the assignment of attributes to the objects of the model.

2.4.3.2 Knowledge-Oriented Modeling of the Business Process

For the modeling of the business process, we have to identify the main relevant knowledge objects according to the IEM generic activity model for knowledge management. Existing process descriptions could be used to save time and effort. The level of the description[92] should be the level on which the operational tasks are performed and the knowledge is used.

We start focusing on a representative product or service as a result of the business process. A product or service is representative if more than 30 to 40 % of the ca-

---

[91] Hedlund (1994); Wilke (1998).

[92] The level of detail should not be exaggerated: Davenport, Prusak (1998), p. 157, mentioned a firm which described "one organizational learning" process, four subprocesses, fifteen sub-subprocesses, and fifty-three sub-sub-subprocesses".

pacity is spent on producing the product or service or if its contribution is of high value to the final product offered to the internal or external client.

From the change management perspective the objects ought to be named with the company-specific terms as this helps to create a common language and common understanding of the process. Our experience also shows that this facilitates the participation of the people involved in the process itself. Due to the high autonomy required by knowledge workers, Davenport et al. stress the importance of their participation for a successful improvement of the process.[93] The IEM methodology has been used in a wide range of projects in industries such as telecommunications and automobile, as well as in public organizations such as ministries, hospitals and the police. Managers and employees from nearly all organizational levels have been involved in describing their processes using the four elements of the IEM methodology.[94]

For data gathering, we use an interview guide, which addresses not only the relevant knowledge objects, but includes criteria to assess the quality and availability of the different knowledge objects. Problems and suggestions are collected to feed the analysis step and the design. The process itself is the guideline of the interview. Nevertheless, we often observed during the interviews that the interviewer has to bring back the attention to the process when the quality and availability of the knowledge objects are discussed. It is possible that the knowledge domain absorbs all attention. The process focus helps to redirect the data gathering and assessment.

Some knowledge management approaches consider these business process models as knowledge assets, which represent the important process knowledge of a company.[95] Generally, this notion is correct. However, a web-based documentation of this process knowledge is not sufficient for extracting value from this asset for the company, not even for a knowledge-based company. "Similarly, many firms viewed the commitment to implementing new knowledge work designs as persuasion rather than mandate. They offered, marketed, or communicated the new designs through education rather than forcing them on those who performed the knowledge work."[96] Often knowledge management approaches with a strong business process reengineering background do not consider this important task of actively marketing and distributing their own modeling results. One of our best practice companies showed how this process knowledge could be trained and transferred with a tool called "Process Rally." Moreover, this tool supports the elicitation and sharing of tacit knowledge of the involved process owners.[97]

---

93 Davenport (1996).
94 Tünschel et al. (1998).
95 Allweyer (1998), p. 45.
96 Davenport (1996), p. 60.
97 Cf. chapter 18.

The outcome of the modeling step is the knowledge-oriented mapping of the business process with the inputs and outputs of the process, the performed tasks and the required critical knowledge domains on an aggregated level. The main knowledge carriers, like organizational units, roles or persons carrying out the specific action, documents, databases or other knowledge related objects, are represented.

2.4.3.3 Knowledge-Oriented Analysis of the Business Process

The links within the IEM generic activity model (Fig. 2.6 to Fig. 2.8) are interpreted as two core knowledge management activities. The link between the resource and the action represents the core activity "apply knowledge" while the link between action and the object product represents the core activity "generate knowledge". The object product shows the knowledge made explicit within the process.

The application of the analysis criteria has shown that there has to be a separate assessment in each identified knowledge domain. Each knowledge domain is assessed in two ways: in the first step we evaluate the knowledge handling from the demand perspective of each process action (Fig. 2.6).

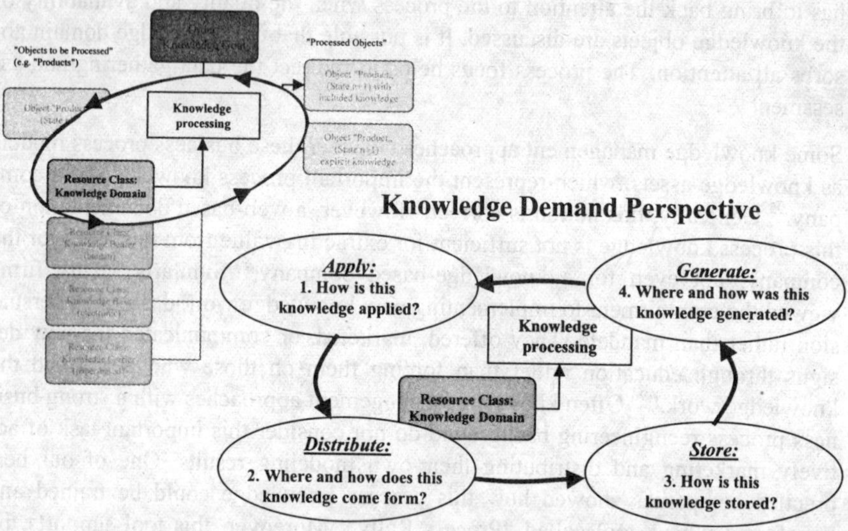

Fig. 2.6: The GPO-WM® Analysis of the Demand Perspective from "Apply Knowledge"

Starting with the core activity "apply knowledge", the quality and availability of the critical knowledge domain is assessed. In the second step, the analytical perspective shifts and the new knowledge is evaluated from the supply perspective, starting with the question: "Which kind of knowledge is generated and how?", following the sequence: "store", "distribute" and "apply" (Fig. 2.7). These two perspectives are complementary. By virtue of the core activity "distribute", the knowledge flows within the business process itself and with other processes are assessed. The assessment of the core activity "store" delivers information about the form in which knowledge is stored and which kind of knowledge carriers are used, like documents or/and databases or people. Following the results from the Fraunhofer survey, these four empirical validated knowledge management core activities are sufficient to analyze the business process and to assure the involvement of process owners from all organizational levels by reducing the level of complexity.

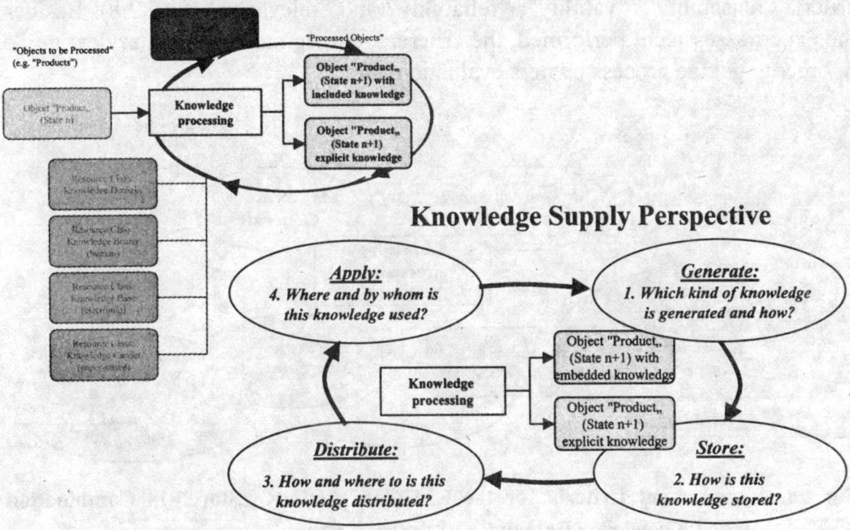

Fig. 2.7: The GPO-WM® Analysis of the Supply Perspective from "Generate Knowledge"

This assessment also creates awareness of the often unconscious handling of knowledge in daily tasks and processes. As we often perform knowledge management activities implicitly, the core process lacks the appropriate degree of integration or connectivity. For instance, new knowledge is generated, though not explicitly, and documented in a database; a database is built but is not used by the targeted user group; knowledge is distributed via intranet though not applied by the staff; knowledge is applied but lessons learned are not generated. These are examples of the successful performance of some individual core activities without these single activities being connected, however. The core process itself is therefore not closed or integrated.

2.4.3.4 Assessment Criteria for Knowledge Handling

One main deficit of existing approaches of business process oriented knowledge management is their lack of empirically validated assessment criteria for the handling of knowledge in the analyzed business process. Based on a survey of KM practitioners and researchers, we have evaluated[98] and defined assessment criteria for every core activity in four categories: a) performance of methods applied by core activities, b) competency of human knowledge carriers, c) quality of knowledge and d) quality of media applied. The 31 criteria are applied as follows: If the knowledge processing business process activity is identified as "store knowledge", than the criteria "targeted" and "reliable" are used to assess the operational method employed in this business process activity to store knowledge. The knowledge stored is described by the object knowledge domain and assessed by the criteria "reusability", "validity", "reliability" and "relevance" (Fig. 2.8). Besides the expert assessment performed, the criteria are integrated in the interview guide in order to get the process owners evaluation, too.

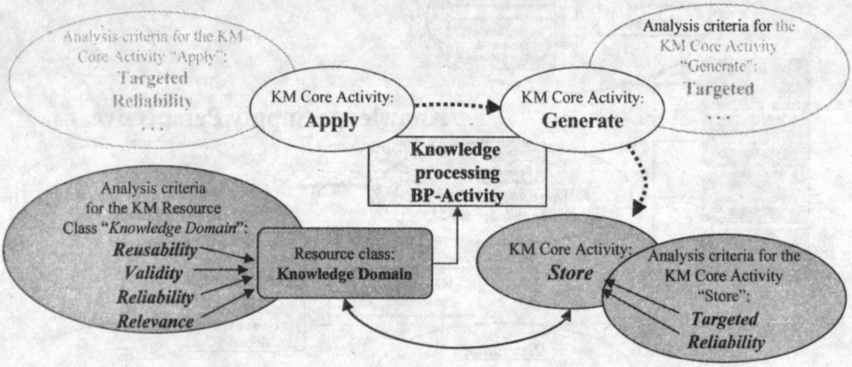

Fig. 2.8: Assessment Criteria for the KM Core Activity "Store" in Combination with "Knowledge Domain"

Resulting from this assessment, the knowledge activity profile and the degree of connectivity per knowledge domain emerge in the context of the business process. Knowledge relations and sources outside the analyzed business process are identified. Strengths like good practices and points for improvement are identified based on the assessment criteria employed and the interviews with the involved process-owners and employees. Together with the information about the general set-up coming from the KM audit, the reasons and causes for the deficits have to been determined.

98 Two criteria have to be met by each assessment criterion: 1) The criteria must be considered as "essentially important" and "important" with an average higher than 4.5 of 6.0. 2) The rate of agreement must have a minimum of 75 %.

2.4.4 KM Solutions

The design of the KM-Solution aims at achieving a closed process of the four core knowledge management activities integrated into the business process to be improved.

2.4.4.1 The Fraunhofer KM Reference Model Supports Holistic Solutions

In our empirical research, we identified five critical success factors for knowledge management initiatives: corporate culture, structures and processes, information technology, motivation and skills and top management support. These critical success factors, combined with the measurement aspect named "control", are integrated, together with the value-adding business processes and the four core activities, into the Fraunhofer Knowledge Management Reference Model.[99] This reference model supports the specification and design of a holistic KM solution, where measures will be defined related to the process elements and design fields. If advanced modeling tools are used to describe the new process design, the outcome of the new design could be simulated with regard to different indicators, such as process duration and costs.

2.4.4.2 Existing Methods and Tools as Good Practices are One Starting Element

The optimization and the design of business processes aim at closing the identified gaps within the underlying core processes and at sequencing the core tasks of knowledge management. One design principle is to use available procedures, methods, tools and results from the process to design the solution. As our projects with different companies have demonstrated, there are already many worthwhile attempts aiming at improving the handling of knowledge. However, these approaches are neither considered nor dealt with as systematic and structured knowledge management.[100] Therefore, in most cases, they have to be discovered, reframed and reused from a knowledge management perspective. The individual or team-based good practice examples identified by the KM audit and the GPO-WM Analysis serves as one good starting point to achieve the commitment of the different stakeholders. If the gap cannot be closed with existing methods, a new solution has to be developed in compliance with the missing or not fulfilled core activity of knowledge management.

2.4.4.3 KM Best Practice Building Blocks

Another good source for KM methods are benchmarking studies and case studies on KM.[101] Within our benchmarking study, we identified around 30 best and good

---

99 Cf. chapter 1, Fig. 1.7.

100 Heisig (1998a); Heisig (1998b).

101 APQC 1996, EFQM (1997), cf. chapter 11.

practice methods, which can be adopted as building blocks into the selected area and process.

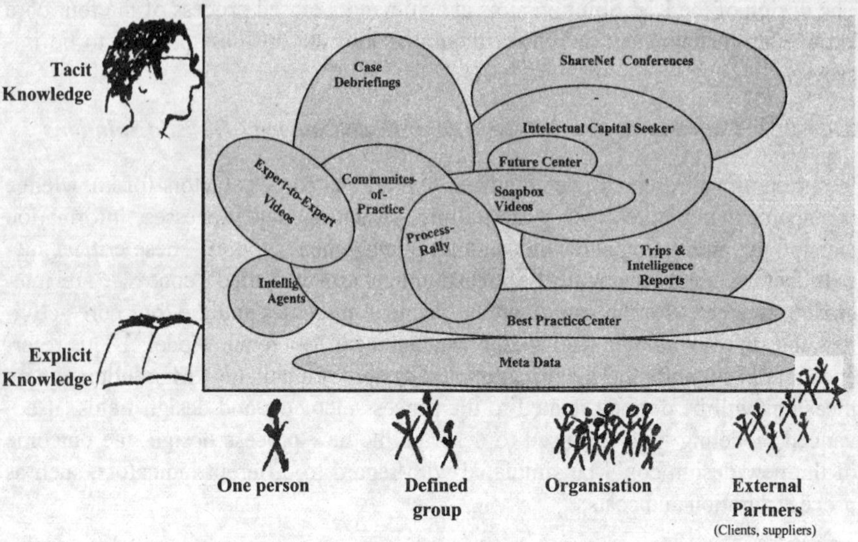

Fig. 2.9: Best Practice Methods as Building Blocks for Knowledge Management

The figure above (Fig. 2.9) shows, on a continuum from explicit to tacit knowledge on the vertical axis, which kind of knowledge is covered by some selected best practice methods. The horizontal axis indicates whether the method addresses a single person, a defined group of people (e.g. department, project team), the whole organization or external partners like clients and/or suppliers.

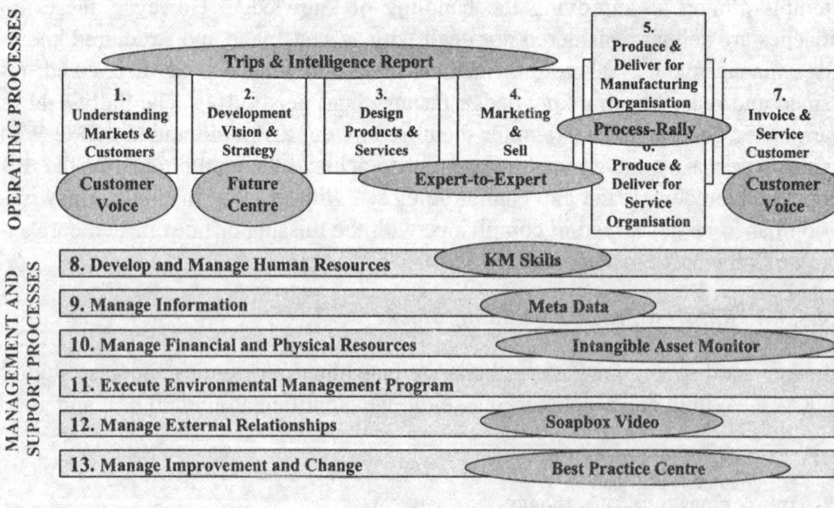

Fig. 2.10: Best Practice Methods Linked to Business Processes

Furthermore, some of the identified best practice methods can be linked with the business processes they mainly support. Fig. 2.10 demonstrates where business processes are supported by some of the methods described in the case studies.

2.4.4.4 Synchronizing People, Processes and IT Smoothly

The term synchronization has been chosen to address the logistic linking of order processes with production processes in the context of process design. Taking into account the barriers of knowledge management, e.g. the lack of time, and the fact that each task has knowledge processing elements, synchronization could serve as the design paradigm for business process oriented knowledge management. The elements of the three layers 'people', 'processes' and 'information technology' have to be synchronized. The integrated core activities of knowledge management ("generate", "store", "distribute" and "apply") are the synchronizing elements linking people, processes and information technology with each other.

The business process tasks itself determines the needed knowledge and the core activity that might be improved by implementing a knowledge management instrument. People, as the bearers of knowledge, bring in their (tacit) knowledge and experience by performing the tasks using the knowledge management instrument. An information technology application supports the business process tasks and the knowledge management activity. The information technology needs to become a smooth support tool. 'Smooth' means in this context that the functionalities to support knowledge management are directly integrated into existing or new applications, which support the business process. The user is not forced to start a special knowledge management application in order to perform his knowledge management activity. The information technology provides the user with functionalities to handle the explicit knowledge stored in internal or external databases or/and to connect them with other people to share their undocumented know-how.

Fig. 2.11 shows two knowledge management core activities (circles in the process layer) integrated into the business process being improved. The first business process task with the inherent knowledge management activity is to 'design a solution'. The knowledge management core activity is to 'apply knowledge'. Therefore, the knowledge management instrument 'community-of-experts' is required in order to bring together the existing non-documented knowledge of the organization to reuse existing solutions. The expert session could be a structured face-to-face meeting or a moderated discussion in an electronic forum via the intranet application IM.NET. Furthermore, the IM.NET provides a search engine to find existing documented solutions in internal and external databases. The search functionality is not a separated application but it is integrated into the document management system, which stores all the documentation including the solutions designed in other projects. The second business process task has been introduced by the knowledge management initiative, because the experiences acquired during the design of the solution have not been secured before. Therefore, a generic lessons-learned-tool has been customized as an after-action-review according to the business process context and regarding the knowledge to be stored. The instrument has been implemented in the lessons-learned-database, which is accessible via the IM.NET, too. Indicators linked to the process and the targeted knowledge elements are defined in order to control the knowledge handling in the process and to monitor the safekeeping of the experiences.

Synchronization is achieved by implementing knowledge management instruments into the business processes that improve certain business process tasks related to the core knowledge management activities. Smooth information technology applications enhance existing information technology tools and support these knowledge management activities inherent to the business process tasks.

This approach towards knowledge management achieves high acceptance from the employees because it starts from the daily working tasks and shows the benefits by facilitating a better performance of these daily tasks. This is one of the biggest incentives an organization could put into place. Reliable indicators give the management the objective tool to provide feedback to the users and contribute to their motivation as well.

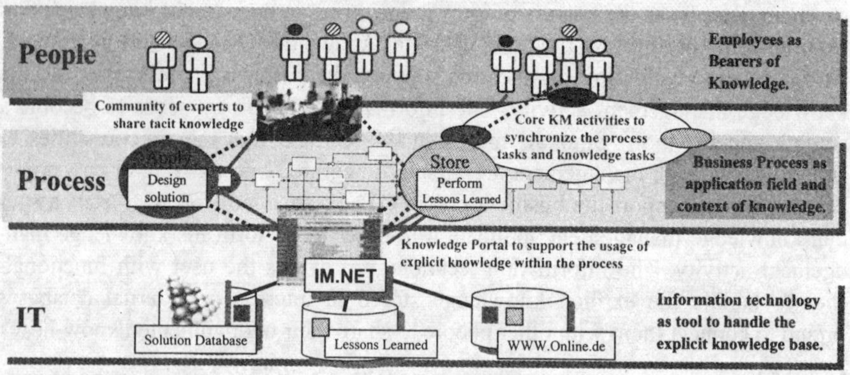

Fig. 2.11: Community of Experts and a Knowledge Portal are the KM Instruments to Synchronize Knowledge Management in a Business Process

2.4.5 KM-Implementation Phase

The third phase of the GPO-WM® Implementation model describes the three implementation steps: detailed planning, implementation of the knowledge management solution and evaluation of the solution. The accompanying activities, such as internal marketing and communication as well as assessment and training of basic knowledge management skills, have to be implemented, too.[102]

---

[102] Cf. chapter 4.

2.4.5.1 *Detailed Planning*

The detailed planning has to coordinate the measures on the three layers 'people', 'processes' and 'information technology'. On the **people** level the managers and employees have to be informed about the changes and the new skills or/and methods will be trained. Therefore, a communication concept and a training concept are required. To achieve the commitment of the management and the employees, they should be involved into the detailed planning of tasks, like the specification of the knowledge management methods and the software support tools. In regard to the critical success factor of corporate culture, adequate measures have to be defined. If the corporate culture has been assessed as an open culture, some "fair rules of the game" have to be defined in order to implement the knowledge management solution. In an open culture, the behaviour of managers often fosters the exchange of ideas and know-how. If the corporate culture is a more rigid one, special change management measures have to be undertaken. The cultural change has to start with the top management and pay special attention to the middle management and informal opinion leaders. If the behaviour has changed visibly and persistently, the main barrier of knowledge management has been overcome. To accelerate the changes in management, behaviour knowledge management related targets should be integrated into the management scorecard. On the employee level, knowledge management related indicators have to be integrated into the annual appraisal system. Last but not least, one low cost incentive measure should be improved. The performance feedback is one of the most expected management tasks by employees but one with the lowest satisfaction figures. Therein lies a great potential for improvement, which most knowledge management projects have not yet addressed themselves to. In case that the management does not want to address itself to the cultural dimension, more detailed organizational rules and responsibilities related to the handling of knowledge have to be defined and implemented.

On the **process** level, existing tasks will be enhanced by process or/and knowledge specific activities addressed to the four core activities of knowledge management. These methods are specified and designed in detail. Furthermore, indicators have to be defined to monitor and control the process and the knowledge usage. These indicators should be connected with the planned software application in order to decrease the required measurement efforts. The indicators should address the quality[103] and validity of the knowledge and information in the database. On the process level, roles[104] and responsibilities for the knowledge usage and quality have to be defined, too. The access concept and data security issues have to be determined from the view of process tasks, too. Modern object oriented process modeling tools provide functionalities to define roles and access rules

---

103 Huang et al. (1999).
104 Cf. chapter 5.

within the process model. The functionalities help the user to transform the model directly into an intranet application (Fig. 2.12).

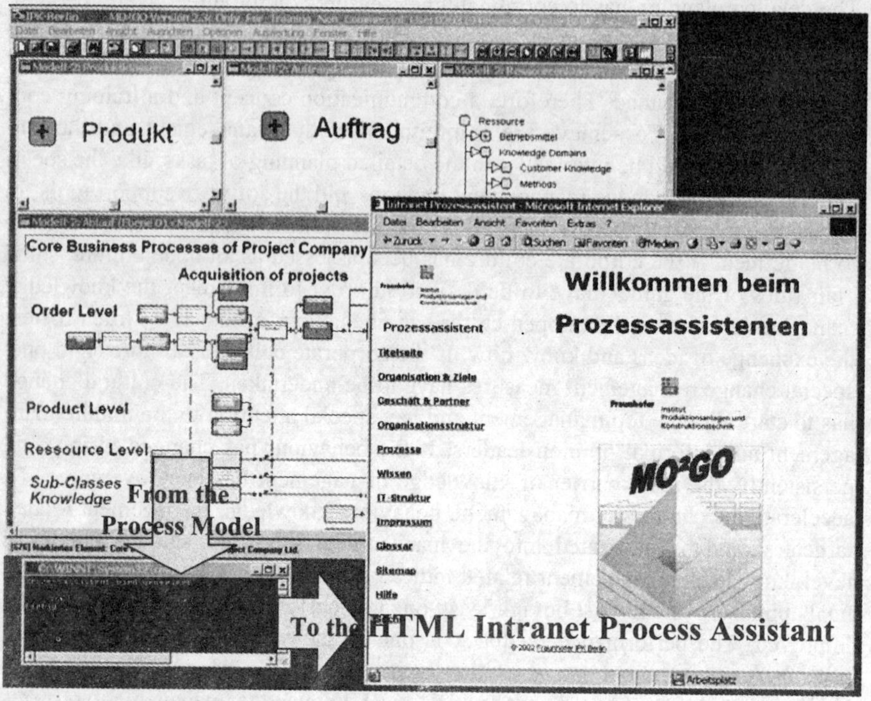

Fig. 2.12: An Intranet Process Assistant is Generated from the Process Model Automatically by the MO²GO-Tool Using the XML Standard

On the **information technology** level, the specification of the functionalities supporting knowledge management tasks are defined from the process perspective and implemented. The end-users should be involved directly into the definition of requirements and the usability testing, because *their* knowledge handling should be supported! Another requirement often made by management is reflected in the following statement of a senior manager of a manufacturing company: "Another IT application will not pass my department door. We already have five different systems to work with. System number six will not be accepted." Therefore, the design idea of one knowledge portal to support knowledge management has been proposed. Unfortunately, the users often consider this application as another one, because the integration into the daily work tasks is not fully achieved. Furthermore, we observed that even the standard functionalities of knowledge management applications are sometimes too demanding. Therefore some companies followed a modular approach. They defined functionalities modules which where put into practice step-by-step. This step-by-step approach should also be used when additional databases are connected to a search engine. In a software company, we discovered in the course of a KM audit that the overall majority of the employees welcomed the search engine very much. But they did not use them because some of the connected knowledge bases were internally not considered as trustworthy.

Another very important task of the detailed planning phase is the communication and internal marketing concept for managers and employees. The communication concept has to address itself to the specific objectives of knowledge management for the individual organization, their strategy, the management and the employees as well as the benefit for all stakeholders. Besides these basic communication aspects, some typical knowledge related attitudes have to be addressed. Not only the typical attitude "knowledge is power", but also some more operational attitudes regarding knowledge handling tasks have to be tackled carefully. The documentation work is often seen as non value-adding work, therefore of less importance and performed not in a very conscientious way[105]. Consecutively, the result is that the stored explicit knowledge is scarce and of poor quality. Therefore, most knowledge is not documented and only stored in brains. Another behavioural aspect relates to the activity of distributing knowledge. In many cases, we have observed that the basic rules of mutual information are not clearly defined or/and not accepted. One department expects to be informed actively by the other, while the other department does not consider it as its duty. Within the process model, the different interfaces and communication rules have to be defined and agreed. The communication task is to inform about new "rules of the game" and foster the commitment.

The question "How can we motivate our staff ?" is one of the most asked questions by managers when they start thinking about a knowledge management initiative. A second concern is addressed to the dimension of roles and responsibilities like access rights to read and write in the electronic knowledge base. Always overlooked and undervalued are the skills and capabilities to handle knowledge more systematically. To overcome this deficit, we defined the Fraunhofer knowledge management skill test[106] and implemented a web-based tool to assess these competencies[107].

2.4.5.2 Implementation

The implementation step covers all activities to put into practice the knowledge management solution, starting with the detailed design or/and customizing of the knowledge management methods, the selection of the basic software application, implementation and testing of the required functionalities up to the redesign of the business process itself with the allocation of roles, responsibilities and indicators.

According to our knowledge of more than fifty knowledge management projects in European companies and more than a hundred published case studies, the pilot

---

105 Lindemann et al. (1998).

106 Vorbeck, Finke (2001).

107 Heisig, Finke, Jaitner (2003) "Knowledge Management Skill Test".

project approach to implement knowledge management in a selected business unit and/or business process has become dominant. Successful knowledge management solutions are focused on the concrete business problem and the knowledge domain and carefully adjusted to the existing cultural framework conditions.

The implementation activities should be used also to increase the involvement of the end-users in order to use their feedback and their ideas to improve the knowledge management solution and to achieve their acceptance and to motivate them to use the implemented solution. End-users have to identify the knowledge management method and solution as "their own" solution. The feedback is also needed to improve the rollout in the whole organization or business area. The first short success stories experienced by the end-users during testing and prototype evaluation could serve as input to the communication measures, too.

During this step, change management activities for managers have to be carried out to create supportive general set-up for the later rollout. The staff should be informed and trained in the basic knowledge management skills required in the organizational units or business process.

2.4.5.3 Evaluation

The most important outcome of the evaluation of knowledge management is that evaluation starts with the first project step. Evaluation often fails because there is no data available about the starting situation and some dimensions, which knowledge management will improve, are never ever measured before. Therefore, a company, which plans to start a knowledge management project tomorrow, should start to gather data today! The problem of balancing costs and benefits is another example for the theoretical and practical problems that arise from measuring intangibles like knowledge.[108]

The current lack of 'hard' figures showing the improvements generated from knowledge management is due to the fact that the majority of knowledge managements projects in the past have been started based on a strategic decision by the top-management. The decisive argument often was: "What will we loose if we do not improve the handling of our knowledge?". In most cases, the expected loss of competitive advantages was one main argument to acquire first own experiences with knowledge management.

In practice, the evaluation of the knowledge management initiative depends on the specific solution that has been implemented. One example is the usefulness of an expert directory in the company-wide intranet. When the project manager was asked about the benefits of the yellow page application, he replied: "How did you calculate the added value of the phone?" This example implies that some know-

---

108 Cf. chapter 7.

ledge management applications will become standard tools, like the telephone, and others not.

2.5 Summary and Outlook

The "*integration of knowledge management into the common business processes*" has been identified by international KM experts as the most pressing theoretical research issue and the approach most likely to be practical for dealing effectively with the practical problems of knowledge management. Consequently, the "*priority on human factors: shift from an IT perspective to a behavioural science perspective*" and a better "*match between the social and technical aspects*" was ranked by the experts as the most important recent theoretical progress in knowledge management. Therefore, "*interdisciplinary and multi-disciplinary approaches, combinations of respective methods and techniques*" are required.[109]

There is no doubt that knowledge management is entering a new phase of theoretical and practical development. After the sometimes naïve expectation that the new IT possibilities can be easily translated into large progress in knowledge management, the shift towards knowledge management approaches which combine the human and process dimension with the information technology is apparent.

The presented Fraunhofer approach of business process oriented knowledge management offers a coherent solution to this challenge. The Fraunhofer approach is build on a theoretical framework with the Integrated Enterprise Modeling method. The IEM method has proven its practical benefits in several projects during the last decade. This basic framework has been enhanced with empirically validated and practically proven knowledge management elements like the four core activities as well as criteria for analysis and design of a coherent knowledge management solution. Based on empirical research, the critical success factors have been identified and transformed into the Fraunhofer Reference Model for Knowledge Management. This reference model combines research approaches and results from different disciplines as required by international experts. Therefore, the Fraunhofer approach offers an integrated method to implement knowledge management to public and private enterprises.

Knowledge management provides a kind of toolbox with several generic methods, instruments and tools to improve the handling of knowledge in daily routine. The proposed method of business process oriented knowledge management helps to analyze the current state, to identify the right points of intervention, to design a

---

109 Cf. chapter 8.

holistic solution, to decide about the most promising application and to implement the chosen solution.

3 The Fraunhofer Knowledge Management Audit (FKM-Audit)

Kai Mertins, Peter Heisig, Ina Finke, Christina Ulbrich

Even though knowledge management is far from being a new discipline,[110] the existing perceptions of knowledge management in European companies show a significant disparity of views. This can be described as follows:

- We have been managing our knowledge at all times. For us, there is nothing new about knowledge management.
- The future of our company depends heavily on our ability to make use of our knowledge in the most efficient way. Therefore, we are urgently concerned about knowledge management.

There is truth in both assumptions. As the intensity of knowledge increases in business processes, the management of this resource is of growing importance,[111] even although the specific application of knowledge in business processes is nothing new. Regardless of whether a company implements new knowledge management systems or maintains its own traditional way of handling knowledge, it is still relevant to ask whether the actual way meets the technical and methodological possibilities and the requirements of competition. Even within a single company, this question will be answered differently. The Fraunhofer Knowledge Management Audit (FKM-Audit) encounters these individual perceptions with a liable analysis.

In the first part of this chapter, the goals to be achieved by the FKM-Audit are specified. In the second part, further approaches aiming at evaluating knowledge management are reviewed. In the third part, the method and the results of the FKM-Audit, which was developed and applied to different industries by the Competence Center Knowledge Management at IPK Berlin,[112] are described.

---

110 Recent surveys by the Conference Board and the American Management Association show that up to 72% of European firms have some kind of knowledge management initiative underway. With regards to global corporations, studies put the figure closer to 80%. Cf. Allee (2000); KPMG (2000).

111 Concerning the increasing importance of knowledge in a global development towards knowledge-societies cf. Drucker (1994), Stewart (1997) and Willke (1998).

112 The Fraunhofer knowledge management audit is an integral part of the method for business process oriented knowledge management (GPO-WM) described in chapter 2.

3.1 Knowledge Management has to Build on Existing Circumstances

The term "audit" implies that a certain field is examined by independent inspectors. By means of a knowledge management audit (KM audit), it is investigated how an organization applies knowledge management within its business processes. In this chapter, the term KM audit is used as a synonym for a survey aiming at the following objectives:

1. **Uncovering strengths and weaknesses within the actual management of corporate knowledge:** the KM audit permits to assess objectively if knowledge management activities are integrated successfully in the business processes; therefore, employees as well as management become aware of the best practices in the field of knowledge management, some of which are already applied in certain areas of the firm but have never been identified as best practice methods; on the other hand existing potentials are uncovered.

2. **Analyzing circumstances, barriers and enablers for knowledge management:** various aspects of corporate culture, leadership, human resource management, information technology, process organization and controlling can be enablers as well as barriers for knowledge management initiatives; it is important that any measure to improve knowledge management takes into account and suits the specific organizational environment.[113]

3. **Increasing awareness for knowledge management within the company:** The involvement of employees in the KM audit and the internal communication of a detailed audit report increase the awareness for the efficient handling of knowledge and the specific potentials and strengths in knowledge management within the firm. The process of sensitizing and involving employees as well as the recognition of the employees needs in concrete business processes, are essential aspects for the success of any project that proposes change. By means of these aspects the audit empowers the acceptance of following measures and avoids that knowledge management is recognized as a demotivating additional burden.

4. **Designing a roadmap for future knowledge management measures:** A knowledge management audit clarifies which measures should be taken, and if any, where the starting point should be. By means of the audit, the existing circumstances and potentials are made transparent and

---

[113] Hylton (2002) states that many of the mistakes of both early and recent adopters of knowledge management can be traced back to the fact that knowledge audits were not originally included in their knowledge management strategies and initiatives.

systematically taken into account when further actions for the implementation of knowledge management are recommended.

5. **Collecting measurable data for the controlling of knowledge management:** It is of growing importance to measure the benefits, which are achieved through KM initiatives. The measurement is often impeded by a lack of data concerning the status quo of the firm's KM practices. The audit can contribute to the gathering of these data, though it differs from methods determining the level of the knowledge inventory at a certain time. However, measurable quantities raised by the KM audit can be integrated within a so-called knowledge-related balance sheet.[114]

3.2 Audit Approaches for the Evaluation of Knowledge Management

Within the last decade, a number of methods for auditing corporate knowledge management have been developed and applied. Some of them are not aiming at the complete range of objectives mentioned above. In the following part of this chapter, a summary of the existing approaches will be given. They are listed according to the increasing degree of which the approaches achieve the mentioned objectives from our point of view.

3.2.1 The Knowledge Audit (According to Liebowitz)

In their concept of the "knowledge audit", Liebowitz et al. (2000), go further into the question of the relevant contents of knowledge: "in order to solve the targeted business problem, what knowledge do we have, what knowledge is missing, who needs this knowledge, and how will we use it?". The knowledge audit comprises, on the one hand, the knowledge already existing within a company and, on the other hand, the current need for knowledge. The questionnaire dealing with this concept is detailed and extensive. A catalogue of questions is drawn up for every relevant content of knowledge. This procedure requires great commitment, as much time is needed for answering the questions.

This knowledge audit designed by Liebowitz et al. (2000) is based on the notion that knowledge has to be seen as part of a company's inventory. If problems occur (e.g. a high turnover of personnel) a knowledge audit can provide an evaluation, looking at the following aspects: what knowledge is needed; what knowledge exists and what knowledge is lacking; and who needs this knowledge and how can it be made available. "A productive audit needs to only concentrate on answering the following questions in order to solve the targeted problem: what knowledge do

114 For different concepts regarding knowledge-related balance sheets cf. chapter 7.

I have, what knowledge is missing, who needs this knowledge and how will they use the knowledge?"[115]

This audit serves to take stock of the existing knowledge, and identify what knowledge is lacking. It will thus derive recommendations for actions capable of solving this specific problem. The current dealing with knowledge and the organization's general set-up are not evaluated.

3.2.2 Knowledge Management Assessment Tool (KMAT)

The Knowledge Management Assessment Tool (KMAT) was developed by Arthur Andersen Consulting in co-operation with the American Productivity and Quality Center.[116]

It is based on an "Organizational Knowledge Management Model": the core activities of the process of knowledge management "share, create, identify, collect, adapt, organize and apply" are supported by four factors, so-called enablers "leadership, culture, technology and measurement".

KMAT strives to achieve two aims:[117]

- to ascertain the position of one's own company with regard to knowledge management in comparison to other companies,
- to evaluate the efficiency of the realization of the knowledge management process (share, create, identify, collect, adapt, organize and apply).

The method of KMAT consists of a questionnaire without additional information about a suitable rollout, e.g. the size of the sample. The questionnaire is highly structured and has a section for each of the four enablers.[118] Within each of these segments four to six statements (for e.g. "employees take responsibility to their own learning", "technology brings the organization closer to its customers") have to be judged on two levels:

- Performance: the person who fills in the questionnaire judges how well he or she thinks the company realizes the given statement. For this judgement a five step scale ranging from "bad", "poor", "fair", "good" to "excellent" is used.
- Importance: the person who fills in the questionnaire judges the importance of a given aspect for the company. Answers are to be given on a five-step scale ("not at all important", "slightly important", "fairly important", "important", "essential").

115 Liebowitz et al. (2000), p. 5.
116 Cf. APQC (2000).
117 Cf. Hiebeler (1996), p. 2.
118 Cf. Arthur Andersen, APQC (1995).

The questionnaire is sent to and evaluated by Arthur Anderson. From the result, it is derived which fields of design should be further developed and how this should take place.

According to Hiebeler (1996), 70 questionnaires were evaluated in 1996. The participating companies represented different branches of industries whose profit ranged from $ 1 million to $ 33.3 trillion.

The described approach leaves much freedom for the realization of the audit. It does support the company to choose well-balanced samples. Following this approach the comparison of results achieved in different companies is restricted. After all, it is left to the user whether he prefers to let the evaluation be based on his individual perception or whether the instrument be applied throughout the whole company or within a management workshop. Since the statements to be judged in the questionnaire are of abstract character, inaccuracies resulting from the individual interpretation must be taken into account even in the case of large-scale sample. The concrete business processes and the organization's general set-up are not recognized in this concept.

3.2.3 Knowledge Management Diagnostic (KMD)

The Knowledge Management Diagnostic (KMD) was developed by Bukowitz & Williams (1999). It is based on a model of knowledge management called "Knowledge Management Process Framework" which consists of seven knowledge management activities ("get, use, learn, contribute, assess, build / sustain, divest").

According to Bukowitz & Williams (1999) the four activities "get, use, learn und contribute" designate the daily routine in dealing with knowledge. By enhancing these activities the company's reaction to the demands of the market is improved.

The three knowledge management activities "assess, build / sustain and divest", are attributed to the strategic planning of the company by evaluating which kind of knowledge will be relevant in the future.

The KMD was designed as a tool for self-evaluation and collects subjective, qualitative data. It thus serves to enable users to determine how well the different aspects of the knowledge management process have been realized in the company.[119]

The interview takes place in a written form, and the choice of the sample is left to the company. Bukowitz & Williams (1999, p. 17), however, encourage the client to involve employees from wide-ranging areas. In addition, the authors recommend the presentation of results in order to trigger a discussion about strengths and weaknesses in respect to the implementation of knowledge management.

---

119 Cf. Bukowitz and Williams (1999).

The questionnaire is divided into seven categories according to the knowledge management process. In every category, 20 statements are given describing possible actions for knowledge management, for e.g. "We build models of our decision-making systems to better understand why things happen the way they do", "People are members of multiple communities, making it easier to transfer knowledge across the entire organization".[120]

The agreement to the statements is measured by a scale ("The statement is strongly / moderately / weakly descriptive of my organization."). Points given for the attributes are summed up for each of the seven categories and compared with the highest possible score.

In the single categories, the supporting factors relating to them, such as corporate culture and systems of information, are questioned on a general level. The specific situation of the company is not taken into account.

3.2.4 Knowledge Audit (According to Pfeifer)

The knowledge audit, according to Pfeifer et al. (2000), focuses on an evaluation of required knowledge at a certain point of time, the carriers of this knowledge, the connections between the knowledge carriers and the need for additional connections.[121]

Furthermore, "it serves as an instrument for uncovering weak points, for encouraging improvements and for controlling the existing measures of knowledge management".[122]

This concept works exclusively with qualitative procedures. "For an exact determination of the individual demands, it is mandatory to talk to those involved in the process and especially important to get a description of the problems and inadequacies of the process. [...] In extensive personal talks [...] the existing behaviour in communication is discussed in detail." In order to extend the existing connections between knowledge carriers by those additionally needed, it is also important to consider whether single persons are capable of promoting an exchange of knowledge: "At this point the employees' personality has to be strongly taken into account."[123]

Owing to the discussions it requires, and the analysis of the structure of communication and the relations of single persons to each other, this concept is time consuming and relatively focused on individual perceptions.

---

120 Bukowitz and Williams (1999), pp. 23.

121 Pfeifer et al. (2000), p. 1275.

122 Pfeifer et al. (2000), p. 1277.

123 Ibid.

3.2.5 Knowledge Management Maturity Model (KMMM)

The Knowledge Management Maturity Model was developed in the Competence Center Knowledge Management of the Siemens AG.[124] It is based on a model of analysis, which is characterized by eight fields of design of knowledge management: strategy and knowledge aims, a company's surroundings and partnerships, employees and competencies, corporate culture and cooperation, leadership and support, forms of knowledge and structures of knowledge, technology and infrastructure as well as processes, roles and organization (Fig. 3.1).

Fig. 3.1: The Eight Key Areas of KMMM[125]

In the model for analysis the demands of each of the above mentioned fields of activities are described. Depending on how the company meets these demands it is assigned a maturity level. These maturity levels are represented in a development model (cf. Fig. 3.2), which describes a company's degree of maturity regarding knowledge management.[126] Accordingly, after the evaluation and by means of the model of analysis, a company is ranked to one of the five levels of maturity. Starting from this attribution, recommendations are made with respect to how the company can improve its dealing with knowledge.

---

124 Cf. Ehms / Langen (2000).

125 Cf. www.kmmm.org, cited 12-17-02.

126 This idea is based on the levels of the CMM (Capability Maturity Model) of the Software Engineering Institute at Carnegie Mellon University. The names of the levels were adopted from this concept.
(Holistic Development of Knowledge Management with KMMM at http://www.knowledgeboard.com/doclibrary/knowledgeboard, cited 1-20-03.)

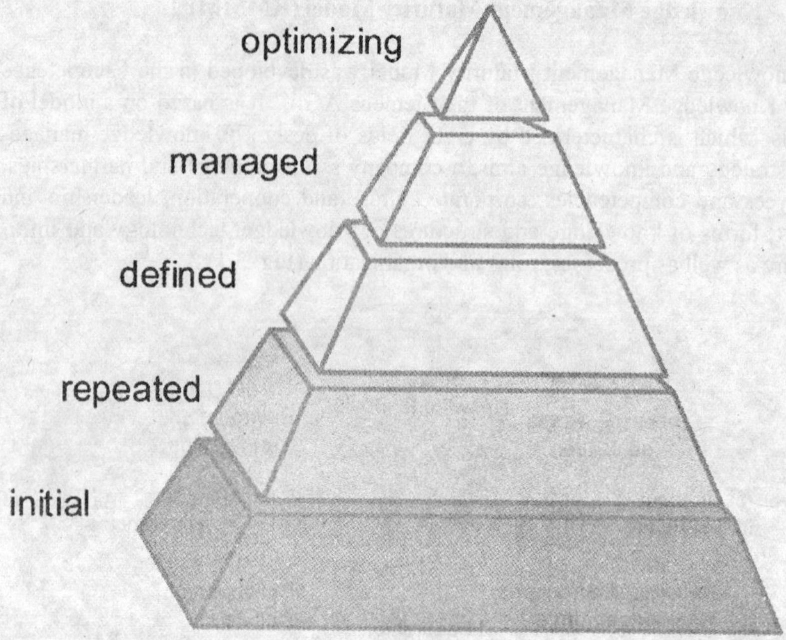

Fig. 3.2: The Five Maturity Levels of KMMM[127]

The model aims at "determining systematically where current knowledge management can be located"[128], i.e. evaluating knowledge management activities. This derives suitable steps for development, which are based on the current status of knowledge management, and thus shows the most appropriate starting point before a KM project actually kicks off.[129]

The company's evaluation is done by means of the 64 subjects while the "detailed structure should not, however, lead to the single aspects being subject to a 'mechanical test".[130] The results are achieved through workshops and interviews with selected members of the organization. It is important for the chosen interview partners to be representative across the company. Before a fixed evaluation, the data must be critically evaluated in the form of a discussion between two consultants.

Qualitative audits are of importance for this concept: "before quantitative surveys take place, the interdependences have to be grasped in a qualitative way".[131] They

---

127 Cf. www.kmmm.org, cited 12-17-02.

128 Ehms, Langen (2000), p. 3.

129 Ehms, Langen (2002), p. 2.

130 Ibid, p.3.

131 Ibid, p. 1.

are used not only for evaluating a company's degree of maturity, but also for the future development of knowledge management activities: "The qualitative – and thus not numerically recordable – results of the KMMM method are vitally important for well-aimed measures of development in knowledge management. They contain useful hints which a company can utilize for improving its knowledge management".[132]

Since the evaluation is performed qualitatively, this concept is especially useful in identifying aspects of knowledge management that are specific to one company. By presenting a profile of the strengths and weaknesses of dealing with knowledge, which allows for attaining "exchange and improved reciprocal understanding about different views on knowledge management problems and solutions",[133] it is possible to heighten awareness for the process of dealing with knowledge.

The following figure gives an overview and characterizes the audit approaches for knowledge management mentioned in this chapter.

Approach	Author	Effort of the Company	Quantitative Methods	Qualitative Methods	Business Process orientated
Knowledge Audit	Liebowitz et al.	low	x		
KMAT	APQC & Arthur Anderson Consulting	low	x		
KMD	Bukowitz & Williams	low	x		
Knowledge Audit	Pfeifer et al.	high		x	
KMMM	Ehms, Langen	medium	x	x	
Fraunhofer KM Audit	Heisig, Finke, Ulbrich	medium	x	x	x

Fig. 3.3: Overview of Different Audit Approaches

132 Ibid.

133 Ibid, p.4.

3.3 Developing the Fraunhofer Knowledge Management Audit

Targeting the aims that were described above, the FKM-Audit method is based on the following objectives:

1. Uncovering strengths and weaknesses within the actual management of corporate knowledge.
2. Analyzing framework conditions, barriers and enablers for knowledge management.
3. Increasing attentiveness for knowledge management within the company.
4. Designing a roadmap for future knowledge management measures.
5. Collecting measurable data for knowledge management controlling.

The Fraunhofer IPK reference model for knowledge management[134] provides the theoretical framework for the design of the Fraunhofer Audit. According to the reference model the method integrates the level of business processes with the level of the design fields for knowledge management.

On the level of business processes the relevant types of knowledge, the demand and the availability of each type are identified. Furthermore the methods of generating, storing, distributing and applying knowledge, which are specific to the processes and the company, are analyzed. Thus the employed activities conducive to knowledge management within the business process are determined systematically.

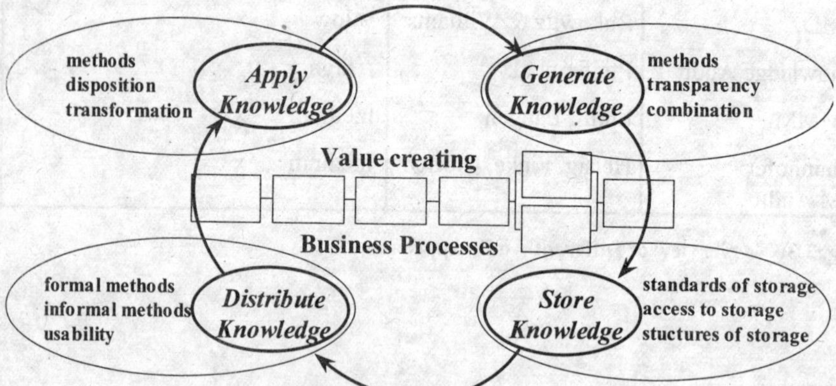

Fig. 3.4: Examples for Process Related Contents of the Audit

134 Cf. chapter 1.

On the level of the design fields for knowledge management the general conditions, enablers as well as barriers for knowledge management, are identified. Fig. 3.5 represents relevant aspects which are evaluated.

Fields of design	Aspects
Prozess organization	roles, knowledge demand
Information technology	benefit, satisfaction
Leadership	feedback, leading by example
Corporate culture	values, social behaviour
Human resources management	motivation, capabilities
Controlling	assessment systems, indicators

Fig. 3.5: Examples for the Relevant Aspects in the Design Fields

The integration of process perspective and fields of design is also realized methodically. Considering the procedure presented in Fig. 3.6, both the processes to be analyzed in an exemplary way and the group of the company's employees to be interviewed are chosen in the focusing phase.

The survey consists of a qualitative and a quantitative part. Key stakeholders of the organizational units, the business process and from different hierarchical levels are chosen for in-depth interviews. Central topics of the face-to-face interviews are the activities that can be attributed to the core processes of knowledge management, and the concrete suggestions of improvement made by the employees to promote the co-ordination or comprehensiveness of the core process. The core activities (generate, store, distribute, apply) and the identification of knowledge are analyzed and evaluated according to the assessment criteria, such as the actuality, availability, transparency and reliability of sources. The way knowledge is handled in a selected business process is grasped and analyzed precisely.

Audit phases	
Initial State (preparation)	Analysis of the relevant documents about processes, procedures and structures (e.g. process model, organigram, job specification, product specification)
Focus Setting	Choosing the target group (e.g. the whole company, a department, a team) and the relevant processes.
Adjustment of Inventory	Customizing the audit to the company's requirements
Survey	Gathering of data: questionnaires for the selected target group and face-to-face interviews with the process owners
Analysis & Evaluation	Analysis of the data; Modeling of the business process for a description of the procedures, creating a roadmap with recommendations for further actions
Feedback Workshop	By means of a workshop, the results are reported back and the suggested measures prioritized (roadmap and action plan).
Project Start	Projects recommended in the roadmap are planned and realized.

Fig. 3.6: Proceeding Scheme of the Fraunhofer Knowledge Management Audit

Besides the qualitative statements given in the interviews, quantitative remarks of the selected target group (e.g. the whole company, a department, a team) are grasped in a task related way by means of a short questionnaire. The results of the quantitative data are compared to the outcome of the qualitative data and illustrated on the level of process. The employees' statements concerning the aspects of the design fields, e.g. with regard to leadership, human resources management and support with information technology, provide a profound basis of analysis. Thus, from the point of view of employees and management, it becomes clear where a company has potential to improve its knowledge management.

The recommendations for further actions, aimed at implementing knowledge management successfully into a company, are summarized in a roadmap. The roadmap illustrates concisely the necessary measures which need to be taken, and serves in the final workshop as a basis of discussions and decisions for the further development of the company.

3.4 Case Studies

The benefits of the Fraunhofer IPK Knowledge Management Audit will be illustrated by using the examples of a medium-sized software company and a research institute, both of whom asked the Competence Center Knowledge Management to conduct an audit.

3.4.1 Knowledge Management in a Medium-Sized Software Company

The company's situation at the starting point is marked by the general situation in the business and its character of being the market leader in that business. Despite the existing competition and the high pace of innovation, the company has a big increase in turnover and a great need for personnel. Due to the actual lack of software development experts, the company has to quickly familiarize new, unqualified employees with their job and at the same time bind experienced employees to the company by implementing incentive schemes. The fast growth of the company necessitates conformist structures whilst the company's innovative character and positive image should simultaneously be upheld. From the management's point of view, knowledge management should help to make full use of the potential for growth and to improve the internal processes. Knowledge is a critical factor for the success of this service company.

The project was started in the middle of May 2000 and within six weeks it was finished with a management workshop. In a half-day workshop, the objectives of the software producer were elaborated on, in co-operation with the internal knowledge management team and the management: Knowledge management is supposed to support the internal change processes and the company's growth. The management expected feedback about the present situation with regard to the existing knowledge handling activities and a competent evaluation and positioning of the company. Concise recommendations for improvement, and the starting point for measures of change or implementation were demanded. The results were to serve as a basis of decision for the implementation of knowledge management into the company. Furthermore, the existing knowledge management project team expected an external assessment on their earlier self-assessment.

Over and above that, the workshop served to determine the relevant business processes for detailed analyses. The client chose two central business processes: on the one hand project development, which is a central process of consulting, and on the other hand innovation and placement of products on the market as an interdepartmental process. While project development aims mainly to safeguard project experience and the familiarized new employees with their job, the quick placement of products is important for the second process.

In preparing the audit, documents were analyzed by the IPK and the short questionnaire was customized to suit the company's situation and wording.

The short questionnaire dealt with aspects of the culture and the need for information and knowledge and was handed out to the company's employees at the end of June 2000, and returned within two weeks. It took each employee a maximum of 15 minutes to answer it. The return rate of 75% illustrates the employees' commitment to the task.

Within two days, the audit interviews were carried out on the spot by one team. Where possible, the half-structured interviews with a representative selection of employees and management were carried out in an on-the-job situation, where the interview partners could relate to sources of information and knowledge they utilized. The interview questions related to the handling of knowledge in the selected processes. By means of these analyses, examples for hindering and promoting factors, which determine concretely the core process of knowledge management (create, store, distribute, apply knowledge), were given. Even the good practices can generally be identified and briefly described.

In the analysis and evaluation phase, the results of the questionnaire were compared to those of the interviews. Examples from the interviews are prepared in a way that illustrates quantitative results from the audit. At the same time, the results concerning the framework conditions, such as leadership, can serve as an explanation for deficit procedures of dealing with knowledge. From the results of the interviews, concrete measures for improvement are derived in order to ameliorate the integration of knowledge management activities in the selected business processes.

The final workshop, which included the management and the internal project team for knowledge management, served mainly to report back on the results and hold a discussion on the short-, medium-, and long-term measures (6 month, 12 month and above 18 month) recommended in the roadmap. In the introduction, the participants were also informed about the general state of the development of knowledge management in the German and European industry. A key element of the workshop was to create a common understanding about potentials and contents of knowledge management in the company. In the roadmap for the introduction of knowledge management, the central starting points, which could be identified in the analysis, were presented and the steps for realization were described. Best practice examples from the Fraunhofer IPK Benchmarking Study (Mertins et al. 2001) and the project experience of the Competence Center Knowledge Management also aided this process. The workshop encouraged discussions and kick starts the introduction and successful implementation of knowledge management.

What did the company learn from the knowledge management audit? A basic result achieved by the audit is transparency of the employees' attitude and expectations towards knowledge management. The prevailing opinion among the majority of management and employees was that an individual as well as a company-wide increase of efficiency could be expected through knowledge management. This is a prerequisite for the successful implementation of change measures within a company. Concrete recommendations from the roadmap were initiated into the company as projects for the internal KM team, and examples of measures derived from the results of the audit were to be presented in the following.

3.4.1.1 The Area of Information Technology

In the area of information technology, transparency could be attained with respect to the usefulness of single applications and how satisfied the employees were with them.

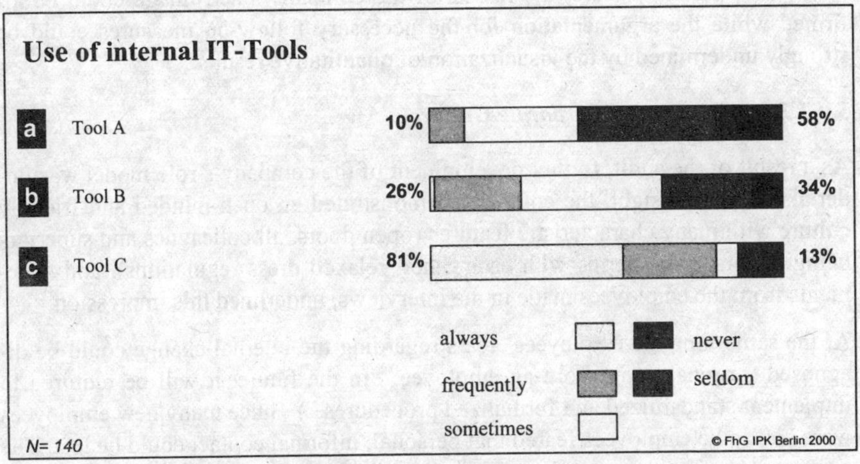

Fig. 3.7: Exemplary Result for the Area of Information Technology: Use

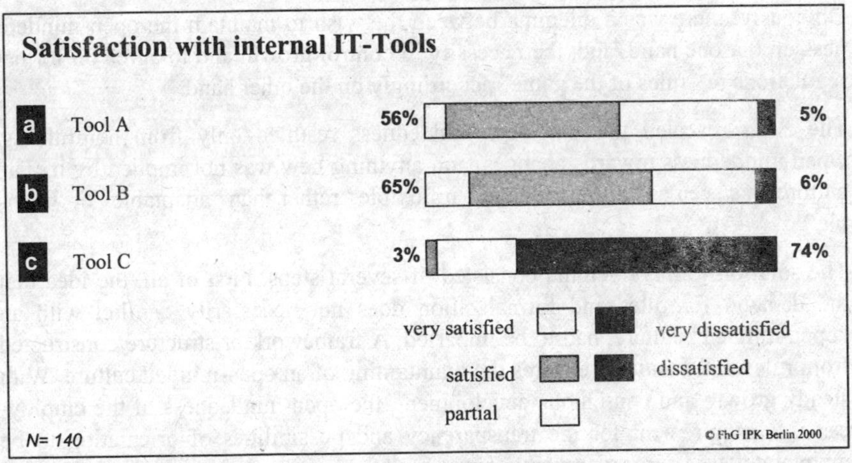

Fig. 3.8: Exemplary Result for the Area of Information Technology: Satisfaction

It became apparent that some of the applications were not even known by those users for whom they had been originally designed (Fig. 3.7), and that some tools were frequently used even though their users were dissatisfied with their performance (Fig. 3.8). Owing to the results of the audit, it was decided that the internal marketing of single applications be optimized and, above all, that the respective target groups be informed which programs should support them at work. In addition, weak points could be uncovered so that well-aimed measures for improvement could be taken (e.g. re-programming of software, integration of programmes, create user profiles). Here, the results of the company's self-image could be affirmed while the argumentation for the necessary follow-up measures could be strongly undermined by the visualization of quantitative results.

3.4.1.2 The Area of Corporate Culture

As a result of the audit, further development of the company's role model was undertaken. At first sight, the company demonstrated an open-minded and friendly culture with many characteristic features (open doors, all colleagues and superiors being on first-name terms with each other, relaxed dress regulations), and statements from the employees made in the interviews, underlined this impression.

At the same time the employees' fears regarding the internal change could be diagnosed (original quote from an employee: "In the future, it will be required to implement standardized and formalized procedures."). Since many new employees were hired, the employees feared that personal, informal contact could be lost. The open communication seemed to suffer slightly from the increasing number of employees.

Obviously, there was a dilemma between the wish to maintain the open-mindedness on the one hand, and, the necessity to control growth and to agree on formal regulations or "rules of the game" accordingly on the other hand.

The audit revealed that the open-mindedness resulted only from helpfulness. Open-mindedness towards changes, and anything new was not implied by it. The culture was seen as "ingrained" and "inflexible" rather than "adaptable" or "flexible".

The solution to this dilemma consisted of several steps. First of all, the idea that the demand for rules and formalization does not necessarily conflict with an "open-minded" culture, had to be imparted. A framework or structure constructed from rules and standards ensures the maintaining of an open-minded culture. With steady growth and continuous development, the open-mindedness of the employees is a requirement for the transparency and possibilities of orientation to be maintained in a large company. Therefore, the employees' open-mindedness is important for potential future changes, however, it must be channeled in a common direction.

Therefore, it was decided to extend the company's existing motto ("We are open-minded!") to "We are open-minded and keep to common rules" or "We are open-minded and promote the exchange of experiences"). A company-wide vision thought to ensure a communication capable of promoting knowledge management in the long term, was concisely developed by means of a workshop.

3.4.1.3 The Area of Process Organization

In the area of the core processes of knowledge management, the results from the audit also led to concrete internal projects. The interviews showed that knowledge gathered from experience was regarded as highly important among the employees to successfully carry out their tasks (more than half of the employees stated that they needed knowledge gathered from experience for more than 60% of their tasks – Fig. 3.9).

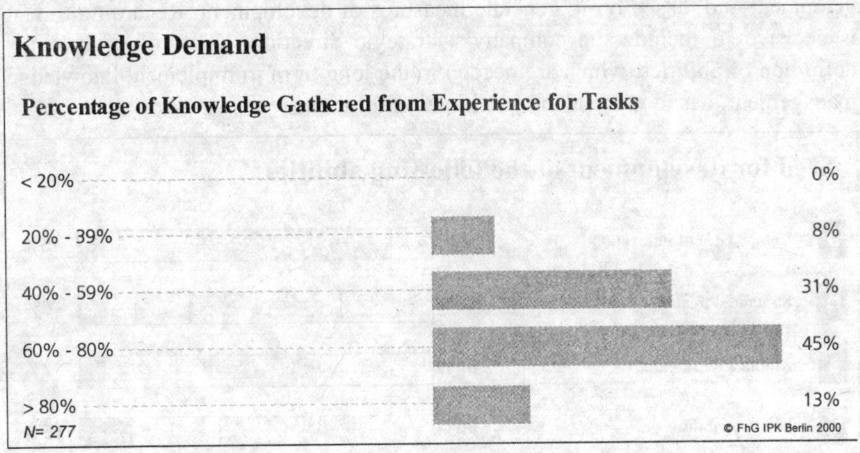

Fig. 3.9: Percentage of Knowledge Gathered from Experience Needed for the Successful Handling of a Task

Since a central issue was that the client makes available the experience of employees gained from dealing with customers (above all for the many newcomers), a systematic project debriefing was introduced. It is obligatory to take place after finishing a project in order to make the relevant knowledge, gathered from the experience of those involved in the project, useable to all interested persons.

There is one last example that illustrates the internal best practice transfer, which can be promoted by means of an audit. In one of the interviews, a programmer presented to us an extensive database, which she had taken care of independently over the years. The data base contained questions and answers for complicated problems of programming. Answers to tricky questions could be found quickly with a simple search request. This best practice uncovered by the audit can be very helpful for other programmers: double efforts can be avoided and the time needed to look for solutions can be reduced.

3.4.1.4 The Area of Human Resources Management

In a successful realization process of knowledge management, the skills required from the employees are increasingly discussed. Besides the framework conditions, such as an open-minded culture or support software, the employee also represents a critical factor of success in the system of knowledge management. Different studies demonstrate that certain competences are necessary for the implementation of knowledge management[135]. The Knowledge Management Audit considers the aspect of abilities in which, via a self-assessment of the employees, the need for development of the relevant competences is being ascertained. From this result, the Human Resource Department can derive the employees' actual need for development and, based on it, generate measures of development. Regarding this, it is decisive to include the company's strategic direction. The strategy-oriented definition of abilities, which are needed in the long term to implement knowledge management, has to precede the qualification for them.

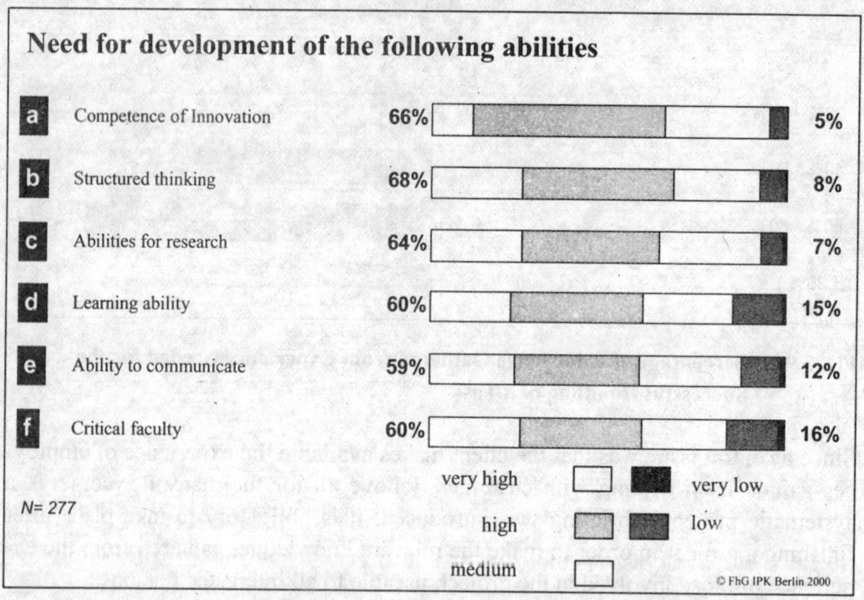

Fig. 3.10: Examples for Abilities Relevant for Knowledge Management

135 Cf. e.g. TFPL Ltd. (1999), Reinmann-Rothmeier et al. (1998).

3.4.2 Knowledge Management in a Research Institute

The second case study concerned a research center with 120 employees. Before the Fraunhofer IPK was asked to do a knowledge management audit, the research center had not explicitly introduced knowledge management. The audit was intended to determine to what extent knowledge management activities were already available and what further actions were required.

The methodical procedure in this case was nearly identical with the previously described case. The return rate of questionnaires in this case was 66%.

The client chose the process of project acquisition and realization as the process to be examined in the interviews, since a need for improvement through knowledge management was dealt with by this process.

Like in the first case, employees and managers stressed the open-minded corporate culture. The great teamwork within departments, seemed to confirm this, and took place mainly on informal terms. The climate between the colleagues was described as "very good" by most of them. However, the lack of formal storage and distribution of knowledge was seen as a problem. Furthermore, there was almost no inter-departmental exchange of knowledge, although it would have been more practical for a more efficient handling of the research assignments.

The analysis of the audit showed that the good cooperation within the departments was accompanied by an open-minded attitude that encourages learning and tolerates mistakes. The strength of teamwork lay in a strong orientation on the informal exchange of knowledge. This way of working, however, brought about difficulties in making knowledge explicit. Some employees did not even have the insight to document their knowledge ("We know what we know."). This coincides with the results from the interviews, which showed that above 60% of the relevant knowledge was stored individually and not distributed in a formal way. Methodical knowledge, expertise, knowledge about products, knowledge about customers and projects were considered the most important contents of knowledge. While methodical knowledge was made explicit comparably well, clear deficits showed up in the areas of knowledge about products and customers. Knowledge about products and customers is, however, both generated and needed on a supra-departmental level. The obvious hypothesis, that the departments considered each other as competitors, could be affirmed through the knowledge management audit. The single departments were independently responsible for their budget and therefore feared to lose assignments or product lines to other departments.

Probably the most important outcome of the audit was that knowledge management, which was considered to be important by the vast majority, could not develop without any additional actions taking into account the framework conditions.

The measures derived from the results took into consideration the need for coordinated integration of knowledge management activities in the business process as well as the identified barriers. The introduction of measures of knowledge management was planned in active cooperation between all departments and all employees. In a workshop, it was defined along the processes, where which knowledge had to be shared, with the internal situation of the competition taken into consideration. In this connection, the rules, which promoted the active cooperation between the departments and sanctioned the disrespect against them accordingly, were defined. In addition, general standards as to how knowledge should be stored in the future were created. A system of incentives for making knowledge explicit was also introduced. Well-documented knowledge ensures, besides the reduced time of research, that new employees are familiarized quicker with their job and also that knowledge is not lost when others leave the company.

These examples show that the result of the audit can initiate the realization of concrete projects, which support and improve the current handling of knowledge. By this, companies are able to include existing activities of knowledge management in a well-aimed way. The roadmap supports the management in making decisions as to how knowledge management should be handled. By demonstrating weaknesses and potentials, the audit gives support to a long-term development of companies.

Moreover, it is very important to consider the economic aspect of audits. It is very important for a company that the audit is processed quickly. It was however possible, to achieve results in a short time and without too much effort.

3.5 Outlook

In a short period of time, the FKM-Audit can provide companies and public administration with a comprehensive assessment of the current handling of knowledge in the selected organizational unit or process. It provides information about existing strengths (internal good practices), potential areas for improvement and the barriers that have to be overcome when measures are taken. In the roadmap, not only short term measures ("quick wins"), but also medium term (up to one year) and long term measures (one to three years) for the implementation of knowledge management are presented.

Above that, concrete alternatives are demonstrated for the analyzed areas. These alternatives comprise aspects concerning tools, methods, qualifications and motivations[136] as well as suggestions for measuring and controlling knowledge management. In this connection, the customers' situation and the market situation of knowledge management are taken into consideration accordingly.

---

[136] Cf. Vorbeck, Finke (2001).

The knowledge about best practices resulting from the first German Benchmarking Study of Knowledge Management and the experience from project work ensure the client that the audit is realized and evaluated on a high level of quality.

4 Motivation for Knowledge Management

Ina Finke, Markus Will

A significant number of companies have by now recognized the potential of improvement that lies in a conscious dealing with knowledge. According to this notion, various projects involving knowledge management have been initiated during the last years, most of them starting with solutions based on information technology designed to support business processes as well as an efficient exchange of knowledge. The implementation of a special technology appears to be a simple and quick solution for pressing problems, e.g. if relevant knowledge is not available for the employees, or if important experiences once made are not being utilized for follow-up projects. Soon, however, many enthusiasts will discover that, by introducing a solution focused on information technology, not all problems can be dealt with in a satisfactory way. Furthermore, new problems occur through the introduction of the IT system itself, e.g. if employees refuse to work with it. It is at this point at the latest that the matter of the employees‘ motivation becomes relevant.

Besides the use of IT systems dealing rather with information than with knowledge[137], companies become increasingly aware of the value of tacit knowledge, e.g. knowledge resulting from experience. With regard to this, there were often demands for a culture of knowledge sharing which could not be achieved by purely technical means. Face-to-face communication is relevant in this case; and the motivation for it, in the nature of things, has to come from inside a person and cannot be established by external factors alone. Especially the exchange of tacit knowledge is dependent on the intrinsic motivation of the employees.[138]

Several questions arise with regard to this: How does knowledge management have to be introduced into a company in order to ensure its success in daily routine? How can the best possible acceptance and motivation on part of the employees be achieved? The following section will establish theoretical fundamentals for our approach to the promotion of motivation in knowledge management.

---

137 Concerning the distinction between data, information and knowledge, cf. Rehäuser/ Krczmer (1996) and chapter 1.

138 Cf. Frey/ Osterloh (2002).

4.1 Motivation and Change towards Knowledge Management

4.1.1 Incentive Systems, Extrinsic and Intrinsic Motivation

Initially, measures of motivation for *knowledge management activities*[139] were based on the development and implementation of classical incentive systems. Incentives both activate motives (= the readiness to behave in a certain way) and have the character of an invitation or challenge, for they bring the employee to carry out certain activities in the intensity desired by the company.[140] One distinguishes between material and immaterial incentives. Monetary gratification, representing the reward for the employees' performance, belongs to the group of material incentives, while immaterial incentives lie in the conditions of performance, such as areas of responsibility, structure of decision, career prospects. The meaning of immaterial incentives steadily increases, since material incentives proved to have only a short-term effect on the employees' motivation. It can be observed – provided that companies concern themselves with measures of motivation for knowledge management – that new, separate systems of gratification were connected with knowledge activities (e.g. the awarding of *Knowledge Miles* or *Knowledge Coins*).[141] Their success, however, has not been satisfactory up to now, as sustainable motivation on part of the employees could not be evoked.

The two central terms of *intrinsic and extrinsic motivation* are also relevant in order to put systems of incentives into practice within a company. There has been much research on the field of motivation.[142] Generally, motivation is defined as the employee's readiness to understand, to perform or to show a certain kind of behaviour. If the result of this behaviour is considered valuable by the acting person, and if, at the same time, the person reckons that the result can actually be achieved by his or her action, motivation is evoked.[143] This approach, which is generally assigned to process-oriented theories,[144] proclaims the usefulness of an activity to be the principal motivation for it. It therefore focuses on the extrinsic aspect, given the fact that it strives to explain the direction of an activity rather

---

139 Knowledge Management activities (KM activities, knowledge activities) include all operations within the business process referring to the four core activities: generation, storage, distribution and application of knowledge.

140 Cf. von Rosenstiel (1999).

141 Frank et al. (2002).

142 For an overview of motivation research cf. Atkinson (1988); Lühker, Vaanholt (1994).

143 This interdependence is dealt with by psychology, especially in terms of "Erwartungs-mal-Wert Theorien"/ anticipation-value theory. This model has been further developed by Heckhausen (1989) and others. This article will not go into depth concerning the distinction between theories focusing on processes or contents.

144 Gebert, von Rosenstiel (1996).

than the nature of it. Intrinsic factors, claiming that the incentive to carry out an action is to be found in the action itself[145], are not considered.

An external incentive for desired behaviour is characteristic of extrinsic motivation. Extrinsic motivation overweighs when employees are only willing to share their knowledge with others if they are being offered a (material) reward. Intrinsic motivation, in turn, emerges from the task itself and presupposes internal incentives for the given activity. Thus, behaviour does not occur for the sake of gratification, but the activity in itself is satisfying.[146]

Intrinsic motivation, however, can be undermined by external gratification. This phenomenon is called *repression-*, or *spillover effect*[147]. Assuming an activity, e.g. the documentation of lessons learned, is carried out for the sake of the activity itself, the introduction of external intensifiers to influence this action can lead to the result that the activity subsequently does not occur unless tied to a reward. This has far-reaching consequences with regard to the employment and the organization of incentive systems in practice. Knowledge management activities with a potential intrinsic motivation are only shown when related to some kind of gratification. If, in turn, gratification cannot be expected (any more), the according activity is not carried out.

External (material) incentives for special activities, such as the documentation of project-bound experiences, imply the notion of "additional work". These basic results were also produced by a recent study of the *Institute for e-management*[148]. It supposes that, this being contra productive to the integration of knowledge management into internal processes, activities involving knowledge management should not stand out or be especially emphasized. This approach is referred to as *thesis of normalization*[149], assuming that no special status should be given to activities involving knowledge management and that they should be assigned the status of any other task or activity. "If activities involving knowledge management are accepted as being average or normal activities, they can be motivated by established strategies. Subsequently, they will lead to average success; they will not be more successful than normal activities, but not less successful either. Thus, it is not new systems of incentives being at stake, but the integration of activities into daily routine in a way that they can be activated with means already established in the company and ready for employment."[150]

---

145 Such as the simple enjoyment of work or the so-called flow-experience. Cf. Csikszentmihaly, Aebli (1999).

146 The shaping of the term intrinsic motivation goes back to Herzberg, 1968.

147 Cf. Frey (1997) and Frey, Osterloh (1997).

148 Frank et al. (2002).

149 Cf. Frank et al. (2002).

150 Frank et al. (2002).

The management of intrinsic motivation is more demanding than that of extrinsic motivation. Strictly speaking, intrinsic motivation cannot be actively engendered; one can merely create the right conditions for its development. A substantial prerequisite is the design of tasks in order to evoke intrinsic motivation.

Concerning the normalization of activities involving knowledge management, the aspect of *task orientation* is given a crucial role. The starting point of an integration of new activities *into* the process has to be the employees' tasks *within* the process. The term *task orientation* describes a state of interest evoked through special features of the task. "There is activity growing out of interest in the task itself, in the problems and challenges it offers. The task guides the person, steers his action, and becomes the center of concern."[151] To understand knowledge management as a working task means to incorporate it into daily routine without significantly adding to the employees' workload. In order to promote intrinsic motivation with the help of task orientation, the following task design criteria are proposed: completeness (including a feedback after a task has been carried out); the diversity of demands; the possibility of social interaction; autonomy; possibilities of learning and development; stress-free adjustability of the task; and meaningfulness.[152] For the application of these principles to the integration of KM activities see section 4.2.6.

4.1.2 Barriers of Change and Change Management

The design and new organization of tasks and processes usually bring about immediate changes regarding the employees' working situation. Mental barriers on part of the employees as well as hindering factors in the their work surroundings make it difficult for them to accept innovations. Thus, the introduction of knowledge management requires *change management* as well. "All measures concerned with the initiation and the realization of new strategies, structures, systems, and modes of behaviour, are subsumed under the term *change management*."[153] This definition shows that change management is a tool for the transfer of innovations, irrespective of their nature, into action.

Change management measures aim at overcoming barriers and blockades hindering the employees. *Barriers*, in our terminology, are defined as obstructive factors in the successful realization of knowledge management in a company. Barriers can be the expression of both a lack of motivation as well as a lack of competence. They can be sorted into three different levels:[154]

151 Cf. Asch (1987).
152 Ulich (1998).
153 Cf. Gattermeyer, Ayad (2001).
154 Cf. Willke (1998).

- individual: e.g. fear of showing oneself up, lack of information, conscious withholding of knowledge, monopolies of knowledge, mental barriers;
- organizational/cultural: e.g. no cooperation between teams, levels of hierarchy, the lack of structured error analysis and avoidance;
- technical: e.g. no IT support, poor quality of data, lacking usability of systems.

Employees are not necessarily aware of the barriers of change, which can also be understood as mental models[155] (such as patterns of behaviour, value profiles, expectations). Mental models do, however, form a number of preconceived ideas persistently and inevitably influencing the employee's judgment of situations, his decisions and his actions. These latent structures can be seen as behavioural predispositions either in a pro-change or a contra-change direction (drivers or barriers of change). A representative analysis of the barriers connected with specific changes in an organization (in this case the introduction of knowledge management) is necessary for a well-aimed definition of measures. The results of the analysis are a starting point for actual change management measures aiming at long-term success, i.e. sustainable motivation. These measures include the dismantling of mental barriers as well as barriers within the work surroundings of the employee. The implications of the thesis of normalization being considered, the barriers impeding the desired behavioural change should be overcome or reduced by way of comprehensive measures of motivation included in the general framework of an organization in a way that allows intrinsic motivation to be evoked. With external incentives alone, a sustainable behavioural change cannot be achieved.

In order to promote intrinsic motivation, there are several factors of acceptance[156] that have to be taken into account in processes of change. If these factors are thoroughly considered in the course of organizing *change management* measures, employees with different motivations can be brought to accept changes. Different types of reactions on part of the employees can be identified as possible ways of behaviour regarding the change objectives, such as spontaneous acceptance or resistance. Different expressions of acceptance with regard to the given change are due to the fact that, depending on the strategy of implementation a company chooses, the four factors of acceptance are activated with different intensity:

- *Sensibility*. The concerned employees have to be thoroughly informed about the objectives and methods of knowledge management, about change measures and decisions, and the implications for their individual performance and working situation.

---

155 Cf. Brauner (1994).

156 Cf. Grewe (2000).

- *Ability*. The concerned employees must be competent enough to cope with the demands of change and fit the requirements of new KM activities.
- *Liability*. The executive personnel must initiate corresponding measures of steering and control, and they must take the function of a *change agent*.
- *Commitment*. The concerned staff has to be involved in the change process with regard to the given conditions.

If – in the course of an introduction of knowledge management – these factors are considered comprehensively, the maximum motivation on part of the persons involved can be achieved.[157] Existing factors of acceptance in a pro-change direction can be utilized as "*drivers of change*". For instance an employee's existing interest in a topic connected with the change objectives or certain abilities already established that fit the new requirements can function as a good starting point for the desired behavioural change on a broader basis. Basically, there are three approaches regarding the change of the employees' way of thinking and acting, with an emphasis on the interdependence of (short-term) *behavioural change* and (long-term) *change of attitude:*[158]

- behavioural change due to incentives,
- change of attitude due to direct and personal experience,
- change of attitude due to socially transmitted experience.

Incentives generally have but a short-term effect. They do not bring about the change of attitude, which is a necessary condition for the long-term internalization of new ways of behaviour[159]. "It must be noted, however, that incentives cause a behavioural change in the first place, which is not automatically accompanied by the desired change of attitude."[160] A change of attitude generally happens on account of personal and positive experience concerning the new way of behaviour.

---

157 Cf. Grewe (2000).

158 Cf. Stroebe (1990).

159 If a person adopts new modes of behaviour (such as the use of new knowledge management tools), there is a discrepancy between behaviour and his previous attitude primarily. Festinger's theory of dissonance (1978) supposes that the individual strives for a solution or reduction of these discrepancies. This human phenomenon is called *strive for consistency* as well. Basically, there are two possibilities: Either a relapse into old modes of behaviour (at the latest when the actual or subjectively attributed value of incentives decreases), or an adaptation of attitude according to the new behaviour.

160 Cf. Mehrwald (1999), p.228.

A substantial means for the achievement of a change of attitude is *participation*[161], providing the possibility of experiencing processes both directly and personally as well as via social transmission.[162] Each person involved, thus, is given the possibility to actively take part in the introduction of knowledge management. This means that, within measures aiming at participation like workshops, introductory events etc., the possibility for personal experience and exchange is provided. Interaction often dismantles fears as well as barriers of reservation concerning the given innovations, for inside the groups, employees work at solutions together.

The following diagram shows the relation between short-term behavioural change due to incentives (extrinsic motivation), whose intensity decreases in time, and intrinsic, sustainable motivation for change, which must be promoted by supporting measures.

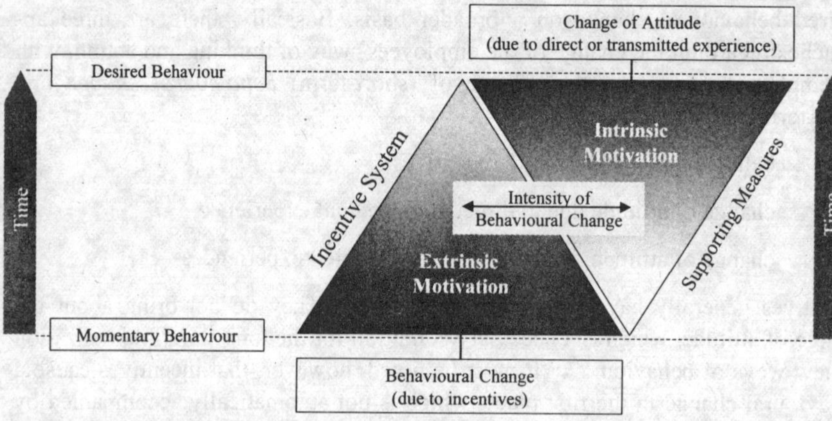

Fig. 4.1: The Relation between Short-Term Behavioural Change and Long-Term Change of Attitude

4.1.3 The Role of Competencies

The barrier of competence is said to be a special case in knowledge management. While introducing knowledge management, it is necessary to refer to the basic competencies in dealing with knowledge. The importance of these competencies is

---

161 Frey, Osterloh (2002), p.221, are talking about the individual participation of each employee regarding the design of his work surrounding in contrast to collective or indirect participation, i.e. worker participation granted by the industrial-relations scheme law.

162 I.e. if a colleague, who is accepted as trustful or as an opinion leader, describes his own experience with a new KM-tool to his fellow workmates.

frequently under-estimated, and failures in the course of the introduction are being reduced to a lack of motivation on part of the employees. Both employees and executives have to be made sensitive for and able to cope with these barriers. The employees' insecurity and their fear of being thought of as incompetent causes reservation and a critical attitude towards all kinds of innovation (mental barriers). Effective dealing with knowledge requires competencies that can hardly be taught explicitly in traditional forms of training. Several studies[163] have concerned themselves with the relevant competencies. According to Mandl (1998), knowledge management is a form of meta-competence including the ability of cooperation and self-management in the first place. By speaking of cooperation, Mandl refers to communication as well as to conflict management and the ability of taking different perspectives on a problem, cooperating and achieving joint decisions. Self-management means for the employee to take his own initiative, to plan independently, to organize, coordinate and to be able to motivate himself.

"Competencies are conditions which are helpful in situations of uncertainty in order to find new strategies of solution in a creative and organized way (competence as a disposition of auto-organization)".[164] Meta-competencies also include the ability of self-reflection with regard to the own way of acting and dealing with knowledge as well as the ability of discussing one's individual mode of communication with others and optimizing it. Meta-competencies are the basis for a continuous improvement of processes and for a culture of knowledge and learning; therefore, they are the basis of the learning organization. Meta-competence can be divided into actual basic competencies of knowledge management,[165] such as knowing different strategies of searching for relevant information, and the ability both to develop them and to ask the right questions. Other basic KM competencies are the ability of structuring one's own knowledge and experience as well as of structuring other information. Furthermore, fluency and concision in both spoken and written (electronic) communication as well as the presentation of information – according to the target group's needs – are important. According to American and German studies on the subject, the exchange of information and knowledge between ‚knowledge workers' happens via verbal communication in fifty to ninety-five per cent of the time spent with communicating.[166] Last of all, a person must be able to take in external knowledge, to understand, accept and apply it. In the near future, these abilities will – apart from central key qualifications such as social, professional, and methodological competence – play a crucial role in the world of employment.

---

163 Reimann-Rothmeier, Mandl (1998).

164 Reimann-Rothmeier, Mandl (1998).

165 Cf. Mertins, Heisig, Vorbeck (2001).

166 Bair (1998), p. 2.

4.1.4 Summary

Approaches for the implementation of knowledge management focusing too much on technical aspects tend to fail in practice due to their lack of consideration of human factors such as motivation and competencies. Taking measures of change management, in turn, the KM project manager can achieve a high degree of acceptance on part of the employees due to participation as well as task orientation. If the process of change is accompanied by personal experience on part of each person involved, the practical realization is likely to be successful. A balanced organization of the factors of acceptance can contribute to the development of intrinsic motivation.

Therefore one must bear in mind that changes of behaviour should not be put down to incentives exclusively. The general organizational framework has to be just as motivating. Each measure has to be designed and organized in a way that each person involved may recognize the importance and use of new ways of behaviour. Only then will employees, on account of intrinsic motivation together with their newly acquired attitude, show the desired modes of behaviour in the long run. The design of tasks and processes (according to the mentioned task design criteria) as well as the employees' active participation in the process of introducing knowledge management can thus cause motivation for a long-lasting change of the previous way of dealing with knowledge, i.e. a sustainable change of behaviour.

4.2 Pro-Active Change Management for Sustainable Behavioural Change

Business process oriented knowledge management[167] is concerned with the requirements of a comprehensive KM system deduced from knowledge intensive business processes of a given company. This section will describe a concept for the successful introduction of such a system, namely the IPK model of intervention for activating sustainable motivation. This model will serve as a basis for change management measures concentrating on the employees.

It has to be noted that this article emphasizes on *how* this introduction takes place from a human behaviour's and a motivational perspective, to achieve the objectives of the given knowledge management project. The objectives proper as well as the tools for achieving them have already been defined in the process-oriented knowledge management approach. The concept of change management proposed here starts with predetermined objectives, which have to be put into practice; but the ways of doing so can differ a great deal comparing the individual cases. KM methods based on face-to-face communication bear a great inherent potential for

167 Cf. chapter 2.

evoking intrinsic motivation (building trust through social interaction) while the introduction of IT-tools into daily routine is more challenging in terms of achieving the employees' commitment. It is necessary to find out about reasonable starting points in order to evoke developments in the employees' minds, apart from changes in the area of technical structures.

While the concept of *participation* concentrates on the employee, enabling him to contribute and re-assess his personal motives by way of including him in decisions and organization of processes, *task orientation* focuses on the change of tasks themselves as well as of areas of responsibility, thus attempting to promote intrinsic motivation.[168] These two approaches are fundamentals of the concept described here. The concept aims at the systematic realization of KM projects – from initial considerations to its integration into the processes of daily routine itself, i.e. its normalization. This form of change management is pro-active, for it analyzes and thus anticipates barriers and overcomes them by way of well-aimed measures.

4.2.1 Analysis of Corporate Culture and Barriers

Each employee is influenced by the corporate culture with a stress on the culture of communication and cooperation in his work surroundings, by the present structure of his personal tasks and by the IT systems employed in the company. This kind of organizational framework has moulded and influenced the employees' mode of behaviour for a certain time, as well as their mental models and their attitudes. Those, however, are the drivers or barriers of change (cf. section 4.1.2). Pro-active change management anticipates these obstacles before they can turn into actual resistance.

The analysis of corporate culture and barriers aims at the description and analysis of psychological and organizational variables intervening in a process. Factors such as experiences/problems with the IT systems employed, the given structure of tasks and communication or the leadership are relevant with regard to this matter. In order to reveal these latent structures (predispositions of behaviour), barrier analysis relies on a combination of research methods: Firstly, hermeneutical talks and observations as well as partly structured interviews (qualitative analysis); secondly, the use of structured questionnaires (quantitative analysis). This research concept can be flexibly included in the business process analysis or the KM Audit as an analysis module, the result of which is a survey of existing barriers on the levels of people, business processes and IT. The survey is completed by a description of causes and interrelations of barriers in the given case as well as proposals for their defeat which, in turn, include the ascertained drivers of change.

---

168 Cf. Kleinbeck (1996), pp. 97, for a definition of the two ways of promoting working motivation: employee orientation and task orientation.

4.2.2 IPK Model of Intervention for Activating Sustainable Motivation

A next step deduces measures from the ascertained barriers and drivers, the measures concentrating on the four *factors of motivation*[169], namely *Sensibility*, *Ability*, *Liability* and *Commitment*. In order to define well-aimed measures, these four factors are linked with four potential areas of intervention accordingly. Combined with the aspects of the analysis of corporate culture and barriers described above, this is illustrated by Fig. 4.2.

In the course of implementing knowledge management into business practice, the interlocking of the areas of intervention has to be the best possible. These measures start with the comprehensive management of communication (transparency). The next stages are measures of qualification (competencies), the adaptation of managerial systems (leadership), and the integration of knowledge management activities including both task orientation and participation (involvement).

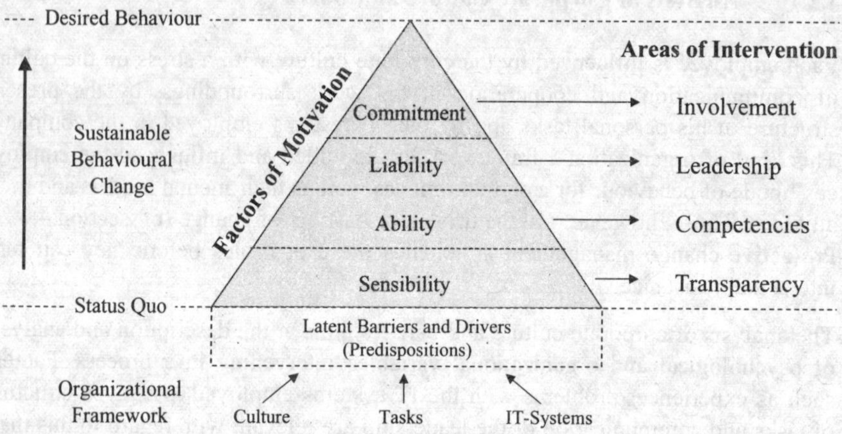

Fig. 4.2: IPK Model of Intervention for Activating Sustainable Motivation

Introductory projects include measures involving all four areas of intervention; they can be subsumed in a change program affecting the whole company, as shown in section 4.3. The introduction of knowledge management can, however, be restricted to individual measures in different contexts, e.g. in small companies, or with regard to special teams or aspects of teamwork only. The areas of intervention would then have a rather modular function. Several given aspects have to be taken into account when defining measures within the areas of intervention,

---

169 Referring to the factors of acceptance mentioned in section 4.1.2, the terminology of the IPK model of intervention will use the term 'factors of motivation', on account of the fact that an employment of the areas of intervention aims at the activation of sustainable motivation

such as the size of the KM project, the extent of change as well as the character of the innovations to be introduced, but, more important still, the ascertained organizational framework conditioning the measures. The results of the analysis of corporate culture and barriers provide those pieces of information that should always serve as the starting point for well-aimed change management measures.

4.2.3 "Transparency" as an Area of Intervention

This area of intervention focuses on the fact that the employees have to be comprehensively and plausibly informed about the introduction of knowledge management. *Sensibility* (perception, to sense, to be aware of), being the factor of motivation served here, is understood to be a basis for all further steps of behavioural change. If an employee does principally grasp – and agree with – the usefulness of the intended innovations, if he does understand how the process of implementation is going to take place, and how upcoming problems and questions can be coped with, he can be expected to support the process of change, or at least not to hamper it in whatever way.

Multi-channeled communication management must both consider top-down processes as well as bottom-up processes and combine them in a reasonable way. First of all, the employees' attention has to be aroused via appropriate, push tools of communication,[170] such as circulars from the management, the company newsletter, e-mail, informational programs, brochures etc. During the course of the project, the achievement of milestones has to be communicated to underline continuity and importance of the project. The *push* tools have to be combined with *pull* tools of communication – such as special intranet services, hotlines, human multipliers of knowledge management (such as key users of IT systems), guidebooks etc. – which enable employees to actively retrieve adequate information about details and backgrounds of the project. Depending on the significance and dimension of the given knowledge management project, the necessity of a professional internal marketing campaign can occur. In order to support the implementation and normalization of KM activities in terms of communication, the communicated objectives of the project should be connected with relevant and up-to-date contents of the respective business strategy.

Persistency and consistency are the main criteria with regard to top-down communication. Recurring basic messages are taken up continuously and integrated into the respective measures of communication. The contents of these official messages have to be consistent beyond all levels of hierarchy. The IPK Benchmarking

---

170 "Push" tools of communication are characterized by an active distribution of messages from a sender to a target group while "pull" tools provide certain information in specific storage devices (encyclopedias, databases, KM retrieval systems etc.), which the recipient can use to actively pull out the needed information just in the moment and context of actual demand.

Study has shown that this kind of support from the top management is important; 33,7% of the companies interviewed claimed that top management promotion and a clear definition of objectives were an important aspect with regard to the project's success.[171] However, one has to bear in mind the possibility of unintended, dysfunctional effects of top management information: Do the actions and decisions of management match with the official proclamations?[172] This fact influences the credibility of the management, and it moulds the reception and interpretation of further management information to come. The analysis of culture and barriers provides substantial indications for this; the ascertained climate in the operative areas leads to conclusions as to whether the messages from above ,fall on fertile grounds'. Contra productive effects, thus, can be anticipated and avoided through professional advisory services in the area of communication.

Furthermore, management creates a methodological and technical framework supporting the dialogue within the organization. At this point, bottom up communication becomes relevant, desirably by way of establishing feedback channels providing direct contact between the employees and the management in case of questions, suggestions or complaints. The anonymity of electronic channels can be used to overcome inhibitions. The milestone messages concerning the project can refer to or be linked with these possibilities of feedback or participation, such as hotlines or the e-mail addresses of responsible persons etc. In the IPK study, open communication was claimed to be an important aspect of managerial systems as well as of the companies' corporate culture.[173] Additionally, one must not neglect strategies of controlling informal communication processes. Rumours can be made transparent, and opinion leaders can be involved as multipliers of knowledge management at an early stage.

4.2.4 "Competencies" as an Area of Intervention

Without competencies, a sustainable motivation of employees to change their behaviour cannot be achieved. This area of intervention aims at the promotion of competencies that are needed for new demands. Learning processes must be well calculated to accompany the process of change. Learning methods, which are included directly in the working process, are known to be more profitable than other traditional forms of job training. Therefore, processes of change also require specific strategies of human resources development. In this context, on-the-job measures are of high importance in cases where rooms for experiments are built and procedures of experience-based learning are triggered, for example through project teams. The above mentioned meta-competencies of self-management and co-

---

171 Cf. chapter 1.

172 In worst case, top-down communication evokes reactions diametrically opposed to the desired effect.

173 Cf. chapter 1.

operation can be developed in the best possible way within such learning-and-working arrangements.

In order to activate competencies, it is of decisive importance that learning and working processes are combined as intensively as possible. The following aspects should be borne in mind for the designing of process-oriented training measures:

- authentic arrangements for learning:
  high compliance with the demands of the work surroundings,
- orientation on practical relevance:
  room for activities, variability and decisions,
- meta-cognition: reflection of one's own way of thinking and learning:
  cooperative learning makes possible the reflection needed to internalize competencies,
- feedback intensity, self- evaluation and task evaluation by others:
  achieved through task design (complete tasks with the possibility of achieving results independently) or supporting measures such as coaching.

If these aspects are being followed in the process of designing measures for exercising or professional development, a long-term success in the development of competencies can be attained. A development, which is oriented more towards the real working process, serves as an important prerequisite for the development of personal and organizational creativity. At the same time it calls for the existence or implementation of a suitable organizational framework within the company, such as the promotion of scopes of action at work and supervision and coaching measures carried out by human resource management experts or specialized trainers. The needed KM competencies are to be promoted in connection with the concrete task. Besides punctual (on-the-job) training measures, the continuous support in cases of problems or questions must be ensured. The users of IT systems, for example, should be given support by key users – a role assigned to an operative employee in each involved department – who have been specially trained on the IT system and are able to link the application of the IT tools to the actual task requirements of his colleagues better than a centralized technical hotline. Those and other multipliers are crucial for successful communication and qualification.

Organizational learning is promoted by individual learning and the development of competencies in the working process. Thus, it is very important to design and to promote adequate learning procedures. However, this requires every employee to be willing to learn, which is determined by the dominant culture and climate. To summarize briefly, every measure for the development of competencies should focus on the relevant basic competencies for the effective dealing with knowledge within business processes so that a further prerequisite for the acceptance of changes is provided.

4.2.5 "Leadership" as an Area of Intervention

This area of intervention deals with liabilities, i.e. duties to be fulfilled and criteria to be met by the employees. Members of an organization can be obliged to accept certain predetermined objectives for behavioural change that have to be clearly set by the top management. Superiors have various means and ways at hand to get changes accepted by acting as a leader – or in this context as a "change agent".

As many studies prove,[174] middle management is in a key position in procedures of change. The head of department must be able to apply to his team the formal requirements of change given by the top management. Furthermore, he is to support the change process as a personal coach of his subordinates.

Formal managerial systems, like *agreements on business objectives* or *employee evaluations*, are to be adjusted by an integration of knowledge management objectives. In every relevant evaluation session, for instance, it should at least be questioned how much knowledge was transferred to colleagues and how much knowledge of others was applied within the own working context, and the result should be included in the evaluation of performance in general. In order to meet the demand of normalization, recourse to existing structures should also take place when systems for giving incentives are designed. Additional efforts, as those made by key users, should be acknowledged and rewarded accordingly. The rewards do not necessarily have to be of financial or other material nature. Social recognition and career opportunities (increase of power and/or reputation), based on clearly defined achievements, can in many cases be more effective. Agreements on business objectives may concern the performance of managers, experts or teams. In agreements on business objectives made with the heads of department, general KM objectives – such as cooperation with other departments, the degree of knowledge reuse (and the increase of efficiency that results from it), or the scope of accessible information which is strategically useful for the whole company or certain business areas – can be laid down for each team. It can be agreed with project managers that the use of the results of precedent projects and the documentation of own project results count for the full achievement of all objectives.

Working closely together with the co-workers as a basis for trust, the superior or head of department has the possibility to give decisive impulses for the necessary changes. Ideally, he does this through consequent empowerment and gives every employee – by means of a general framework that supports changes – room for personal development. Knowledge of informal structures enables him to react to

---

174 The superior or manager of department fulfills the function of an interface between the formal organization and the natural social system. He is a representative of the management and, as well, included in the informal communication process of his team. By this, he ensures an important transfer in both directions. On the key role of the intermediate management see also the "middle-up-down" concept of Nonaka, Takeuchi (1995) and Berner (1999).

barriers of change in bottom-up direction, e.g. by putting those in favour of changes into the position of a key user or an ‚internal change agent'. This can happen through giving social incentives, such as the recognition of innovation. The communication tasks of encouraging new, desired behaviour and convincing its opponents with consistent and rational arguments complete the scope of tasks of the "change agent". Information developed precisely for executive personnel, such as the handbook "How to manage KM barriers",[175] can help him to see potential barriers and give him recommendations as to how sustainable changes of behaviour can be achieved.

4.2.6 "Involvement" as an Area of Intervention

As described in the first section, a sustainable behavioural change through intrinsic motivation can only be attained if employees have the chance to make good experiences or to profit from the experiences made by colleagues, which were socially transmitted to them, so that their attitude can adjust to the new behaviour. At this point, involvement comes into play. The necessary acceptance and personal commitment can only be expected from an employee who is involved actively and in "bottom-up" direction in the process of change, who concerns himself with project goals and the requirements of knowledge management and works on the best solutions together with colleagues. The second aspect of involvement relates to the concept of task orientation that was mentioned in the first section.

Sustainable Motivation through Participation

According to the participatory concept of organization development[176], the starting point for changes in the organization should be the individual employee or a team. Direct employee participation and the possibility of communication on the job increase the perceived self-determination and responsibility of the individual employee, as well as his willingness to cooperate in the team, giving rise to trust and satisfaction.[177] Trust, loyalty, a good work climate and satisfaction are again essential prerequisites for a culture of knowledge sharing.[178] Especially in the case of tacit knowledge, the willingness to share knowledge and apply knowledge of others depends on trustful relations.

It is important for an employee concerned with the realization of KM demands to have a certain extent of freedom to act, which enables him to learn through experiences in connection with a concrete task and to make his own decisions.

---

175 Bachmann et al. (forthcoming 2003).

176 Cf. e.g. French, Bell (1994).

177 Significant relations between these variables have been examined in the Worker Representation and Participation Survey (WRPS) 1994, cf. Frey, Osterloh (2002), pp. 226-234.

178 Smith (2002).

This demand is fulfilled in the best possible way through "bottom-up" elements within the implementation process, because the team has to find the best solution in an autonomous and cooperative way. However, the users of an IT system must also be involved in the process of its design and implementation, even if it takes place as a full-scale measure within the company. Therefore certain requirement details for designing the knowledge management IT system should be developed in cooperation with the future end-users.

Employee participation does not take place in just one event, but in a process which organization development (OD) divides into the following phases:[179]

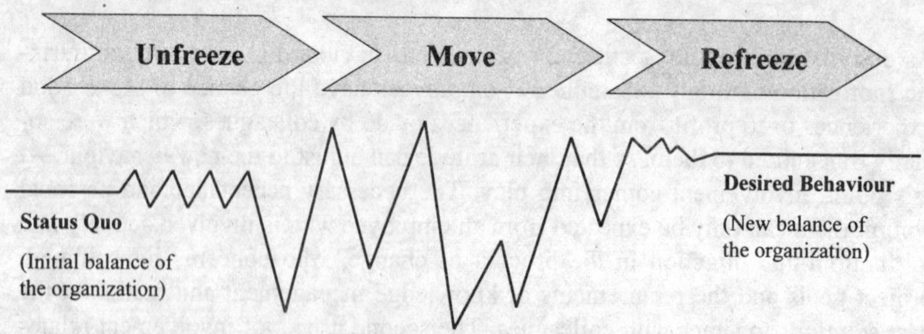

Fig. 4.3: Phases of Organization Development

From the standpoint of communicational psychology, attention and interest for knowledge management must be triggered. This can even be initialized by taking the measures already described in the section about "transparency". Starting from these measures and the strategic KM objectives, the operative teams have to discuss the realization details in relation to their own tasks. In order to optimize the dealing with knowledge, the previous work processes and flow of information must be critically questioned. Scenarios dealing with real or anticipated problems of the organization that must be prevented from being realized or visions that have to be achieved can be an asset. In any case, the interest and the need for changes must be triggered ("unfreeze"). The OD methods at hand are, for instance, "open space" or future-workshops ("Zukunftswerkstatt").[180]

In a next step, the employees must be involved in the actual process of change. According to the OD concept of Kurt Lewin, the group – as a medium for learning and reflection – plays a decisive role in this connection. Apart from that, practical experience is the most important source for the change of behaviour and attitude. The employees should be encouraged to work on solutions together with their colleagues, to test new procedures and methods on the job, and to reflect and im-

179 Here, reference is made to the classical OD model of Kurt Lewin.

180 Concerning the methods of organizational development, cf. Comelli (1985).

prove them continuously in frequently held workshop settings. The strategical demands of knowledge management are to be included in the discussion by a suitable moderator as operational goals and not as instructions. In order to actively prevent burnout symptoms, all ineffective procedures, analyzed by the team, must be eliminated and the remaining tasks be optimized together with the affected colleagues. It is important that the new behaviour is not only a result of an external incentive. Through trying out and reflecting individual and collective behaviour, procedures, which cause this behaviour to be up valued, internalized and, in the end, intrinsically motivated, come into force ("move").

Finally, the optimized procedures including knowledge management activities must be stabilized. The "best practice" methods, which were determined empirically, must be woven into the daily routine. At this point, the phase of normalization starts ("refreeze"). For that purpose, the manager has to integrate the team-specific details of the KM realization into his/her managerial system and adjust his behaviour to the collectively defined KM objectives. Decisions of the team on the operative realization of KM demands have to be retransmitted to those responsible for the project to finalize process documentation and job descriptions. By this, bottom-up implementation was transferred into formal structures and therefore completed.

Sustainable Motivation through Task Orientation

By virtue of a *task-oriented integration of KM activities* (normalization), it is important not only to give incentives for individual KM activities, but also to relate KM activities reasonably to the context of the tasks in question. Depending on the extent to which changes take place – i.e. the scope of new KM activities – the supplementation of existing task structures has to be guided by the criterion of *completeness*, including aspects of planning, executing and controlling in order to evoke intrinsic motivation through task design. This means that KM activities cannot be inserted just where they seem to fit best at a certain moment, but that the team's core assignments must be newly structured and expressed in alternated job descriptions and responsibilities, if necessary. The employee's scope of action must be designed in a way that employees can comprehend the purpose of individual activities in connection with the business objectives of certain areas or the whole organization (*meaningfulness*). In this context, it must be ensured that the employee gets a *feedback* of the success or failure of his performance. This may arise from the task itself (direct visibility of success), but may also be done by the superior, or by colleagues, for instance through the acknowledgment of expertise by other experts or by means of feedback functionalities for used documents and corresponding author rankings in technical KM-systems. "*Autonomy*" as a task design criterion can be fulfilled by giving opportunities for self-regulation and self-management (*stress-free adjustability*) as well as delegating the responsibility for the results to the employee performing the task.

The possibility for *social interaction* is of high importance for KM activities. The optimizing of the flow of information and the culture of cooperation within existing teams can be essential for the intrinsic motivation needed for the realization of KM activities in the operative business – especially when the transfer of tacit knowledge is concerned, which depends on trust and mutual understanding. This can only be achieved through social interaction. As measures for the task-oriented design of KM activities through the extension of social interaction, experts can be granted additional time to participate in a *community of practice* or other supra-department and supra-organization networks can be built up. From a motivational perspective social interaction refers to immaterial incentives, such as the recognition of the employee's expert status and appreciation of his or her special knowledge or competencies. This can be connected with granting individual possibilities of learning and development, such as special training measures on the employee's field of interest or empowering and encouraging the respective person to take over more responsibility, for instance in coaching or mentoring assignments. Moreover, the informal process of communication which takes place e.g. in coffee-corners, should be supported, because the importance of informal exchange of tacit knowledge cannot be underestimated.

All task features together allow for an essential element of behavioural change. Through various aspects, the employees are offered different ways to make experiences (social interaction, employees having to face the consequences of their decisions etc.). Such scopes of action and spaces for development are decisive for learning new behaviour in respect to handling knowledge.

4.2.7 Summary

The starting point of pro-active change management is always the individual in his work surrounding, including the existing motivational mechanisms and barriers of change resulting from it. The fields of intervention presented above are to be understood as a start for measures intended to activate sustainable motivation. Ideally, all factors of motivation and the corresponding areas of intervention have to be considered and included in a comprehensive change program. In reality, however, in order to develop economically reasonable measures, the focus has to rest on certain points, depending on the general organizational framework and on the extent to which changes actually take place. The business process oriented knowledge management analysis and the analysis of corporate culture and barriers provide the necessary information on where to place emphasis on. The area of intervention "competencies", for instance, can be neglected by reason of a lack of new skill requirements, since, for example, the support for new IT functions can be covered by brief instructions and by multipliers (e.g. key users). Small KM projects may, depending on the need, focus on certain areas of intervention and take calculated individual measures.

4.3 An Example of Putting Pro-Active Change Management into Practice

In this section, an example for putting measures for sustainable behavioural change into practice will be given based on a KM introduction project, which took place in a big company. The program for process-consolidation, which came into play in our project, serves as an example as to how various measures can be linked in order to activate the central factors of motivation.

4.3.1 The Starting Point

In a big German industrial company, the executive board intended to implement a new knowledge management IT tool within the sector of quality assurance. Various departments from quality assurance, product development and manufacturing to sales and service partners, being dispersed in a globally decentralized structure, had to work together in applying the new KM system. Over the years, the staff had worked with self-made IT tools and had not received the relevant information from outside, which got it used to protecting itself and its surroundings and routines against other departments. Living in a bureaucratic, top- down corporate culture, the staff experienced that communicating only good news and never by-passing official reporting processes was the way to get appreciation from colleagues and superiors. Fast and target-oriented interaction between the different teams or departments was often hampered due to internal competition and politics.

The implementation concept aimed at motivating and rewarding the involved staff for sharing knowledge throughout the company by using the KM system frequently in their daily routine. Detailed documentation bringing a benefit to those potentially receiving knowledge were the desired results of behavioural change, i.e. to generate relevant information on product defects, and store this information properly by means of the new IT system.

4.3.2 The Comprehensive Change Programme

The introduction of knowledge management in the company was linked to the selected process of quality assurance (QA). For that reason, the concept of change management was introduced as a program for process consolidation, intended to ensure a process-oriented introduction and to make a connection with the employees' areas of responsibility. This specifically designed change program was roughly divided into three phases and comprised in the form of packages of measures: "marketing", "consolidation" and "realization".

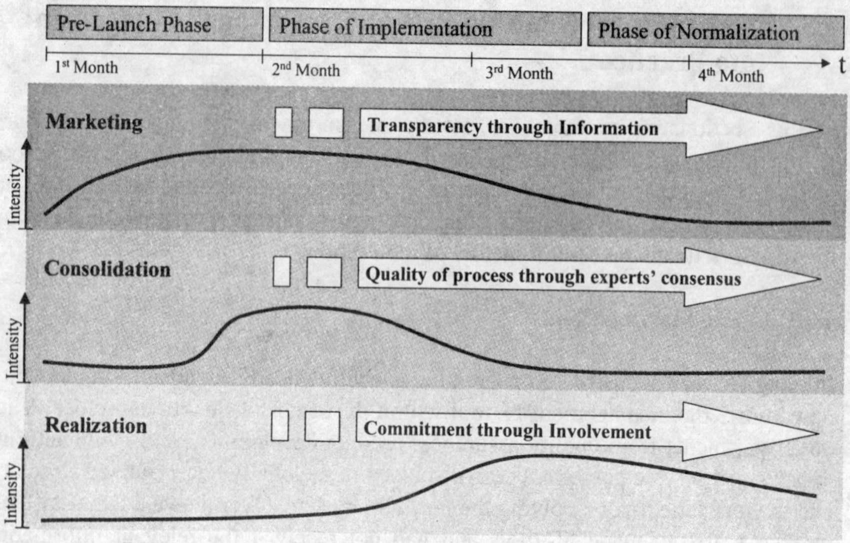

Fig. 4.4: Change Programme Roadmap (Business Example)

The actual working procedures and the existing KM activities of the business process in question were examined by means of interviews with employees and managers. This phase of analysis already proved to contribute in a decisive way to make the interviewees sensitive for changes. In order to prevent any rumours, the background of the KM project must be made transparent in the interviews. Persons with a key function within the relevant areas of the company were informed in advance and won to support the project as promoters, which added to the existing possibilities to control informal communication.

The analysis of processes and barriers led to a potential of improvement on:

- the level of process: the instructions on processes applying to the whole company were not put into practice consistently at its base,
- the level of people: errors were not being dealt with effectively and the inter-department communication was insufficient,
- the level of IT: different and partially proprietary IT systems in different areas.

After evaluation of the interviews and the questionnaires used, the results of the examination and the potential for optimizing the business procedures derived from them were presented to the concerned superiors and the top management. By this, the change program was initialized.

4.3.3 “Marketing“ as a Package of Measures

Before the official implementation phase took place, department managers and other persons with key functions were informed about the change program that was planned. Backgrounds, goals and the *procedere* of the knowledge management project were made transparent. The awareness of the employees for the challenge of a successful introduction was increased (pre-launch phase). The board of management, heads of department and other top managers were required to be able to give information about the project whenever and wherever they communicated and to link it positively with relevant issues. In order to do so, it is necessary to provide management briefings and to win top managers as promoters. Interviews with the chairman of the board, in which he emphasized the optimization of business processes and knowledge management as the relevant fields of action to achieve the company's strategical goals, were published in various media. In addition, the responsible managers of the involved staff units were used as promoters in the way that they promoted and sensitized for the planned process-consolidation in their presentations and visits to sites abroad.

The marketing activities proper started with a kick-off event. In introductory events, the strategical KM goals – the pilot project taking place within one quality assurance process and the new IT system – were presented to the different areas of the company as well as to the different target groups. Selected manager demonstrated the usage of the new IT-system in a practical context, e.g. searching for a popular document using the new tool. In this connection, info sheets and give-aways were distributed in order to increase the mid-term attention. For the further dissemination of the over-all strategy, presentations given by the relevant top managers were taped on video and put in the intranet.

After this far-reaching information campaign had been initialized, the target groups had to be contacted more directly and informed about what they had to expect. This was realized mainly through a multiplier approach involving key-users of the new IT-system to spread the message, enabling other end-users and act as a first level support. Feedback channels in connection with the package of measures “consolidation” were used for further problem solving. The employees were, however, to be informed continuously about the general progress of the project and the next concrete steps in order to get the issue and its importance into the employees' minds. This was done by means of a project-related intranet page on which the most important results of the analysis phase were summarized and on which milestone reports about the ongoing process were to be published. Furthermore, the existing employees' magazine was, from time to time, used to publish reports about the ongoing project as well as introductive activities. Funny incidents and solved problems within the phase of implementation, presented as highlights or success stories, are capable of leaving a long-lasting, positive impression on the employees.

By means of special *pull* tools (e.g. a steadily updated intranet page) and the feedback channels, the project had to be transposed into the normalization phase. In this connection, a continuous improvement has to be followed, while this improvement process should always be accompanied by communication. It was, for instance, planned to change the intranet page into an online process-assistant from which the employees could get all the information related to procedures which they need for the job, including support for the KM instruments, such as guidelines for methods of knowledge capturing, storing and retrieval.

4.3.4 "Consolidation" as a Package of Measures

Top-down distributed process documentations and instructions (including the added KM activities), which are in effect throughout a whole company, contain the risk of not being accepted by the employees as a basis for their work. Individual interpretations may lead to the process not being realized homogenous in the different areas. However, consistent application of KM-tools within clearly defined process steps is a prerequisite for an efficient exchange of knowledge throughout the company.

In order to increase the commitment for a consequent process realization, a participatory process consolidation was recommended. It was carried out in three stages: the employees of the staff unit, who are responsible for a company-wide design and control of the respective process (process owner), were included in the process consolidation. In workshops, not only a common sense about the detailed course of process and the KM instruments to be included, but also a common language among the employees was achieved. This enabled the employees of the headquarter to take over their championing of the project and to represent in a credible and consistent way the goals and methods of the relevant quality assurance process in all sites and plants. The results of the workshop and discussions in working groups subsequent to them were summarized in an interactive process manual.

In the second stage, the plants' departments for quality assurance, which had to fulfill the function of an interface at the particular site, were included. Here, the organizational structures and procedures specific to the plant were registered together with the QA multipliers and according to the company-wide general conditions laid down in the process manual. These were the foundation for the actual design of the process and the realization of the project at the concerned site.

In the third stage, it was called upon the QA multipliers to discuss the changes with the involved departments and to notice their detailed ideas and demands. To ensure homogeneity of processes, a colleague from the staff department, who was assigned a championing function for the respective site, gave them support or settled inconsistencies which arose in certain fields together with his colleagues in the headquarter, if necessary.

These communication structures and feedback-channels for this consolidation process can be made clear with the following figure (Fig. 4.5):

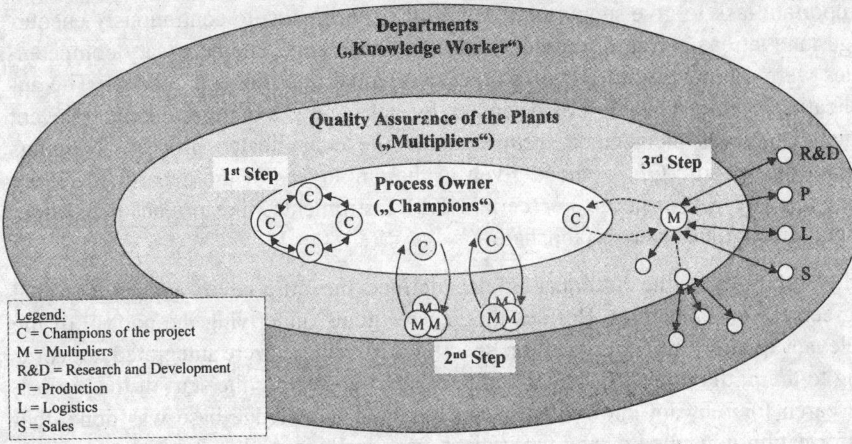

Fig. 4.5: Stage Concept for Participatory Process Consolidation

A critical point with regard to this procedure was the mixture of top-down and bottom-up processes intended to involve the employees. Due to the size of the company and the prevailing business culture, a participation of *all* involved persons was not feasible. Therefore, the competent, strategic instructions given by the experts of process and their communication designed for the target groups were important on one side. On the other side, the centrally developed instructions had to leave enough space for adjustability for the sites and different departments to include and fulfill their needs. Without this differentiated development of the process manual, the acceptance of the consolidated process could not have been attained, which had been shown by the analysis of barriers that revealed earlier experiences of the affected employees with top-down instructions from the headquarter.

4.3.5 "Realization" as a Package of Measures

The transition from implementation phase to normalization phase was supported by a series of workshops in which, depending on the demands and the general organizational framework in the individual areas, different methods of organization- and team development come into play. Task-oriented analyses of the flow of information and exercises for the improvement of communication and cooperation in the team serve as examples. By this, the status of the introduction of knowledge management is being reviewed punctually over a long period of time, developments are being consolidated and new measures launched.

In daily routine, the innovations – such as the use of a new IT system – must be tested and realized. For this, the necessary qualification must be taken into consideration and e.g. contact persons for technical-methodical support must be made accessible. In this phase, the key users and the manager of department have the important task to give support and room for experiments, to continuously encourage innovations and to discuss questions and problems. The necessary competencies were outlined by means of a process-oriented qualification concept. The application of the new IT tool designed for retrieving and investigating relevant knowledge was included in lectures about the consolidated process. Trainings about the specific QA methods given to the employees of the departments were extended by KM aspects. For certain KM instruments, like project debriefing, methodical guides were established.

The staff unit in the headquarter also defined the criteria for a project-related agreement on objectives. Besides the general goals underlying the control of the relevant quality project procedure, specific KM criteria were integrated. According to them, the project manager had to assure that, before the start of the project, research for relevant knowledge in the provided knowledge base was done, that information generated during the project work was inserted in the relevant operative data systems and that, at the end of the project, a densified version of the newly produced knowledge was fed to the knowledge base in the form of a final report. These criteria contribute to the evaluation of performance and influence the way incentives come into effect. Apart from that, the flow of information and the cooperation within the project team can be evaluated with the help of specially developed questionnaires and contribute to the degree of aim achievement.

A system for incentives that could be used, among others, was a cafeteria system, according to which the employees could, depending on the number of collected bonus points, select from the offered gifts. More important in this connection is, however, the possibility to present – in the case of above-average performance – the results of the mentioned quality assurance project in front of the company's head of the board. Exceptional performances of the project team in relation to efficient dealing with knowledge could furthermore be awarded by publishing success stories in the employees' magazine and, thus, lead to a motivation for further KM activities. Here the principles of feedback and recognition of achievements and expertise as immaterial incentives were put into effect by making individual and team achievements public throughout the company.

4.4 Outlook

In order to ensure a successful introduction of knowledge management, a sustainable behavioural change on part of the affected employees must be ascertained through intrinsic motivation. The described model of intervention of the Fraunhofer IPK aims at the best possible interplay of the four factors of motivation: *sensibility*, *ability*, *liability* and *commitment*. An analysis of the given organizational framework in relation to (latent) barriers and drivers of change is the essential foundation for pro-active change management in practice. Experiences made with strategies for the introduction of knowledge management, which were too much based on technical solutions, show how important it is to focus on the individual. Here, the described model for intervention is considered to be a beginning and intended to steer research and practice in this direction.

In order to fulfill the demands of normalization – and by this the integration of knowledge management into daily routine –, the organizational-psychological concept of task orientation offers important criteria for a restructuring of task areas and responsibilities capable of activating motivation in the process of implementing knowledge management.

The earliest possible involvement of employees can promote the necessary acceptance and commitment of employees in the organization (*see* "*involvement* as an area of intervention"). If the other areas of intervention (sensibility, ability, liability) are also designed in a way in which the employee is being encouraged and qualified to actively take part in the organizational changes, a sustainable behavioural change on part of the individual employee is achieved.

5 Role Models, Human Resources and Strategy

Arne Jaitner

A fundamental consideration when designing a KM system is the definition and assignment of roles. So far, there is no consistent approach to answering questions such as: What are the tasks of a knowledge worker? Can the IT department perform KM tasks? Do we need a Chief Knowledge Officer (CKO)? Based on the Business Process Oriented Knowledge Management (GPO-WM) approach the following article contributes some arguments to the definition of KM roles. First, it reflects on the relation of business processes and strategic orientation (5.1), then discusses the close ties between approaches to knowledge and knowledge sharing (5.2) and finally provides elements for the definition of roles based on business processes and strategic positioning (5.3).

5.1 Business Processes and Knowledge Strategy

5.1.1 Knowledge as an Object Class

The Business Process Oriented Knowledge Management approach offers an access to defining roles based on the knowledge demand of the employees' daily activities. The modeling of business processes based on integrated enterprise modeling (IEM) provides a model of the activities – the "Actions" that combine the three object classes "Product", "Order" and "Resource". According to the overall concept, knowledge can be modeled as a sub-class of the superordinated class "Resource".

As a "Resource", knowledge is a prerequisite for an "Action" that follows an "Order" to bring forward an according "Product". The sub-class "Knowledge" can be linked to the resource classes that include specially trained personnel and IT-Systems.[181]

[181] Cf. chapter 2.

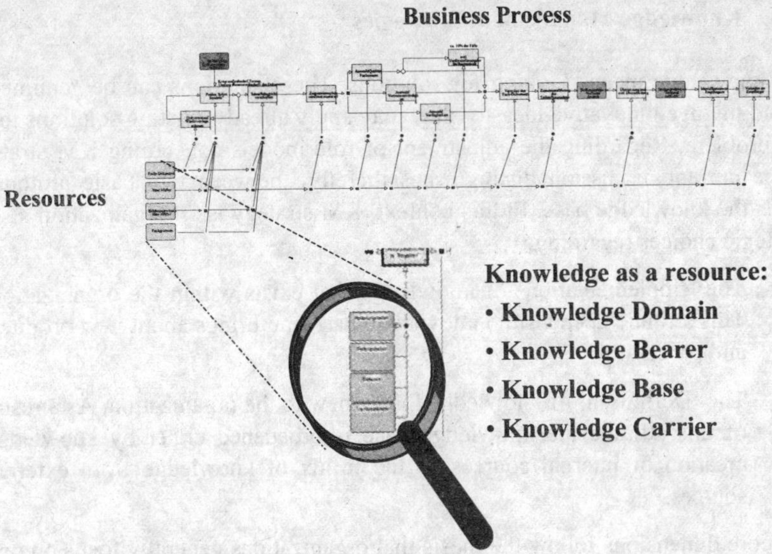

Fig. 5.1: Knowledge Described as a Resource

Describing knowledge as a resource offers an integrated approach towards defining KM roles. At the same time, however, it is necessary to specify – in the context of this method – which ways of sharing knowledge fit the actual knowledge demand in order to produce a more precise description of the roles in question. Regarding the relationship of personnel and IT – and, subsequently, the forming of role models – the strategy of the organization and the chosen approach to KM has to be taken into account.

In sociology, roles are defined as a set of expectations, e.g. attitudes, beliefs and assumptions that are associated with a certain position. In our context roles consist of operations that are fulfilled within the context of the business process in order to provide an "Order" or a "Resource", to carry through an "Action" or to deal with a "Product". A person may play several roles; at the same time a role may be played by several persons. Roles are used to allocate, standardize and control tasks.[182]

A role model describes a set of roles and their relationships. These roles have been clustered because they are interconnected or otherwise related to each other. A person may participate in several role models; a role model may be filled by several persons.

---

182 Cf. Kosiol (1968); Rühli (1993).

5.1.2 Knowledge Management Strategies

Using knowledge means looking for solutions. These solutions can be genuinely new and involve innovative ideas or they may apply already existing solutions to a given problem. Regarding the adjustment of role models concerning KM strategies the current discussion distinguishes therefore between two basic attitudes towards the knowledge base. In this context, KM strategy is an organization's set of strategic choices regarding:

- The 'Implementation' of knowledge that exists within the organization. This attitude deepens the knowledge base and brings about new products and processes.
- The 'Definition' of knowledge that is new to the organization. As a result of this attitude the knowledge base is broadened either by knowledge creation of internal sources or the influx of knowledge from external sources.

These core dimensions follow the thesis that organizations generally focus on one of two learning processes, single-loop and double-loop.[183] Organizations, which develop their knowledge within an existing frame of reference, promote single-loop learning. Double-loop learning, on the contrary, demands the questioning of the current frame of reference. By challenging existing assumptions and reference models, new knowledge can be created within the organization. It has been pointed out that most firms do not successfully follow both core dimensions. This is mostly due to limited resources within the firm, but also to different organizational cultures and structures.[184] The differences between these cultures have been characterized as those of "exploiters" and "explorers".[185]

The KM strategy of a firm can be described as the set of strategic choices towards the core domains of Definition and Implementation. The Definition of a solution will usually rely on given predispositions and the Implementation of a solution may still involve a new thought. Still, our distinction offers the possibility to make clear that there are two basic types of knowledge demand. Also, the Implementation of a solution necessarily follows its Definition and not vice-versa. Typically, the relation between Definition and Implementation can be visualized as shown in Fig. 5.2.

---

183 Argyris and Schön (1978).

184 Cf. Lorsch (1985); Levinthal and March (1993); Volberda (1998).

185 Bierly and Daly (2002), p.278.

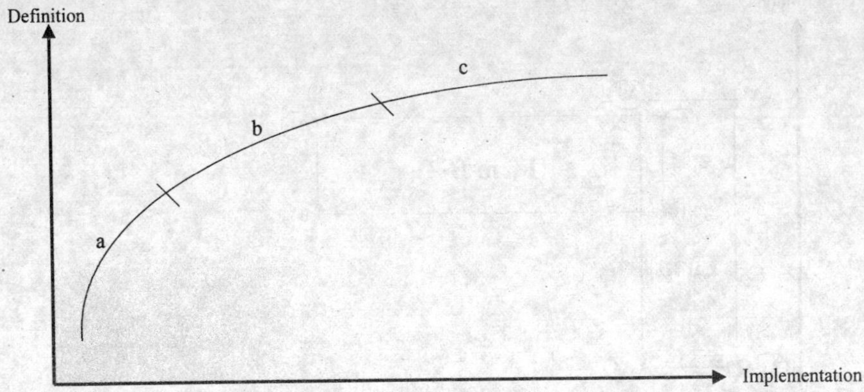

Fig. 5.2: Typical Development of Definition and Implementation of Solutions

While the first step of the Definition of a solution implies little or no Implementation (a), the innovative power essential for the Definition decreases sharply once the solution has been successfully distributed (b). As the solution becomes an established product, the process of Definition comes to a halt (c). The end of the curve signifies the end of the Implementation step in a saturated market.

The distinction between Definition and Implementation leads us to a fundamental categorization of firms: those that are mainly involved in the Definition of solutions, and those that are busy distributing them on the markets.[186] Let us call the first type A, the latter type B. As pointed out above, Definition and Implementation can hardly ever be completely separated. Any type A firm needs to go along with the Implementation process at least in order to make their products known to the firms of type B. On the contrary, type B firms need to redefine those solutions to fit them into their portfolio and adapt them to their customers' needs. This relation can be visualized by the social learning cycle described by Boisot.[187]

186 The term 'firm' seeks to imply that we are talking about commercial organizations; it does not exclude that both types coexisting in one firm. E.g. R&D departments will usually belong to type A while marketing staff will follow patterns of type B.

187 Boisot (1999), pp. 253.

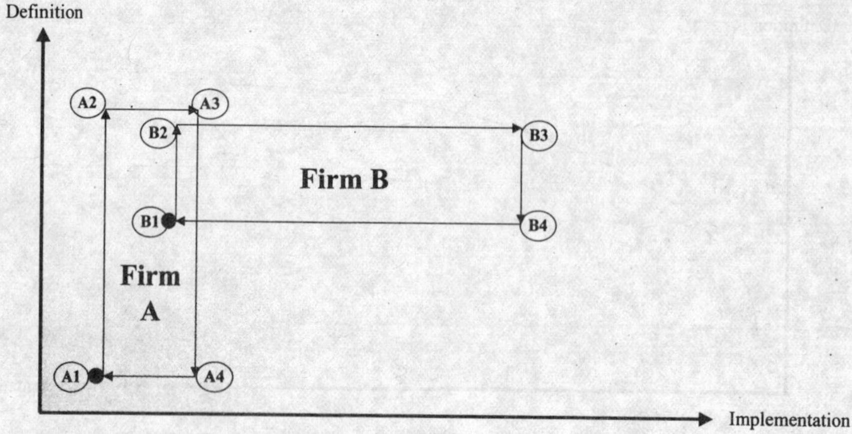

Fig. 5.3: Types of Firms Regarding Solutions

Fig. 5.3 transforms the curve of Fig. 5.2 into two separate development cycles. The type A firm starts at the left bottom corner (A1, marked with a dot) with building the Definition of a new solution (A2), then transforms the relevant know-how to firm type B (A2 to A3) and returns to the starting point to develop a new product (A4). The type B firm starts its cycle (B1, also on the lower left corner) on a higher competence level regarding the Definition of solutions since it can rely on the research provided by A. For this reason B will not reach the level of Definition ability achieved by A (represented by A2, A3) but certainly goes a longer way when it comes to the abilities of Implementation (B3). When the Implementation comes to an end, the Definition ability necessarily drops (from B3 to B4 and A3 to A4 respectively) because the solution is well established and there is no more specific advantage compared to the market. For the same reason the Implementation abilities are reduced to the starting points (from B4 to B1 and A4 to A1 respectively) since there is no use for further distribution.

The graph clarifies the importance of developing strategic orientations in order to deal with solutions. When the knowledge of firm A regarding the solution has been transferred to firm B there is a steep decline in Definition ability (A3 to A4) and also a reduction in Implementation ability (A4 to A1). This happens because the explicit knowledge regarding the solution in question is sold and available 'on the market'. Cashing in on the defined solution allows firm A to leave the implementing business to somebody with more competence in this field and also saves time and resources in order to develop a new solution that has to be ready when the Implementation cycle is finished. Firm B, on the contrary, does not need to build up much Definition abilities – but certainly needs a new product when the market for the old one is saturated. Both firms benefit therefore by concentrating on their core competencies. It is clear that the build-up, the maintenance and the application of this know-how require different knowledge structures and, what is more, different ways of dealing with knowledge. But let us first have a look at the strategic implications of this model.

5.1.3 Strategic Orientation

The differentiation between Definition and Implementation has a wide range of effects on the organizational demand for human resources. The Type A organization goes a long way when it comes to the Definition of solutions, but builds up comparatively little ability in the Implementation of solutions. Thus, a well-skilled research force that is capable of establishing industry leadership and preventing others from imitating the solutions defined will be crucial for its success. On the opposite, B will try to build up a flexible and extrovert workforce that is very efficient in adapting to insourced solutions and capable of reaching out to potential clients. This extrovert respectively introvert behaviour corresponds to the distinction of market and resource orientation.

The classic market structure theory explained the strategy of an enterprise and its success with the given market conditions. According to this theory a firm's potential rentability depends on the strength of its competitors. The enterprise has to develop and maintain competitive advantages in order to work profitably. Its strategic behavior has to adapt to the market structure of the respective sector of industry.[188] The market-oriented approach was decisive for the strategic disposition of enterprises from the 1950s to the 80s. In the 90s the flexibility and adaptability of human resources gained increasing importance for the strategic positioning of firms. The market oriented approach was confronted with a resource-based view that considers internal structures and processes instead of products and markets as the decisive strategic features.[189] This perspective takes the importance of information technology and its influence on the economic dynamics into account and argues that competitive advantages rely on the development of intellectual capital rather than on optimizing existing techniques and procedures. This step implies a 'farewell' to the optimization of processes as introduced by Taylor. Taylorism has, in a way, become a victim of its own success since the competitive advantages that had been brought about by Taylor's approach have been more or less thoroughly realized.[190] While there is, historically speaking, a shift in the main perspective from a market based to a resource based view, market orientation remains of great importance for firms whose economic potential depends mainly on their positioning in the market. This is the case with firms that are mostly occupied with Implementation as described for the type B firm. On the other hand type A firms have to rely for economic success to a much greater part on the intellec-

---

188 Cf. Porter's three generic types of strategy in Porter (1980), p. 62.

189 Cf. Wernerfelt (1984) pp. 171. One of various compilations of the 1990s is Shristava (1992).

190 On the development of Taylorism cf. Jaitner (2001), pp. 304.

tual capital and the excellence of their employees for the Definition of solutions. Our model relies on cooperation and the sharing of abilities: firm B may follow the resource orientation to some extent, firm A to a much greater extent. Since the implementation activities of firm B depend on the contacts with firm A, innovative ideas are clearly of the utmost importance for both of them. As a result the strategic orientation mark a tendency rather than being mutually exclusive.

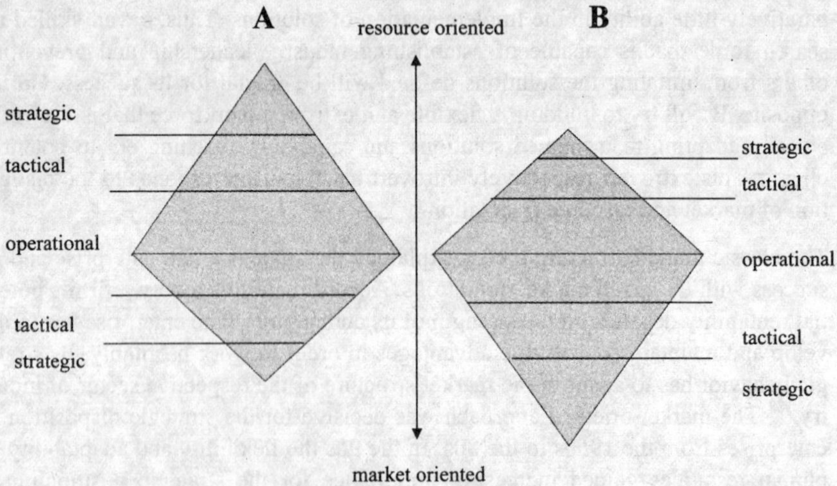

Fig. 5.4: Types of Firms Regarding Solutions

Fig. 5.4 visualizes the resource orientation of firm A: the strategic and tactical dispositions in this field are clearly more prominent than those of firm B. On the contrary, firm B puts stronger emphasis on the strategic, tactical and operational market activities. At the same time these orientations remain relative; both firms maintain both orientations while their activities follow specific needs. This has significant impact on developing human capital. Therefore, it must be a central target to recruit, develop and challenge individuals in a way that fulfills the strategic demands of the organization. In order to specify the human resources and role models, the categories of Definition and Implementation will be analyzed regarding their use of personnel and IT.

5.2 Codification vs. Personalization

5.2.1 Roads to Sharing Knowledge

Recent experience has indicated that the mobilization and conversion of tacit knowledge through IT-Systems is only successful on a rather limited scale.[191] Many enterprises, which had hoped to develop an IT-tool to collect and publish their company's knowledge internally, were disappointed with the results. Strategy-oriented consulting firms are a good example for this development because the innovative power of their human capital is their driving force. While the implementation of IT-solutions delivers fast and reliable first views of comparable projects, problems and situations, the clients of consulting firms clearly expect customized solutions. IT-solutions tend to offer off-the-shelf consulting approaches that are rarely accepted by demanding customers. This is the reason why for example Arthur D. Little follows the principle "No 'one size fits it all".[192] In an effort to define the strengths and limits of IT-based KM, the approaches of Codification and Personalization have been identified.[193] Codification stands for a fast, reliable, reuse-oriented and IT-based approach. According to GPO-WM it comprises electronic or paper-based knowledge carriers as knowledge resources and is fit for answering a large number of similar questions quickly, simultaneously and at low cost. The quality of the answers, however, depends solely on the information stored in the available database. Personalization emphasizes individual expertise capable of providing in-depth analysis and creative problem solving.

This approach is comparatively expensive because it consumes working time of highly qualified and thus expensive staff and often requires additional travel costs. A decisive advantage of the Personalization approach is that face-to-face contacts allow a much more effective communication. Dropping a document in a taxonomy or retrieving it is clearly a much more restricted way of communicating than to sit down in person, in a videoconference or at least on the phone and talk a problem through. But why is that so? First, the questions asked and the answers given will be more specifically targeted and detailed. The step from explicit individual knowledge to explicit common knowledge is therefore optimized by both the sender and the receptor of knowledge. The reduction of complexity that takes place in personal communication can hardly be simulated. It would take an enormous database to be able to store and retrieve that one specific piece of information from a wide range of possibilities. Second, and even more important, Personalization facilitates the first step mentioned above – from tacit individual knowledge to explicit individual knowledge. Tacit individual knowledge is by definition non-explicit. It requires questions and communication to make us realize that

---

191 Cf. Heisig (2000), p. 6.

192 Cf. chapter 12.

193 Cf. Hansen et al. (1999).

we know what our colleagues need to know. This does hold true not only for knowledge that can fairly easily be made explicit but also for 'soft factors' which can only be described vaguely – but may still be a most valuable piece of information. This applies for example to preconscious and nonconscious contents that can be implicated by metaphors or examples. Last but not least personal communication offers a variety of communication channels, such as body language, mimicking or the modulation of the voice for making forms of tacit knowledge explicit that are not available to IT-systems. A major drawback is the distribution of personalized knowledge. While codified knowledge is readily available, personalized distribution is restricted to one or two receptors. In order to reduce the effects of this disadvantage and to possibly achieve a snowball-effect, knowledge sharing mechanisms need to be installed and promoted.

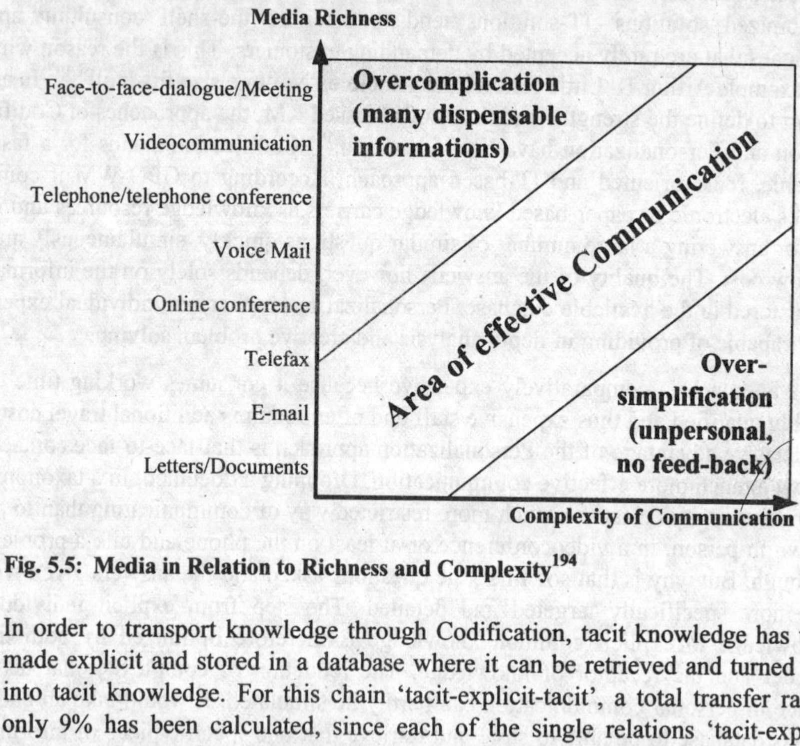

Fig. 5.5: Media in Relation to Richness and Complexity[194]

In order to transport knowledge through Codification, tacit knowledge has to be made explicit and stored in a database where it can be retrieved and turned back into tacit knowledge. For this chain 'tacit-explicit-tacit', a total transfer rate of only 9% has been calculated, since each of the single relations 'tacit-explicit' amounts to a transfer of about 30%. By comparison, the step 'tacit-tacit' allows – via socialization – a transmission of 80% of the original knowledge.[195]

Apparently, the approaches of Codification and Personalization imply different levels of explicability. While neither tacit nor explicit knowledge can ever be

194 According to Pribilla et al. (1996), p. 21.

195 Lee (2001), pp. 406.

transmitted to the full extent, Personalization allows a more thorough transfer of both types of knowledge.

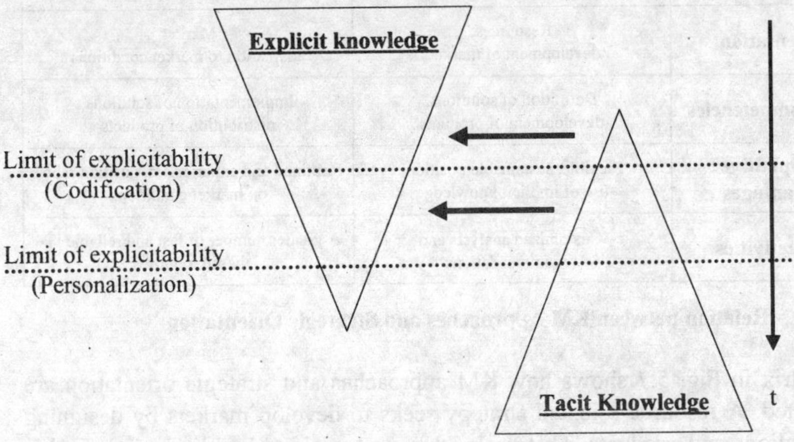

Fig. 5.6: Limits of Explicitability

Since Personalization of knowledge implies significantly higher internal costs than Codification, the knowledge demand of a given enterprise has to be analyzed thoroughly to decide which approach to choose. Generally speaking it makes sense to integrate rather basic instruments such as questionnaires or checklists into IT-tools, while more complex competencies regarding innovative and specialized solutions should be shared personally. The more generic a service, the more economic the codified approach; the more customized a service is, the more the successful knowledge transfer relies on personal contact. A distinction between generic and complex services is therefore the prerequisite to deciding how much working time, travel expenses and communication costs should be invested to access tacit knowledge.

5.2.2 Knowledge Management Approaches and Strategic Orientation

The distinction between the Definition and Implementation of solutions has led us to the description of two types of firms with different strategic dispositions. There was a correlation between the Definition of solutions and a resource oriented strategy on the one hand, and between the Implementation of solutions and a market oriented strategy on the other hand. How can this be related to the two basic KM approaches – Personalization and Codification – discussed above?

A resource-based strategy depends heavily on the personal exchange between employees. New ideas call for the sharing of the contents base of knowledge. Conversely, the Implementation of mature solutions is to a great extent a matter of routine that has to be supported by reliable and readily available information.

	Personalization	Codification
Orientation	Resources; development of markets	Market; adaptation to market conditions
Core competencies	Definition of solutions, development of products	Implementation of solutions, distribution of products
Competitive advantages	Flexible, sustainable position, use of implicit knowledge	Routined and experienced, position relies on market conditions
KM activities	Customized analysis and recommendations	A greater number of fast and reliable IT-based inquiries

Fig. 5.7: Relation between KM Approaches and Strategic Orientation

The matrix in Fig. 5.7 shows how KM approaches and strategic orientation are interrelated. A resource-oriented strategy seeks to develop markets by designing new solutions and products. The aim of this strategy is to build up a sustainable knowledge structure that allows a flexible reaction to new technologies and concepts. In order to make best use of the intellectual capital available, knowledge management has to exploit implicit knowledge wherever possible. This fact, and the need for detailed and customized knowledge material require a Personalization approach.

The implementation of solutions as well as the distribution of goods depend strongly on the market conditions. A successful positioning in the market requires experienced and routined employees who are able to quickly retrieve reliable answers to unexpected questions and problems – if possible even from a client's office. It can be expected that the same questions will be asked many times by different employees. The solution for those specific knowledge needs will most likely be a codified data pool that allows IT-based inquiries.

5.2.3 Knowledge Sourcing and Core Processes

Personalization and Codification have been identified as main roads to sharing knowledge. They offer different advantages and drawbacks and clearly fulfill different needs. The fact that both approaches have been in use for several years means that there is no single 'best way' to KM. Let us therefore take a closer look on how these two paths of access to knowledge relate to the core processes modeled by Fraunhofer IPK.

	Personalization	Codification
Activities	**Providing individual customized advice considering specific tactical and strategic conditions**	**Providing fast, reliable and high-quality information by reusing codified knowledge**
Generate Knowledge	Reflexive-creative process using personal expertise	Production of documents on exemplary cases
Store Knowledge	Discursive description based on notes, minutes, project reports	Preparation and adaptation of documents (e.g. liner notes, anonymisation, adding attachments)
Distribute Knowledge	Establishing and maintaining personal contacts and networks; telephone calls, informal meetings	Assessement and categorization according to the taxonomy, providing links for cross references
Apply Knowledge	Adaptation of given experiences on up-to-date challenges	Reuse/ copy and paste of e.g. exemplary case studies, presentations, checklists, questionnaires

Fig. 5.8: KM Approaches and Core Processes

First, let us remember the main characteristics of Codification and Personalization. Whereas Codification provides high-speed access to previously stored information, Personalization offers advice based on an in-depth analysis of the customer's needs and options. Generating knowledge under Codification means the preparation of documents well before the actual demand arises, while Personalization is in a position to use the most recent developments in the relevant fields as a background for personal expertise. Codified knowledge is stored in a database that allows fast and reliable access but a restricted complexity of communication. Personalized knowledge may rely on notes and reports but is mainly reproduced from the professional background of the advisors; therefore it will include inherent preconscious or even unconscious messages, details and hints. By including for example gestures, body language and the modulation of the voice Personalization offers a much wider range of communication channels to transport complex messages than could possibly be represented in a database. Codification calls for ways of distributing knowledge that stick very closely to the taxonomy agreed upon. Search engines are only of limited usefulness if the knowledge searcher is looking for an overview of solutions to a certain problem. Personalized distribution of knowledge relies on personal communication, which allows for the specification of demand and supply to a much more detailed degree than taxonomy could possibly offer. When it comes to applying knowledge it becomes clear that Codification offers out-of-the-box 'convenience information', while Personalization only uses fresh ingredients for a customized advice. Also, the use of codified knowledge is a tool or a first step in order to get an overview of a certain field or to come up with an idea for further proceedings while personalized knowledge transfer is already an integral part of the solution process. Ultimately, the difference between both approaches will be reflected in the price to be paid. Codified information is ready-made and therefore reasonably priced if a frequent use can be assured. Personalized expert advice, on the contrary, will come with a steep price tag, which includes not only working time but also communication cost for phone calls, videoconferences and possibly flights and hotel rooms. Because of the specific advantages and disadvantages, both approaches will seldom be used exclusively; the categories of Definition and Implementation are however strategic dispositions that call for a preference of Personalization or Codification respectively as a path of access.

At this point, it becomes clear that we are not only talking about specific knowledge needs but also about very different kinds of employees and KM personnel. Let us take a look at the role models that accompany the described KM approaches.

5.3 Modeling Roles

The GPO-WM perceives knowledge management tasks as part of a business process. When improving a business process there is no basic distinction between KM and 'regular' tasks – on the contrary, KM tasks are designed as an integral part of the process. From this point of view the introduction of roles should avoid the creation of new positions exclusively for KM purposes. KM tasks are rather to be integrated into the job description of existing positions. The roles categories listed in the following chapters characterize functions that will be generally performed by existing personnel that are mainly compensated for non-KM tasks.

5.3.1 Types of Roles

When designing processes, an appropriate match of responsibilities with the right personnel is crucial. In order to define roles, the following types can be differentiated:

Strategic Roles

Basic tasks are the strategic and tactical alignment of the KM process, the description of responsibilities and the definition of roles including their support and controlling. These tasks have to be performed according to a long-term strategic disposition as well as for mid-term tactical management, e.g. for the support of temporary task forces with special knowledge demand. A further aspect is the embedding of knowledge management processes in the current organizational culture. This process has to respect the given cultural and motivational conditions; in most cases, change management measures have to be initialized, supported and controlled. Further activities should include the implementation of activities, which will measure and continually improve the acceptance of KM tools and contents. Strategic Roles are of the highest importance during the conception and implementation of KM processes; the respective workload decreases when the system has gone productive. The neglect of cultural and motivational aspects during the introductory step of KM necessarily leads to a lack of acceptance of the project – and often to its failure. Once the idea of KM has been rejected by the workforce, there is little sense in a relaunch. As a consequence, quite a number of KM projects of the first wave were not successful since they often provided just an additional database. This proved definitely not sufficient for the integration of KM into the current business processes.

Once KM has been implemented the main tasks of Strategic Roles are reduced to controlling, developing and improving the ongoing processes. According to this profile, Strategic Roles should be separated either horizontally (in a higher hierarchy, e.g. as a coordinating manager) or vertically (in a separate organizational unit) from the Operational Roles.

Operational Roles

The owners of Operational Roles are responsible for the generation, storage, distribution and use of knowledge. As opposed to the long- and mid-term strategic and tactical alignments, the precise definition of these core activities is subject to short-term adaptations. Operational Roles include many tasks that require overview knowledge on the respective products and markets as well as profound insight into the processes involved and their mechanisms of knowledge transfer. While the controlling of core activities and the processes mentioned is situated on the level of Strategic Roles, Operational Roles are still decisive for the success of KM. Generally, there is a danger of employing an insufficiently qualified workforce in this field because the tasks to be performed tend to be underestimated by management. Since the owners of Operational Roles are usually to be recruited from the operational business, it is crucial to promote Operational Roles to an extent which makes clear that KM is a core target of the enterprise and not inferior to the operational business. Incentives as well as temporary assignments can be helpful in increasing the acceptance of this point of view. The prestige of Operational Roles depends directly on the motivation for KM in the introductory step. If KM cannot be rooted in the corporate culture at this early stage, it will be close to impossible to recruit adequately qualified and experienced personnel for operational KM tasks.

Technical Roles

The development of IT-systems for KM (e.g. implementation of data bases, intranet structure, linking of external sources) as well as the technical procedures for publishing content are tasks performed by IT-specialists. Moreover, Technical Roles usually include a formal or informal support line concerning the use of IT-systems. Since the tools in question (e.g. for the design of intra- or internet pages) are increasingly easy to handle, more and more technical tasks are performed by the owners of Operational Roles.

5.3.2 Generic Role Models

The requirements placed on knowledge workers in a firm are defined by their role in KM. There are different findings about possible roles in KM processes.

North[196] comes to the conclusion that knowledge workers in knowledge-oriented firms can be divided into five groups:

- Visionaries (top management),
- Information managers and infrastructure managers (operative management),
- Knowledge engineers and entrepreneurs (middle management),
- Knowledge practitioners (operative employees), and
- Support colleagues (back office, secretary).

The roles and tasks are determined by existing organizational concepts or management approaches. North refers to the perspective of Nonaka and Takeuchi[197] who suggest a "middle-up-down" approach in which the key role in organizational KM is assigned to the middle management. According to analyses of "top-down" and "bottom-up" approaches, Nonaka et al. conclude that a third way, the "middle-up-down" approach, is most appropriate for supporting the generation of knowledge. The top management develops a vision while the middle management devises concrete concepts, which the employees are able to understand and to implement. The key position lies on the middle level since this is where the discrepancy between visions and reality can be solved. These role models are to a great degree consistent with the generic descriptions of the TFPL-study[198]: An initial strategy team (Chief Knowledge Officer), a planning team (Chief Knowledge Team), an implementation team, knowledge practitioners and knowledge workers. There is a fundamental concordance with the four role models distinguished by Davenport and Prusak:[199]

- Chief Knowledge Officer (CKO),
- Manager of Knowledge Projects,
- Knowledge Management Worker and
- Knowledge-oriented Personnel.

---

196 North (1998), p. 126.
197 Nonaka, Takeuchi (1995), p.130.
198 TFPL Ltd. (1999), pp.59.
199 Cf. Davenport, Prusack (1998), pp. 107.

The case studies in the context of our study at the IPK confirm these role definitions although our terminology is somewhat different. In brackets you find role descriptions that have been subsumed under the category in question:

Coordinator
(Chief Knowledge Officer-CKO, Knowledge Coordinator, Practice Leader, Director of Knowledge Management, Director Intellectual Capital Management, Director of Learning)

This position is a senior management role model performing many tasks described for the strategic role type. It describes a leading position within the organization closely related to the management roles for human resources and corporate information. Core functions are the promotion of a knowledge culture and the design and implementation of a KM infrastructure according to the strategic disposition of the firm. The Coordinator is generally responsible for the creation of economic benefit from knowledge use. An extrapolation estimated 25 positions in the year 1997 and 50 positions in the year 2000 worldwide, with about 40% filled with female managers.[200] The high proportion of female contributors is probably characteristic for this position since it is usually only a temporary appointment (for this reason the positions estimated for 1997 are already partly included in those for 2000). The Coordinator is of particular importance during the conception and implementation step of a KM project because his or her leadership qualities and competencies are indispensable for the organizational development and change management. After the implementation step, however, the role owner is often assigned another position within the organization. Because of the degree of specialization, the leadership qualities required and, last not least, the temporary appointment for the implementation step the Coordinator role is the only one for which a – temporary – creation of a new position seems justified.

Knowledge Project Manager
(Knowledge Advocate, Knowledge Network Program Manager, Manager of Knowledge Projects, Knowledge Program Manager, Internal Marketing Manager)

This role is also dedicated to strategic functions – including a strong focus on tactical questions. Knowledge Project Managers are concerned with the development of project goals, recruiting and coordination of knowledge teams and day-to-day project management. Although leading a team of knowledge management workers is a permanent task, Knowledge Project Managers experience the highest workload during the implementation step. During this time other tasks should be adequately reduced. In the operational step, KM management positions are mainly responsible for overseeing the availability of knowledge resources. This task can be performed as a part of the daily routine whereas it will often be combined with related activities (e.g. management of internal communication, change management or IT projects).

---

200 Earl, Scott (2000), pp. 196.

Knowledge Management Worker
(Intellectual Capital-Seeker, Knowledge Steward)

This role includes the broadest variation of activities: apart from the procurement and scanning of potential knowledge content it deals with the categorization and structuring of the knowledge base. It implies wide areas of the operational and peripheral areas of the Technical Roles. While the procurement of content requires high social competence, the structuring of the knowledge base rather requires logical-abstract reflection. Eventually, the owners of these roles have to dispose of varied and specialized operational knowledge to categorize the potential codified content or partners for the personal approach according to their relevance and affiliation. Since this role has to combine a thorough operational background with management abilities, experienced employees are recruited and particularly trained for these positions. The GPO-WM approach guarantees that Knowledge Management Workers will remain a part of the operational routine, the tasks off this role should not amount to more than a third – the IBM case study i.e. speaks of 10% – of the overall workload. Incentives are frequently awarded in order to motivate suitable personnel for this role.

Knowledge Oriented Contributor
(Knowledge-oriented personnel)

Last but not least, KM depends on motivated contributors: it is they who supply the results of their daily work as potential knowledge content, and are crucial for spreading the stored knowledge within the organization (especially with regard to personalized transfer). Other than the roles mentioned above, Knowledge Oriented Contributors are not compensated for the handling of knowledge. Therefore, it is even more important to convince them of the importance of KM for the development of the organization and its culture and to invite them to co-design the knowledge base by sharing their experiences and tools. They are the main addressees of the motivation campaign during the implementation step.

5.3.3 Specific Conditions of Generic Role Models

Apparently, the generic role models will be adapted to the requirements of the processes and not vice-versa. This procedure should take two specific conditions into account. First, there is a difference between the personalized and the codified approach regarding the distribution of the tasks described in Fig. 5.7.

	Personalization	Codification
Activities	**Providing individual customized advice considering specific tactical and strategic conditions**	**Providing fast, reliable and high-quality information by reusing codified knowledge**
Generate Knowledge	Knowledge Oriented Contributor	Knowledge Oriented Contributor
Store Knowledge	Knowledge Oriented Contributor	Knowledge Project Manager Knowledge Management Worker
Distribute Knowledge	Knowledge Oriented Contributor For the first contact: Knowledge Management Worker	Knowledge Project Manager Knowledge Management Worker
Apply Knowledge	Knowledge Oriented Contributor	Knowledge Oriented Contributor

Fig. 5.9: Involvement of Roles Regarding KM Activities

As Fig. 5.9 points out, Knowledge Oriented Contributors (the shaded boxes) operate within a personalized approach involved in all activities – including those of storing and distributing knowledge. On the other hand, the codified approach generally requires a stronger involvement of KM specialist roles as represented by the non-shaded boxes.

Role models	Concept	Introduction	Operation
Coordinator	Codification: ++ Personalization: ++	Codification: ++ Personalization: +	Codification: +0 Personalization: +0
Knowledge Project Manager	Codification: ++ Personalization: ++	Codification: ++ Personalization: +	Codification: + Personalization: +0
Knowledge Management Worker	Codification: + Personalization: +	Codification: ++ Personalization: +	Codification: ++ Personalization: +
Knowledge Oriented Contributor	Codification: 0 Personalization: 0	Codification: + Personalization: ++	Codification: ++ Personalization: ++

++ = very important; + = important; +0 = inferior importance; 0 = little importance

Fig. 5.10: Reduced Involvement of high-Ranking Roles after Implementation

Second, Fig. 5.10 above displays a significant difference in the need for personnel during the course of the implementation. While highly qualified roles such as Coordinators and Knowledge Project Managers are indispensable for the concept, most of the burden of the operation is on Knowledge Management Workers, and, above all, Knowledge Oriented Contributors. The result is a reduced involvement of high-ranking roles as the operation of the KM system draws closer. Knowledge Management Workers are of a higher importance within the Codification approach both for the involvement of roles regarding KM activities and during implementation. Independent of the approach, however, the decisive role for the operation remains the Knowledge Oriented Contributor. It therefore appears appropriate to allow a significant time budget for all Contributors involved in KM activities as well as to provide incentives for generating, storing, distributing and applying knowledge. These findings support the attempt of the business process oriented KM approach to generally integrate KM and non-KM tasks within the business process in question. As a result, full-time positions can be avoided at least on a long-term view.

On a short-term base it can be discussed against the background of Fig. Fig. 5.9 whether the roles of Coordinators and KM Managers shall be filled on a full-time basis for the concept and part of the introduction. However, the tasks described for the operation are to be integrated into the work descriptions of the respective process owners. Fig. 5.9 and Fig. 5.10 emphasize that the motivation of Knowledge Oriented Contributors and, to a lesser degree, of Knowledge Management Workers are decisive for the success of KM projects.

5.3.4 Defining Roles

For the definition of KM roles, three steps are suggested.

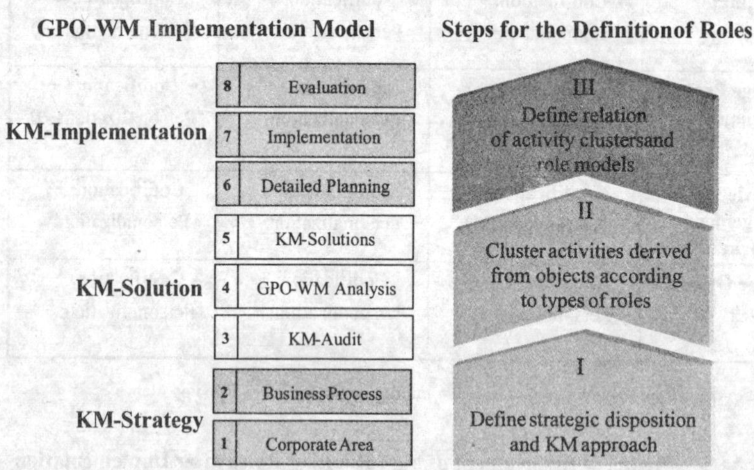

Fig. 5.11: GPO-WM Implementation Model and Steps for the Definition of Roles

First, the basic strategic distinction between Definition and Implementation leads to a decision on the use of a personalized or codified approach to KM. Second, an analysis of the strengths and weaknesses in the current use of knowledge gives an overview on the framework conditions of the business processes in question. This understanding allows the sorting and clustering of activities derived from the objects „Order“, „Action“ and „Resources“ regarding strategic, operational and technical types of roles. Third, the definition of the role models is based on the findings regarding strategic disposition, KM approach, framework conditions and process analysis.

The first step is dedicated to the discussion of the strategic disposition of the organization. This means in our context a discussion about the organizational mission regarding Definition and Implementation and the orientation regarding resource- or market-orientation.[201] This step is crucial both for the definition of roles and the proper setup of the KM design. This step relates to the phases 1 and 2 of the GPO-WM implementation model.[202] The definition of the strategic disposition will mostly include a series of workshops in order to discuss the current activities, a SWOT analysis (strengths, weaknesses, opportunities, threats) and the development or adjustment of a strategic business plan. Since the KM design needs to be adaptable to the further development of the organization, both current activities and future perspectives should be covered by the analysis of the strategic disposition. This includes the selection of the precise business processes that are to be improved by KM methods and tools. The processes in question should be knowledge intensive and complex enough to expect a significant improvement by applying KM tools; also, the approach for their improvement should fit the overall strategy of the firm. As a result, the relation of personalized and codified knowledge management according to the present and future strategic positioning will be defined. The specific advantages of both approaches should be discussed according to the specific strategic needs of the organization using the concepts of Definition and Implementation respectively as well as the distribution of solutions and products. Also, the evaluation of the intellectual capital given will provide valuable hints for a further development of the organization.[203] Step I is designed to set the strategic long term goals and policies regarding the use of knowledge into a relation to the extent of Codification or Personalization of the planned KM activities.

The second step is dedicated to analyzing the business processes selected regarding their relevance for the definition of roles. In the course of the KM-Audit the framework conditions of the processes in question will be examined. This concerns especially the methods and tools used to support the current knowledge use.[204] The modeling of the process itself provides an overview of the knowledge tasks – the ‘Actions’ based on ‘Orders’ – performed in the process. Also, the rele-

---

[201] Cf. chapter 5.1.2 in this article.

[202] Regarding the GPO-WM implementation model cf. chapter 2.

[203] Cf. chapter 7.

[204] Cf. chapter 3.

vant knowledge objects – the 'Resources' – used to perform these tasks are described. If necessary for a deeper understanding, in-depth interviews will be conducted with the process owners responsible for key actions, orders and resources. This GPO-WM analysis is followed by the design of KM-solutions to be implemented. The choice of these solutions follows the results of the analysis as well as the audit and the strategy discussion. The implementation of KM-solutions implies a decision on the remodeling of the processes, the support facilities as i.e. IT-architecture and hardware and last not least on the KM-tasks to be performed. In order to receive a qualitative and quantitative overview on the strategic, operational and technical tasks the KM-activities implicated by the remodeled knowledge objects are sorted and clustered. An acquisition process may i.e. consist of two order objects, which imply strategic tasks, two actions that require operative activities and three knowledge resource objects that are transmitted codified – and therefore imply technical support. The sorting procedure may well assign two or more activities to one knowledge object – the resource objects mentioned i.e. are well to require both operative and technical activities. The clustering forms strategic, operational and technical groups of activities that require comparable competencies.[205] Part of this step is a check whether the business processes modeled and the activities clustered fit the present and projected strategic disposition of the firm defined in step I. If not, adjustments in the process design should be made before proceeding to step III.

The last step works out the eventual definition of the role models using the findings of steps I and II. The activities clustered according to the types of roles are now set into relation to the generic role models. In the context of the detailed planning, duty lists will be prepared on the basis of the activity clusters. The generic role types will be used to assign qualitatively appropriate duties to each role. The guiding principle for formulating this list of duties for the role models will be the requirements of the processes. Eventually, the qualitative findings of the definition of the relation between KM activity clusters and role models are transformed into quantitative results. Using the methods of human resources quantification, the duty lists for the various roles provide the basis for an estimation of the actual human resources demand. Change management measures have to be defined and put into effect together with the human resources quantification in order to secure the acceptance of the KM approach. The human resources planning and development will have to rely on the results of the strategic analysis conducted in step I both for the adjustment of the current task profile and the creation of new positions. The implementation puts the remodeled processes into effect; a thorough evaluation allows to readjust decisions on the previous phases if necessary.

At this point, it should be emphasized that in order to support KM, it is not absolutely necessary to create special teams or to appoint single persons for the roles in question. Alternatively, it is – with the possible exception of the management roles

---

205 Cf. chapter 4.

during the implementation of KM measures – possible to train and motivate every single employee to integrate KM into his daily routine. To make employees aware of their own responsibility in creating a knowledge-oriented company is one of the main tasks of change management measures on the KM-implementation level.

5.4 Outlook

In the course of abandoning the "one size fits all" attempt to codify the collected knowledge of an organization, its strategic disposition comes into view. It is widely acknowledged that there are several ways to manage knowledge. And the field of possibilities becomes wider: neither will firms striving to offer customized services stop developing specific KM solutions nor will organizations seeking to rapidly and economically answer a large number of similar questions stop optimizing their IT-systems. There are few areas where this diversification becomes more obvious than for designing role models. Where there is no generic approach, there will be no generic role models.

The methodic consequence will have to be an even more thorough examination of the business processes and their underlying strategic assumptions. This allows to analyze and describe the objects defined for the processes towards role types as well as towards the strategic disposition and the KM approach chosen. The results of this procedure lay the foundation for the description of roles. Independent of a decline of the buzzword KM the integration of strategy, business processes and the assignment of tasks and roles will remain a bare necessity under the conditions of resource-based organizations.

6 Knowledge Management Tools

Kay Alwert, Ingo Hoffmann

The value of information technology for organizational knowledge management becomes especially evident in the areas of networking and communication. It is not the complicated computing algorithm, but rather the linkage of heterogeneous data and information sources and the exchange between experts that represents the actual advantage. The most typical knowledge management tool in an organization is the portal to which employees and partners, as well as suppliers and customers, are given access. By the use of browser-based tools, information, applications and services can be made accessible from any location and recent updates can be distributed quicker to the employees. An additional aspect explaining the popularity of internet tools is that this software and this form of information distribution even caters for mobile employees. The company portals function as the main entry to the employee's business applications and information as a whole. As a universal tool, the computer is used to handle diverse tasks of numerous employees in organizations. However, generally work and the exchange of knowledge still take place in discussions, meetings, and so forth, as well as on the computer.

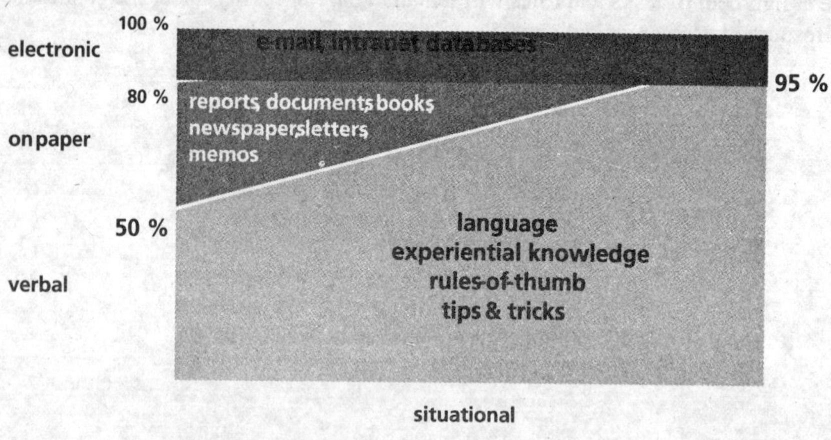

Fig. 6.1: Verbal Communication Dominates[206]

Not all knowledge can or must be saved explicitly and directly in the computer system. The transfer of information and knowledge still occurs primarily through

---

206 Bair (1998), p. 2.

verbal communication. Fig. 6.1 shows that between 50% and 95% of all exchange is verbal, depending upon the situation. While the preparation of a project report is situated rather on the left side of the graphic, meetings and conferences are situated more on the right. The richness and effectiveness of face-to-face communication should not be underestimated.[207]

When planning IT support in knowledge management an effort should be made to try to avoid that the technical realization ousts the measures independent of IT, e.g. the structuring of the knowledge base as to its content, the planning of its integration in the organizational procedures[208] and the measures concerned with the promotion of knowledge exchange and its acceptance among employees.[209]

This chapter is addressed to people concerned with the selection and implementation of Knowledge Management Tools. For this purpose, a brief introduction to KM-Tools and Fraunhofer IPK's approach to these issues will follow. Subsequently, we will elaborate on the field of structuring knowledge and information, which is of increasing importance. We will also show some methods designed for the structuring of knowledge. In the third section, we will illustrate the technological side of KM-Tools and briefly introduce some Knowledge Technologies. All aspects will be brought together in the fourth section, presenting a method for choosing an appropriate KM-Tool and explaining it exemplarily. In the annex, the reader finds a list of KM-Tools available on the market, which cover a wide range of functionalities. The list is supposed to serve as a possible starting point for a company-specific evaluation and selection.

6.1 Definition of KM-Tool and the IPK Approach

First of all, it is necessary to define what type of knowledge is being targeted with knowledge management tools and what the meaning of KM-Tool is in this context. A broad definition of a knowledge management tool would include E-mail and computer-videoconference systems. However, we consider this kind of software to be the basic technology constituting a possible building block for a knowledge management system. This technology only becomes part of a knowledge management tool when the content structure and defined processes for the handling within the knowledge base are included.

Accordingly, **KM-Tools** are tools which automate and support the handling of the electronically available knowledge base, or, to quote Ruggles: "Knowledge management tools are technologies which automate, enhance and enable knowledge generation, codification and transfer."[210]

---

207 Cf. also Fig. 5.5.

208 Cf. chapter 2.

209 Cf. chapter 4.

210 Ruggles (1997), p. 3.

The **knowledge base** as a whole can be described as the amount of problem-specific knowledge existing within an organization[211]. Hereby, knowledge is carried by various knowledge bearers (persons) and knowledge carriers such as documents and databases, which among them present context-sensitive, verifiable associations or sub-groupings. This definition of a knowledge base also comprises elements, which are not electronically available, e.g. employees and documents on paper.

According to our model of knowledge management, information technology is only one of six main design fields (cf. Fig. 6.2). Knowledge management tools that offer computer-based methods and techniques to enhance the handling of knowledge in organizations fall into this category. These tools are solutions that can handle the complexity of a knowledge base and cover the entire range of the core processes of knowledge management (generate, distribute, store, apply).

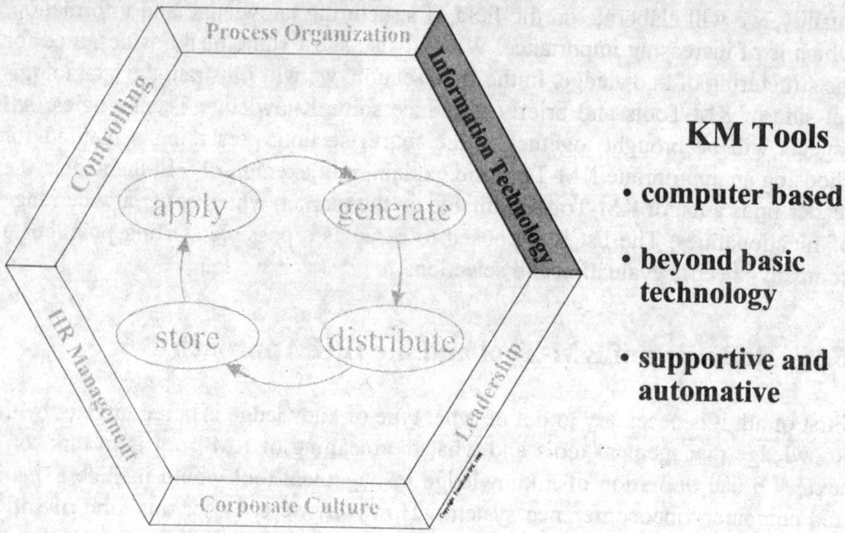

Fig. 6.2: IPK Approach of KM-Tools

Before we describe special KM-Tools and technical functionalities, we shall settle the question how successfully a KM-Tool can deal with tacit knowledge. Processing explicit knowledge with a software tool requires the existence of information in an electronically processable form. By definition, tacit knowledge does not exist in an explicit form. Therefore, tacit knowledge must be handled in an indirect way. Information about tacit knowledge is made explicit and is subsequently made electronically available. This information about the hidden elements of a knowledge base is also described as meta-knowledge, which can be administered

211 Gabler (1994), p. 38.

electronically in the form of meta-data and meta-data structures. Some application for automated expert location are already in use.[212]

The most common way to deal with tacit knowledge is Yellow Pages, which refer to the owner of the tacit part of the sought knowledge. Documents on paper are also not directly accessible by way of a KM-tool. Here, as in the case of people, meta-data about the factual objects can be made available electronically. An electronic library system serves as an example for this. The availability of non-electronical knowledge carriers is thus not being directly improved by KM-Tools. Yet the described systems, via access to meta-data, make it possible to do research on those carriers and to locate them.

6.2 Structuring Knowledge and Information

When looking at the architecture of knowledge management systems frequently used nowadays, it can be seen that they offer various functionalities for an appropriate infrastructure or components for the controlling of organizational procedures. A knowledge management system, however, consists of technology, organizational processes, knowledge content and the resulting structure. Only an analysis of all four components together leads to a solution that responds to the circumstances of the organization in an optimal way. In this context the knowledge structure represents the transition between content and technical solution. By the way of structuring, the possibilities of later use as well as the visualization and therefore the interface between system and user are being influenced. This means that the potential benefit for the user and the basic architecture as well as the effort needed for the installation, the initial feeding and updating of the system are being anticipated. It is therefore necessary to take account of this important aspect at an early stage when dealing with KM-systems.

Knowledge structuring offers further advantages, which are of importance for a comprehensive consideration of knowledge management. Common structures, once made explicit and being communicated, allow for greater transparency. The structure determines the language for the exchange of knowledge and therefore has far-reaching, cultural side effects on the organization as a whole.

6.2.1 Definition of Knowledge Structure

Various articles and books approach the challenge of structuring knowledge and information bases using results from computer science and its technical possibilities. This, however, is only one aspect of structuring. IT tools and their various functionalities can support the structuring of knowledge and information sources, e.g. through procedures of information retrieval, data and text mining, procedures

---

212 Cf. chapter 13, Aventis.

for content analysis, automatical clustering etc. An essential aspect of knowledge structures is, however, that they are based on experiences made by experts and that they create a mutual understanding between persons. Computer-generated structuring is but one part of this. Yet it should not substitute the entire process of achieving a consensus between employees, experts and the executive personnel, a process that also promotes acceptance.

It is the aim of knowledge structuring to put the organizational knowledge base into an action-based, accepted context by means of connecting, modeling and visualizing relations between the elements of the knowledge base. In this regard, knowledge structure can be defined as the explicit representation of verifiable, action-based relations within the knowledge base.

6.2.2 Search Strategies and Knowledge Structure

In order to find single elements of knowledge and information in a stock of data without being dependent on chance, various concepts and strategies[213] can be used. These strategies and concepts result primarily from the respective context of search, that is, from different cases of application and their objectives. The following paragraph briefly shows two typical cases of application, which, in their oppositeness, demonstrate the whole spectrum of a search.

- Key-Word based Search:

  The user knows what he is looking for and is familiar with individual, precise search words and meta-data, such as the author of a document and the period in which it was created. He tries to formulate his search entry as precisely as possible, using different key words – probably combined via logical operators (or, and) – which are fed into a search engine. The search engine scans the entered key words and retransmits the results, which it might possibly arrange according to further meta data (e.g. the date of creation). Ideally, the searched file is among the results. Frequently, however, this kind of search does not lead to the desired result. This occurs if the correct keywords are not known and, subsequently, an immense amount of results or no results at all are retransmitted. In an ideal case, the request is formulated in such a precise way that only the one sought file is retransmitted. Search requests that are less precise lead to no or too many results.

- Browsing:

  The user does not exactly know what he is looking for and starts the search process with a general request, in order to subsequently start a new request based on the results achieved from the first. It is helpful for the

213 Ellis (1989), pp. 318-338.

user if the knowledge structure of the respective area is visualized. These visualized structures are summarized under the term knowledge maps (cf. chapter 6.3.3 in this section). Not every sort of structuring, as little as the way in which structures are being visualized, is appropriate for every case of application and every user, as the difficulties of browsing show. As a result, the benefits of explicit structures depend on the respective cases of application.

In practice, the procedure of browsing is used much more frequently. According to a survey of the Delphi Group, almost 70 % of the asked people prefer information browsing.[214] This is due to the fact that in the most cases of search for knowledge, the precise content of the search result is not a priori known. In the browsing process, the identification of the search result is much simpler, since the context of thematically related knowledge is also being presented. Thus, the user can "feel his way" towards the result. An additional positive aspect of this approach is the learning effect attained by the use of explicit knowledge structures during the course of research.

Both concepts, however, use structures in one way or another in order to bring a determined or undetermined search to success. In the case of "browsing", the method, which is applied, is called **pre-structuring**, since structures are determined before the search. Yet, there is also a form of structuring inherent to "key-word-based search", namely the structuring of files according to key-words, dates of creation, file size, author, etc. – thus, by means of meta-data attributed to the files. This method is called **post-structuring**, because an explicit structure is not created until after the search.

The relation between searching and filing is of great importance as well. In general, it can be assumed that an increasing effort with regard to the structuring and, subsequently, the filing of knowledge elements leads to an improvement in the results of retrieval. Therefore, the requirements of future users and their efforts in using the system must be taken into consideration when choosing the method of structuring.

6.2.3 Methods of Structuring Knowledge and Information

In the course of structuring knowledge and information bases, there are certain inevitable steps that have to be taken, such as determining the method of structuring and its actual establishment with regard to contents. Some methods of structuring relevant in recent discussion shall be introduced here.

214 Cf. Delphi Group (2002), p. 4.

Semantic Network

Semantics is an important term concerning modern approaches to structuring knowledge and information. Several practical approaches in the fields of information sciences and informatics are based on ideas involving the notion of semantics. Semantics, according to Morris, are concerned with the relationship between the sign and the signified (an object or state of affairs the sign refers to), i.e. with the meaning of signs[215]. In information sciences, semantic networks were introduced in the field of artificial intelligence first. They originate from cognitive psychology, which is a branch of psychology dealing with processes of consciousness, and particularly with those of perception. The establishment of semantic networks was modeled according to the structures of human memory. Concepts and the associative relations existing between them are in general a constitutive structural element of knowledge. Within semantic networks, the connections between concepts are made clear by way of linking them via so-called associations, thus creating a network of concepts and relations. Concepts are frequently called knots, while the connections between them are described as edges. With regard to their operational integration into a computer system, the different types of associations are restricted to a reasonable number[216]. Typical associations are:

- "is a",
- "belongs to," or,
- "is linked with".

Many approaches applied in the process of knowledge structuring draw on the theory of semantic networks. These are, e.g., mind mapping, thesauri, taxonomies and topic maps, but also ontologies and process models.

Thesaurus

The principle of thesaurus originally belongs to the field of library and information sciences and works on the basis of a structured collection of concepts and their designations[217]. It provides a standardized vocabulary, which improves processes of retrieval as well as the establishment of indices. Two essential features are characteristic of thesauri: firstly, there is an unequivocal reference between concept and designation; and secondly, the relations between the concepts are made obvious. Frequently, a concept can be verbalized in various ways. Therefore, one designation, the so-called descriptor, is chosen as a representative for the concept. Other designations for the concept such as synonyms are called non-descriptors.

---

215 Cf. Morris (1977).
216 Cf. Helbig (2001), p. 20.
217 Cf. Westphal (1998).

All thesaurus entries that can be involved in the establishment of an index are descriptors, while non-descriptors make a reference to the given descriptor by way of respective links. Among other criteria a descriptor has to describe its non-descriptors in a way as comprehensive, unequivocal and exact as possible. It further has to correspond with the terminological conventions of the (scientific) field that it belongs to, and it has to be easy to remember as well as uncomplicated. A descriptor is mostly presented as a singular noun[218], and it should be avoided to make use of adjectives, adverbs, verbs or abbreviations as descriptors. A thesaurus makes use of both hierarchical and associative relations. Hierarchical relations within the thesaurus are an expression of conceptual priority or subordination, while associative relations point at different descriptors.

Topic Map

Topic maps, being a relatively recent standard dating from the beginning of the 1990s[219], serve as a tool for search and navigation within a knowledge base[220]. A topic map can be understood as a sort of semantic net stretched over the actual information resources. It allows direct access to the sought information, which is saved in documents. The information resources do not have to be homogenous and its sources can have different formatting (such as PDF documents, MS word files, e-mails, text files etc.).

A topic map consists of topics, associations, and occurrences. The central element of a topic map are the so-called topics. Within the scope of the topic map, they can represent any object of the real world. One or several topic types can be assigned to one topic map (cf. Fig. 6.3), while the types are, in turn, themselves defined as topics within the topic map. References to elements of the knowledge base are called topic occurrences. An individual topic can be connected to any number of knowledge resources. A topic occurrence would be, e.g., an article about a certain subject matter within a document, as well as a picture, a video, or similar references. The actual advantage of topic maps as a way of structuring becomes obvious in the use of topic associations. Associations establish connecting lines between the topics. Associations, like topics, can be assigned different association types.

An innovation of the original standard, ISO 13250, led to the development of the XML Topic Map Standard, XTM, which helped to realize an improved utilization of topic maps in connection with internet technology.

---

218 ISO – 2788, p.6.

219 Cf. Widhalm et al. (2001), p.2.

220 The first definition of a Topic Map Standard was introduced as ISO 13250 in autumn 1999 by the ISO JTC1/SC34/WG3 task group of the W3 Consortium.

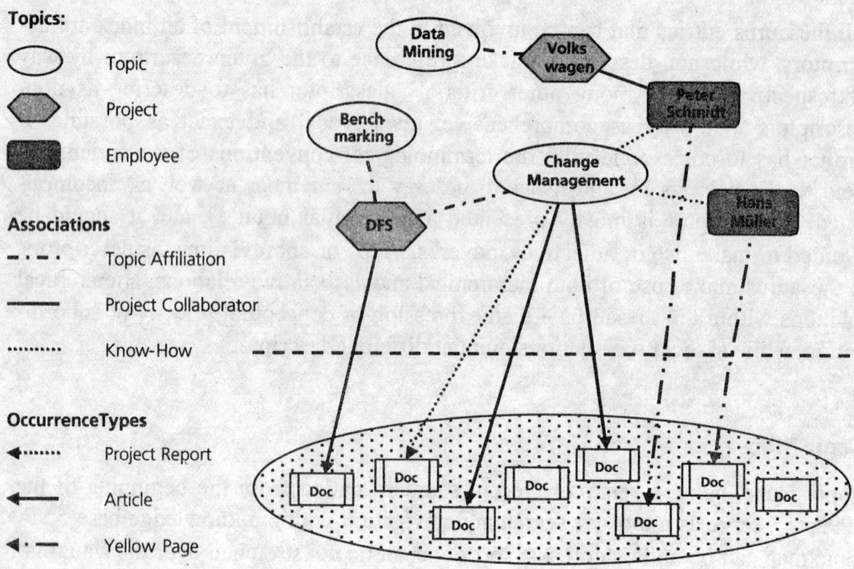

Fig. 6.3: Topic Map Diagram

Taxonomy

Taxonomy designates a classification system organizing concepts within a hierarchy. In this case, hierarchy represents a mere relation of class and authority, that is, an "is a"-association. The concepts should be unequivocal and are supposed to only occur once within a taxonomy. Exemplary taxonomy design also excels by completeness, i.e. the field it deals with should be fully covered. Taxonomies are frequently used in the field of biology. The term taxonomy, which was originally very strictly defined, has recently become more flexible and now includes general hierarchical systems of concepts as well[221]. This is due to an increased use of taxonomies in rather abstract branches of the economy and sciences, which, in the nature of these things, do not deal very well with such unequivocal "is a" – relations defined by biology.

Ontology

During the last few years, ontology has become one of the central key words in the field of information management and knowledge management respectively. The term itself is more than 2000 years old and designates a branch of philosophy dealing with nature and the organization of reality. In the early 1990s, however, it

221 Cf. Delphi Group (2002).

was taken up again by information sciences. The most common definition of the term is based on Gruber. "An ontology is an explicit specification of a conceptualization."[222] This fairly vague definition is further specified by Studer. Accordingly, the term ontology can designate a consensus within a group (of persons) with regard to a formal model within a special field of application called domain[223]. The relevant terms of the given field of application are commonly registered and sorted by way of a classical taxonomy, thus producing three constitutive elements of the ontology, concepts, relations and attributes. In contrast to a mere taxonomy, an ontology excels by providing possibilities of further specification of the concepts described. This is achieved by adding rules (axioms). Axioms are employed in order to model relations or conclusions. These conclusions, or rules of inference, are often self-evident to the human user, but for computer-based processing they have to be indicated explicitly. The definition of rules of inference can thus help to uncover knowledge which is only tacitly present in the ontology, provided, of course, that the rules of inference are properly interpreted by the respective applications[224].

Process Models for the Structuring of Knowledge and Information

(Fraunhofer IPK Approach)

One of the aims of knowledge structuring is to place the knowledge at the employee's disposal at the exact point of time when it is needed to carry out a given activity. Therefore, the employment of processes can be regarded as an adequate basis for the structuring of knowledge and information. A process model highlights and describes those connections between single activities that are relevant for the change of a given state. These process descriptions represent an important and explicit component of knowledge, i.e. the know-how. In addition to the mere process description, Fraunhofer IPK also assigns the different types of knowledge resources to the single activities of the process. Knowledge, according to the common perspective of business administration, is therefore understood as a resource relevant for success within business processes.

In order to utilize process modeling methods for the structuring of information and knowledge, the modeling procedure concentrates on those knowledge and information sources required by the process on the one hand and the results of knowledge work, the knowledge products on the other. Knowledge, therefore, is basically divided into two categories by way of a fundamental act of structuring, namely the knowledge about a process, the 'know-how', and the knowledge within a process, the 'know-what'.

---

222 Cf. Gruber (1993).

223 Studer (2001), p.4.

224 Cf. Studer (2001).

The IPK Procedure Model of Business Process Oriented Knowledge Structuring

The knowledge structuring procedure according to the IPK approach can be integrated as a module into the implementation model for business-process-oriented knowledge management[225]. It is therefore important that the process-oriented way of knowledge structuring supports the four core activities of knowledge management (generation, storage, distribution and application of knowledge) and includes the requirements of future users at an early stage.[226]

There are five main steps in the procedure of business-process-oriented knowledge structuring:

- modeling the selected business process[227] with the necessary knowledge domains, bearers, carriers and bases including the existing knowledge structures,
- defining the users' requirements for the knowledge structure derived from the business process,
- defining and pre-structuring the relevant knowledge as to the main knowledge objects and their relevant metadata; the themes in the evaluated knowledge domain and the connections between all these elements,
- defining and specifying the structure in consensus with leaders, stakeholders and experts,
- implementing the knowledge structure through the definition of roles and processes of maintenance.

Starting with modeling the business process, both the structures employed by the organization in documents and IT systems as well as mental models of the employees are being gathered in interviews and workshops. From the users' requirements and the analysis an initial knowledge structure emerges. The consensus about the final knowledge structure is achieved in an iterative way using workshops and involving users and management (cf. Fig. 6.4). In order to visualize the developed knowledge structure (cf. chapter 6.3.3), different methods of knowledge mapping, such as topic maps, knowledge carrier maps and process assistants, can be employed.

225 Cf. Fig. 2.3.
226 Cf. chapter 2.
227 Cf. chapter 2.

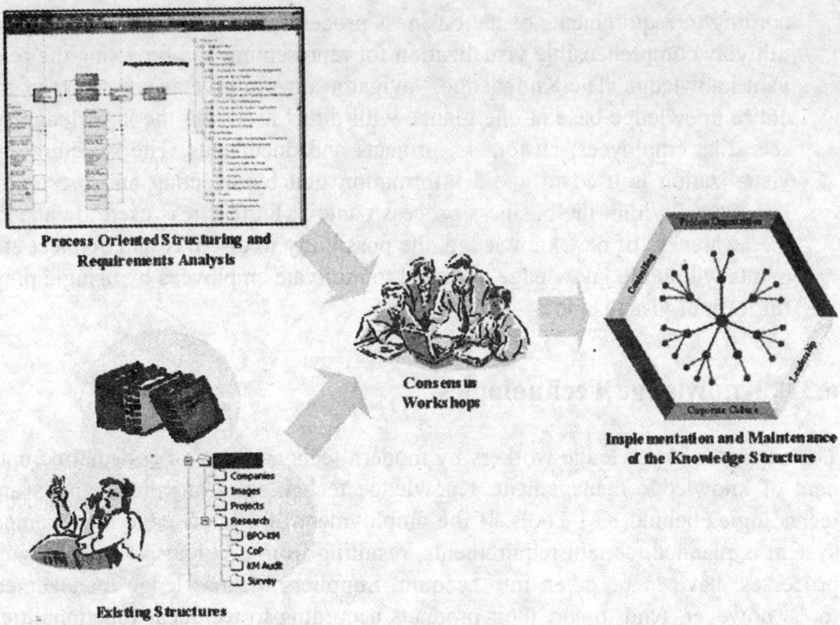

Fig. 6.4: The Principle of Process-Oriented Development of a Knowledge Structure

Following this approach, three different types of knowledge structuring and their visualization have been developed by the Fraunhofer IPK and are currently applied and tested in organizations.

- **The Web-Based Process Assistant:** The Process Assistant is a KM tool that presents knowledge and information according to the organizational business processes. The user navigates through single steps of his work while being supported by relevant connected knowledge and information elements (e.g. forms, important further readings, tips and tricks from experts, contact to experts and customers, work descriptions, web links etc.). The Process Assistant is XML-based and can be automatically generated from the process-modeling tool MO²GO.[228]
- **The Fraunhofer IPK Topic Map Engine:** The IPK Topic Map Engine serves as a tool for research and navigation within the knowledge base. The Topic Map Engine visualizes the semantic network of the relevant concepts of an organization. It allows to navigate from concept to concept in an associative way.
- **The Knowledge Navigator:** The Knowledge Navigator is a KM tool, which allows the handling and organization of complex knowledge structures ac-

[228] Cf. Fig. 2.12.

cording to requirements of the business processes. The Navigator uses an intuitively comprehensible visualization for representing and browsing the relevant knowledge. The Knowledge Navigator aims to give an overview of the entire knowledge base at one glance with direct access to the knowledge assets, like employees, customers, projects and documents. The structure and visualization is used to avoid information glut by selecting and presenting knowledge within the business process context. Further it is used to enhance the awareness of new knowledge, the possibility to discover and connect elements within the knowledge base and to motivate employees by using a playful form of visualization.

6.3 Knowledge Technologies

The support of knowledge workers by modern technology is an essential component of knowledge management. Knowledge technologies are the methods and technologies behind KM-Tools. If the employment of a knowledge management system is planned, certain requirements, resulting from problems in current work processes, have to be taken into account. Suppliers of knowledge management tools, however, tend to sort their products according to technical functionalities, which will be called technology enablers in this chapter. In the process of selection, these two attitudes have to be connected. Finally, we want to elaborate on potentials of visualization of knowledge and point out several possibilities of this important cross-section-like function.

6.3.1 Various Classifications of Knowledge Management Tools

The comparability of the KM-tools that are available on the market is not a simple issue. In order to get an overview of the various systems, the classification of tools can be helpful. In the following we would like to introduce some methods of classification. We start with a brief historical overview and refer to some approaches for the classification of KM-tools. In many cases, a historical classification can explain the special use of a certain product or how the manufacturer understands its use.[229] The following historical roots are relevant:

- tools which have been developed out of classical information archives or the retrieval of information,
- solutions from the field of communication and re-activated concepts from the field of artificial intelligence come into play in the analysis and automated classification of documents as well as in automated searches,
- tools for modeling explicit knowledge structures such as business processes modeling tools and Topic Map Tools, and finally

---

229 Cf. also the very comprehensive study by Maier (2002).

- software that attempts to combine several techniques and to support different tasks in knowledge management equally.

The next approach is an example of a well-structured architectural model. The classification of tools beyond an abstract model of system development can help to classify existent technology. Tools that are on the market can be compared with each other and their strengths and weaknesses can be evaluated on different levels of the architectural model. In addition, this kind of model can help to specify new and individually adjusted solutions. On top of making individual solutions possible, this model displays how knowledge management can be formed independently of one tool or of one company.

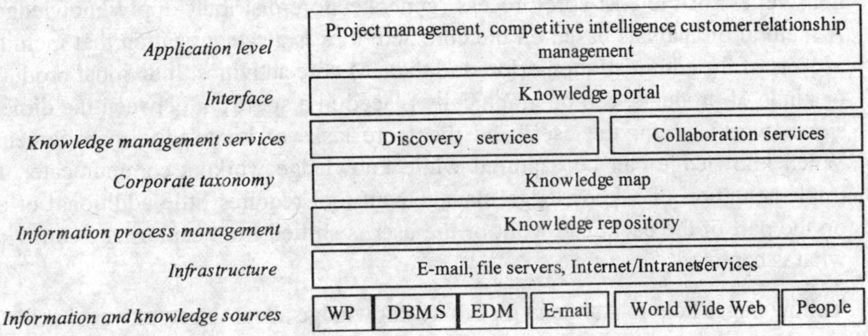

Fig. 6.5: Ovum KM Tools Architectural Model[230]

Individual systems cover these levels with differing degrees of intensity. Sometimes, not all elements are at hand. The initial level of the model consists of information and knowledge sources. These are delivered to the upper levels by means of infrastructure. On the next level, the information management process level, the access control to the knowledge repository is handled. The knowledge structure, here called corporate taxonomy, defines important knowledge categories within the company and establishes the interface between the data level and the knowledge and information level. On the next level, services for the application of knowledge are provided, e.g. in the form of visualization tools, and for collaboration, e.g. in the form of collaborative filtering. The user interface is described as a 'portal' that gives the user access to the knowledge base. Portals often provide functionalities to design a personalized interface with direct access to the individually required information and knowledge sources. From applications such as project management, CRM or competitive intelligence the user can access the relevant knowledge through the portal.

A further possibility is to categorize according to the functionalities from which knowledge management systems are constructed. Most knowledge management

230 Woods, Sheina (1998), p. 8.

tools use existing technologies to provide a collaborative framework for knowledge sharing and dissemination. They are implemented by use of e-mail and groupware, intranets, information retrieval and document management systems. Applications from data warehousing to helpdesks can be used to improve the quality of knowledge management[231].

Finally, classification according to the core tasks of knowledge management systems can be applicable. Tools can be examined regarding the support they provide for the tasks of generating, storing, distributing and applying knowledge. According to the IPK approach, a solution based on knowledge management should give support to a company's dealing with knowledge in a way that ensures the comprehensiveness of the KM-core process (generate-store-distribute-apply knowledge). A method of analysis based on the core activities is a categorization that includes aspects of the administration effort and the KM core activities. Individual products or kinds of products, can be graphically placed in a spectrum between the dimensions 'workload for the user' and the 'core tasks of knowledge management'. When knowledge can be captured while knowledge workers communicate, the implementation of a knowledge management tool requires little additional effort on the part of the user. The work of the user is shifted from capturing to utilizing what is captured[232].

Yellow pages help to store and distribute knowledge about the employees' skills.

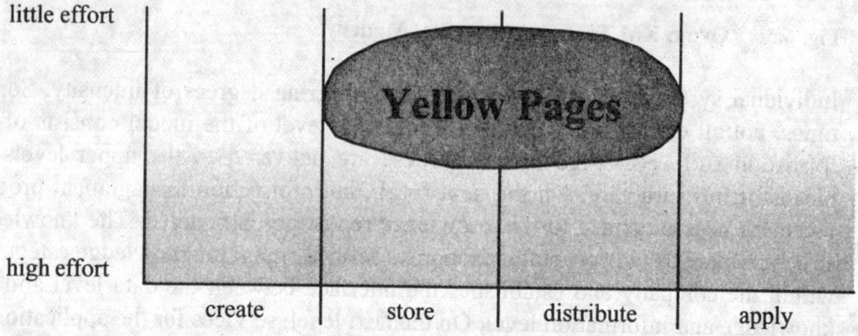

Fig. 6.6: IPK Categorization with an Example

Therefore, this tool would be placed in the figure as shown (cf. Fig. 6.6). If properly implemented, little effort is required from the user to up-date the skill profile. Some applications provide semi-automated functions for this task.[233] Horizontal categorization into the four core activities does not take place with the same continuity as vertical categorization.

231 Woods, Sheina (1998), p. 36.

232 Bair, O'Connor (1998), p. 23.

233 Cf. KnowledgeMail in Chapter 13.

The overall advantage of this categorization is that it connects the functionalities of KM-Tools directly to the core activities of knowledge management performed by the end-users. It also provides a means of analyzing strengths and weaknesses showing which core activities are supported and which not. It illustrates that the storage process is supported by several functionalities or tools, whereas the creation function is neglected. This analysis identifies shortcomings at an early stage, thus providing a valuable basis for decisions as to the selection of tools.

Integrated in our knowledge management approach, the results supplied by the model also become relevant for other areas. For example, if there is no suitable tool and defined process for storing knowledge, employees cannot be expected to be able to exchange relevant project information systematically and to retain it for future reference. If the exchange is not supported technically, the analysis of the employees' behavior may provide a tool to support these activities.

6.3.2 Technology Enablers

In order to combine the users' view on KM-Tools with the view taken by knowledge managers, it is necessary to tie the technical functionalities offered on the market to the four core activities. These technical functionalities are also called technology enablers.

The following classification of tools and technology enablers is in accordance with the core tasks of knowledge management in Fraunhofer IPK's approach. The table is an aid to see on which core task the main focus of a knowledge enabler is placed. The transitions between the single core activities are sometimes blurred, so that some functionalities contribute to more than one core activity. A knowledge management system should provide a seamless support for all four core activities.

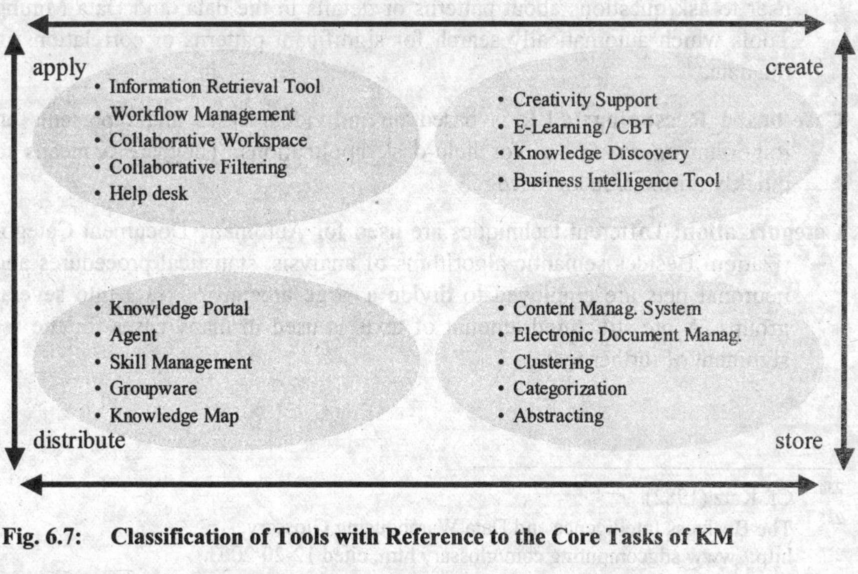

Fig. 6.7: Classification of Tools with Reference to the Core Tasks of KM

We assigned tools and technologies into the category of "creating knowledge" if they help to generate new ideas and insights or if they teach the user something new. Storing knowledge in documents involves marking it with additional information (meta data) for retrieving it. Tools, which support this process, are put in this category. It is part of the distribution of knowledge to find its right addressees. How does the tool support the delivery of the right knowledge to the right people? Here, user profiles with a description of the knowledge needed or sought can be used. With respect to software tools, storage and distribution are often very similar. The application of knowledge is supported when the tool helps users to find the right documents and files and then enables them to easily apply that knowledge to the actual work to be done. We identified 25 basic KM-Technologies that are listed and briefly explained in the following list.

Abstracting: Automatic Abstracting or Summarization of Texts tries to find the most relevant parts of a document to build a summary that provides a clue to the relevance of the material. It is a valuable method to determine whether or not a user really wants the document.[234] Abstracting can be improved if meta-data of the text is available.

Agent Technology: Software agents based on the essentials of artificial intelligence enable the user to let the system independently search for information according to a personal profile and to use various sources and other agents.

Business Intelligence Tools: Software that enables business users to see and use large amounts of complex data. Three types of tools are referred to as Business Intelligence Tools: Multidimensional Analysis Software (OLAP – Online Analytical Processing) that gives the user the opportunity to look at the data from a variety of different dimensions; Query Tools allowing the user to ask questions about patterns or details in the data; and Data Mining Tools which automatically search for significant patterns or correlations in the data.[235]

Case-based Reasoning: CBR is based on individual cases and represents an important approach e.g. for help-desk applications. They are a means to quickly utilize case knowledge.

Categorization: Different techniques are used for Automatic Document Categorization. Besides semantic algorithms of analysis, statistical procedures and neuronal nets are employed to divide a large amount of texts into several groups. A pre-structured amount of texts is used in many cases for the assignment of further texts.

---

234 Cf. Katz (1982).

235 The Business Intelligence and Data Warehousing Glossary, http://www.sdgcomputing.com/glossary.htm, cited 12-20-2003.

Clustering: Clustering of data is a method by which large sets of data are grouped into clusters of smaller sets of similar data[236]. A matrix is created, which contains values and reports the distance between two documents. Documents are close if they contain the same concepts. Clustering of Text Documents and Automatic Text Classification are related to each other.

Content Management System (CMS): CMS is a tool that enables a variety of (centralized) technical and (decentralized) non-technical staff to create, edit, manage and finally publish a variety of contents (such as text, graphics, video etc.), whilst being constrained by a centralized set of rules, processes and workflows that ensure a coherent, validated website appearance.[237]

Collaborative Functionality: Collaborative Functionality helps to retrieve knowledge by focusing on the meaning and value given to information through collaborative use.

Collaborative Filters: Collaborative Filters track the popularity of concepts by paying close attention to evaluations by experts and communities of practice. These filters define what is likely to interest a user based on what is significant or reliable for other users and based on user profiles. They can alert the user whenever a new document is added or changes are made to existing ones.

Collaborative Workspace: A Collaborative Workspace is a shared electronic area, where group members can work collaboratively. A workspace is organized into separate but interdependent areas, each one providing structure and functional aspects to perform activities related to a learning task. The structure and organization of a collaborative workspace depends on the group, the learning task and the nature of the collaboration[238].

Creativity Support: Tools, which support the employee in the creative process of generating ideas, in brainstorming or in the development of innovations, can be subsumed under this concept. Examples are Mindmapping Tools, Brainstorming Support and tools for visualization.

Data Analysis: Pattern recognition, classification and forecasting are the techniques used for data analysis. Data analysis is a method useful for generating new knowledge.

---

236 Center for new Engineering, http://cne.gmu.edu/modules/dau/stat/clustgalgs/clust1_bdy.html, cited 12-20-2003.

237 Content Management Systems Europe, http://www.contentmanager.eu.com/history.htm, cited 12-20-2003.

238 Barros, Verdejo (1998).

Data Warehousing: A modern database where data and information are stored. Connections that are not readily apparent can be uncovered with the use of Data Mining and OLAP. These techniques are part of data analysis.

E-Learning / CBT: Abbreviation for "Computer Based Training", comprises learning systems, which are based on computer and multimedia technologies[239]. They support just-in-time trainings in which employees can acquire knowledge and skills suitable in the best possible way for their needs.

Electronic Document Management: Documents are a central means of storing and spreading knowledge. Procedures for using and maintaining such documents, such as to check whether an update is overdue (document lifecycle), can be easily implemented in knowledge management systems.

Groupware: Groupware is another basic technology that is used for knowledge management systems[240]. Groupware offers a platform for communication and collaboration between employees. Ideally, groupware consists of software for e-mail, scheduling, document management and workflow automation.

Information Retrieval Tools: By means of Information Retrieval, tasks reaching from text searches to the automatic categorization and summation of documents can be solved. Advanced search algorithms use Thesauri and Text Mining to discover contexts that could not be found with simple queries. Semantic text analyses can also be implemented.

Knowledge Discovery: Discovery of new links and patterns in the knowledge base which are not evident. Technologies used for this purpose are e.g. Data Mining, Text Mining und Data Warehousing. The mining procedures work with statistical procedures based on rules.

Knowledge Map: Knowledge Maps are systematically developed[241] navigation tools, which may represent an organization's explicit and tacit structures of knowledge[242]. They can refer to databases, documents and knowledge gath-

---

239 ARCHmatic-Glossar und -Lexikon, www.glossar.de, cited 12-20-2003.

240 Cf. Tiwana (2000), pp.80.

241 Cf. Schüppel (1996), p. 214.

242 Cf. Eppler (2003), Grey (1999), http://www.smithweaversmith.com/knowledg2.htm, Bergmann: http://www.flexible-unternehmen.de/fb990708.htm, Guretzky: http://www.community-of-knowledge.de/cp_artikel.htm?artikel_id=39, cited 12-20-2003.

ered by experts and teams[243]. They are either signposts containing meta-information or refer to it.[244]

Knowledge Portals: Knowledge Portals are web-based applications that offer a single access to various knowledge sources. They can be customized by the user in order to serve the specific knowledge needs. They are well suited for the distribution of knowledge throughout all levels of an organization.

Semantic Functionality: Semantic functionality helps the user to recognize different concepts with similar meanings or similar concepts with different meanings in different contexts. Also to quickly see similarities among concepts and to make sense of vast quantities of information objects based on their content. This is done by means of clustering, automatic categorization, semantic networks, company dictionaries and thesauri, linguistic analysis and data extraction, rule-based systems and pattern identification.

Skill Management: Skill management deals with the maintenance and documentation of employees' skills. Clearly, it is necessary to predict skills needed for the future and to take measures to acquire them. Yellow pages, which contain the names and competency profiles of company experts, help to get into contact and increase networking capabilities of new employees.[245]

Text Analysis: There are linguistic and statistical approaches to automatic text analysis. Text analysis is employed for categorizing, classifying and clustering documents.

Text Mining: Text mining is concerned with the task of extracting relevant information from natural language text and to search for interesting relationships between the extracted entities.[246]

Workflow Management System: The business processes of a company are a major source of knowledge. In addition, the integration of knowledge management into business processes is an important factor for success. Workflow Management Systems automate the communication and information processes within an organization by the means of defined rules and standardized documents.

---

243 Stewart (1997) S. 119,
Bergmann op. cit.: http://www.flexible-unternehmen.de/fb990708.htm,
Guretzky: http://www.community-of-knowledge.de/cp_artikel.htm?artikel_id=39, cited 12-20-2003.

244 Cf. Davenport/Prusak (1998).

245 Cf. Weber (2000), pp. 161.

246 Cf. Fraunhofer AiS – Textmining, http://www.ais.gmd.de/KD/textmining.html, cited 12-20-2003.

6.3.3 Visualization

Although the psychological findings about the cognitive effect of visual forms of presentation are still incomplete, it is relatively certain that visual information is easier to memorize than verbal information[247]. It can be perceived more quickly and completely than texts, series of numbers or acoustical signals[248]. In cognitive psychology, this process is even described as a visual encoding of information and the subconscious establishment of cognitive maps[249].

As an instrument of knowledge management, knowledge maps serve to visualize complex structures of knowledge. By means of Knowledge Maps, the organizational knowledge is made available for all members of the organization and can be multiplied through a large-scale use within the organization[250]. Knowledge Maps provide an opportunity for users to see connections between seemingly disparate items. Enhanced by this view users can draw conclusions, interpretations, and decisions[251]. Our natural abilities to recognize visual and spatial clues and patterns are supplemented by means of knowledge visualization tools[252]. Visualization helps to retrieve knowledge by

- helping the user to easily navigate through large quantities of information,
- presenting knowledge in a way that helps users to see similarities and connections between knowledge and information objects,
- presenting knowledge in an interactively accessible way.

This is done with computer graphics (2D and 3D), animation and metaphors for navigational views for the representation of information[253]. Even though, according to our benchmarking study for knowledge management[254], visualization by means of knowledge maps has only seldom been used up to now, it is an instrument, which increasingly gains importance. The instrument of knowledge maps causes a certain fascination on behalf of managers. According to a survey by Romhardt, many managers are sure that knowledge maps will be used more frequently in the future[255].

---

247 Cf. Issing (1988), p. 544.
248 Cf. Fischer (1998), p. 29.
249 Probst, Raub, Romhardt (1998), p. 277.
250 Sveiby (1997), p. 45.
251 Bair, O'Connor (1998), p. 23.
252 Rao, Sprague (1998), p. 70.
253 Bair, O'Connor (1998), pp. 25.
254 Mertins et al. (2001), pp. 97.
255 Romhardt (1998), p. 125.

6.4 Choosing the Right KM-Tool

It is not easy to choose the right KM-Tool for employment in an organization. Like in the case of choosing software tools in general, several criteria have to be borne in mind. The following questions give a short survey of the criteria that are of influence and have to be considered:

Technical Framework[256 257]:

- Is the existing infrastructure homogenous or heterogeneous?
- Does the organization develop or buy applications?
- Which requirements have to be met by its performance: response time, amount of data, amount of users?
- Scalability: how adjustable is the system to infrastructure and possible growth?
- Can a new system co-exist with the existing infrastructure or is it even compatible with it?
- How long are systems kept operational?
- What kind of technical development can be expected? How long does the chosen technology comply with the market standard (independence from platform)?
- How well developed or proven is the infrastructure on the one hand and the new system on the other?
- What sorts of skill levels on behalf of technical resources (e.g. administrator, knowledge engineer or software developer) are needed?
- How aggressively does the organization adopt new technology?
- How much effort does the implementation of the different systems require?
- What is the supplier's market position?
- What kind of support is provided?

256 Cf. Oracle (2002), p. 3.
257 Tiemeyer (1997).

Functionalities Framework:

- What functionalities must the tools have in order to respond to the needs of the organization?
- Which functionalities can be covered with the systems that are already in existence?
- Can missing functionalities be supplemented by additional systems?
- Where do functionalities exceed the desired effect and where are they insufficient?
- What is required by the user and the knowledge engineer with regard to the handling of the functionality?

The tools offered on the market mostly combine different functionalities. In addition, they differ a lot from each other as to their nature and quality. Important criteria for comparison between KM-Tools are:

Economic Framework:

- How much does the purchase of the system cost?
- How much effort does the initial configuration of the system demand?
- How high are the annual costs for licenses?
- What budget is needed in order to maintain the system (e.g. support, hotline assistance, training) and to follow the development of the system (e.g. upgrades)?

Apart from these, the general software requirements, such as protection against unauthorized access to data, multi-lingual applicability, mobility, suitability for handicapped users etc. have to be taken into consideration.

In the following, we want to introduce a method that supports the systematical process of choosing knowledge management tools. The first step to be taken is the "requirements analysis". In this regard, the Fraunhofer IPK concept recommends an audit in order to find out the present IT situation, the general set-up and cultural aspects concerning the dealing with knowledge and information.[258] In a second step, an analysis of the requirements focusing on the business process takes place. This analysis is based on the understanding that dealing with knowledge is part of the daily routine. Here, the present situation is further specified by means of process-oriented expert interviews. The requirements for a KM solution from the perspective of the system's future users are being examined and analyzed. The result of the analysis is a requirements document, which contains the most important requirements for the system and the general set-up, which has to be borne in mind.

258 Cf. chapter 3.

This is a simple example of a profile of requirements generated from a requirements document:

- create transparency regarding employees and competences,
- survey and localization of the organization's knowledge topics,
- ensure the validity of knowledge objects.

From these requirements, those functionalities must be derived which are able to fulfill the requirements that result from the present situation.

In the given example, these are the KM functionalities of

- Skill Management / Yellow Pages,
- Knowledge Map (visualized structures),
- Electronic Document Management (Document Lifecycle).

Fig. 6.8 shows a graphical evaluation of the analysis phase. The various functionalities are listed along the horizontal axis, while the spectrum between the technological state of the art and fundamental technologies is depicted on the vertical axis. The broken lines in the lower part of the individual columns depicting the functional aspects reflect the present situation respectively. The upper broken lines represent the results from the requirement analysis. The space between them reflects the discrepancy between the present situation and the desired situation. Of course the requirements can undercut the current inventory as well, signaling that a present functionality has already been acquired.

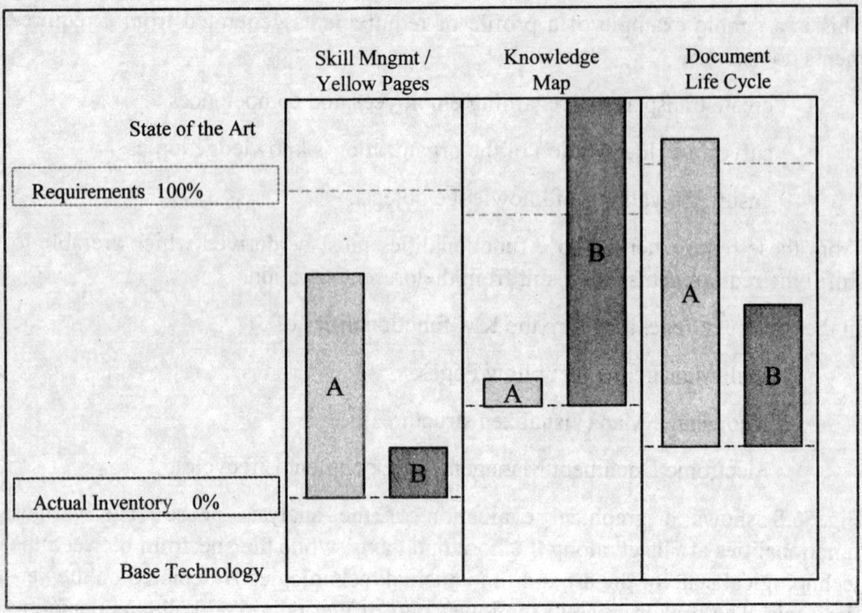

Fig. 6.8: Comparing Tools

On the basis of ascertained functionalities as well as the respective present situation and the requirements waiting to be fulfilled, the relevant software tools on the market can be screened, and tools providing the required functionalities can be selected. Most of the common tools offer several functionalities. Thus, one tool will generally cover more than one column. According to the example we have chosen, the figure juxtaposes two different KM-Tools selected on the basis of previous screening. Tool A represents a classical document management system, with an additional skill management module. Structures, however, are merely depicted by way of a classical explorer structure. The strengths of Tool B lie in structuring and visualization, and it is able to alert the users if the document has changed. However, in order to keep track of the employees' competencies it is necessary to create skill profile documents. When the functionality does not reach one of the broken lines, the tool is not adequate with regard to the ascertained requirements. It therefore has to be dismissed completely, or else the requirements would have to be adapted accordingly. If there are several tools in the figure, like in our example, the differences between them, concerning the actual requirements within the organization, can be recognized without difficulty. Thus, their strong and weak points, as well as the possibility of tools complementing one another, can be made transparent.

Concerning the selection of an adequate KM-Tool, one does not only have to take account of individual functionalities and of their factual results with regard to the requirements, but one must also consider the general economical set-up, services, the supplier's market power etc. A value benefit analysis can help considering all criteria of selection (see list above) and their importance with regard to the identified requirements.

Fig. 6.9 shows an example of value benefit analysis. All criteria of selection are listed in a chart (vertical list); and their importance with regard to the requirements of an organization is ascertained. The tools are transferred horizontally into the chart and subsequently are appointed evaluations in numbers from 1 to 100 with regard to whether the given criterion has been fulfilled. After that, the individual evaluations are multiplied by their factor of importance, and then, the individual tool columns are added up, with the result being a specific score or value for each given tool. If this form of analysis is established for every tool and every criterion, the values of the individual tools can be handily compared. The highest score or value, therefore, designates the most adequate product with regard to the requirements and criteria.

Value-Benefit Analysis					
		Degree of Performance			
		Tool A		Tool B	
Criteria	Weighting	%	pts	%	pts
Skill Management / Yellow Pages	25%	75	19	20	5
Knowledge Map (visualised structure)	15%	10	2	100	15
EDM (Document Life Cycle)	15%	100	15	50	8
Performance	10%	80	8	70	7
Maturity	5%	100	5	90	5
Cost of Purchase	15%	20	3	35	5
Cost of Operation	15%	20	3	20	3
		Value of Benefit			
Sum	100%		54		47

Fig. 6.9: Example of Value–Benefit–Analysis[259]

In this example, some criteria for the evaluation of tools have been selected and their importance with regard to the objectives of the organization has been determined. After performing the evaluation of both tools by every criterion the aggregated result shows that Tool A is more appropriate for this organization than Tool B.

For further readings, there is a list of tools and market leaders in the annex, giving an overview of KM-Tools available on the market.

6.5 Outlook

There are already many specific KM solutions on the market. KM IT solutions, however, are frequently just a collection of methods and various functionalities which are, by now, quite easy to handle. It is the selection of an adequate tool, the elementary structuring, and the introductory measures it requires, which determine

259 Tiemeyer (1997).

the actual cost of implementing a KM system. It has become obvious that a company's corporate culture and the barriers it brings about as well as the system's potential acceptance by future users are critical factors for success. The future challenge therefore consists in the development of methods, which support the interplay of these technologies and their required integration into the organization's business processes while simultaneously including the employees.

Moreover, in recent years knowledge management has increasingly concentrated on mobile access to huge and widely spread knowledge sources such as those existent in international organizations, on the internet, and in topic networks. The idea of utilizing this globally available knowledge by way of quick access to the sources independent from geographical factors is, of course, fascinating. Therefore the current developments focus on standardizations based on natural language structures as well as so-called semantic based knowledge technologies. These methods and technologies, which can be subsumed under the term semantic web, are supposed to allow for automated acquisition and exchange of knowledge as well as to provide information about exchange languages such as XML and standards based on it.

The integration of the current context of knowledge acquisition and exchange in automated processes is one of the major challenges ahead. Only if these contexts are taken into account, there will be adequate results responding to the users' demands.

Automation on the basis of actual contexts of work as well as standardizations simplifying the exchange of knowledge are therefore the most urgent tasks of computer-aided KM. Additionally, KM has become an activity operating beyond the boundaries of individual organizations, and even beyond national boundaries. The support from information technology is thus necessary and will continue to play a decisive role in the future of KM.

6.6 List of Knowledge Management Tools

The following knowledge management tools were selected due to their leading market position, their innovative approach, and their wide use. The tools we have included were listed and investigated in the relevant professional literature. Our evaluation is based on information from those professional sources and from our own research. Beside our research from the first edition the following sources were used: Meta Group[260], Delphi Group[261], Vision Project Report[262], Business Communities[263]. The List of web addresses is updated last on 17th of December 2002.

We grouped the KM-Tools as followed:

1. Complete KM suites
2. Learn and teach
3. Ontologies
4. Portals
5. Process-oriented KM
6. Search engines / categorization tools / intelligent agents
7. Skill management
8. Toolkits for developing individual solutions
9. Virtual Teams / Collaboration
10. Visualizing tools

Every tool belongs to one group. Although some tools could be included in more than one group, the main emphasis of the product was decisive.

260 Meta Group (2002).
261 Delphi Group (2002).
262 Vision Project (2002).
263 Bullinger et al. (2002).

Complete KM Suites

Complete KM suites comprises a rich environment for communication and good integration of existing knowledge sources as well as powerful search engines combined with a portal and yellow pages.

Company	Product	Web
Autonomy	Active Knowledge	http://www.autonomy.com/
Blue Angel Technologies	MetaStar 7	http://www.blueangeltech.com/
BOV AG	Bazaar42	http://www.bov-ag.com/produkte/con_bazaar42.asp
Comma Soft AG	Infonea	http://www.comma-soft.com/
Documentum	Documentum 5	http://www.documentum.com/
Empolis GmbH	e:kms	http://www.empolis.com/products/prod_emp.asp
Gauss AG	VIP Enterprise 8	http://www.gauss.de/
Hummingbird	Hummingbird KM	http://www.hummingbird.com/products/km/overview.html
HyperWave	eKnowledge Infrastructure	http://www.hyperwave.com/e/
OpenText	Livelink	http://www.opentext.com/
Verity	K2	http://www.verity.com/products/index.html

Learn and Teach

In this category we grouped tools that help to create new knowledge with enriched presentations. Users can learn from others and create new knowledge for themselves.

Company	Product	Web
Chromedia GmbH	MediaNaut	http://www.medianaut.com/
IBM	Learning Space	http://www.lotus.com/lotus/products.nsf/fa_prohomepage
HyperKnowledge	HyperKnowledge	http://www.hyperknowledge.com/

Ontologies

A number of companies have products that employ ontology-based technologies. Because of the high relevance of this technology for the field of upcoming semantic web developments, these tools are grouped into an own category.

Company	Product	Web
Autonomy	Active Knowledge	http://www.autonomy.com/
Blue Angel Technologies	MetaStar 7	http://www.blueangeltech.com/
Aidministrator	Sesame	http://www.aidministrator.nl/
akados	KAMATO	http://www.adakos.de/KAMATO/Features.htm
Applied Semantics	Auto-Categorizer	http://www.appliedsemantics.com/
Applied Semantics	Metadata Creator	http://www.appliedsemantics.com/
Applied Semantics	Page Summarizer	http://www.appliedsemantics.com/
Empolis GmbH	k42 Knowledge Server	http://www.empolis.com/products/prod_k42.asp
Intellidimension	RDF Gateway	http://www.intellidimension.com/
Intelligent Views	K-Infinity	http://www.i-views.de/web/
Interwoven	MetaTagger	http://www.interwoven.com/products/content_intelligence/
Ontoprise GmbH	Ontobroker	http://www.ontoprise.de
Semagix	Freedom	http://www.semagix.com/products/overview.shtml
Semantic Edge	Knowledge Tool Suite	http://www.semanticedge.de/solutions/en_solut_knowledge.htm
Semtation GmbH	SemTalk	http://www.semtalk.com/

Portals

The idea of a portal that makes all knowledge available at one time and place is found in many of the tools we investigated. Most of the complete knowledge management suites were implemented as portals. In this group, tools focus on a user interface that grants access to a multitude of different sources and search engines. Software like the Knowledge Navigator of the Knowledge Raven Management GmbH supports the user additionally with active and selective modules.

Company	Product	Web
BEA Systems	WebLogic Portal	http://www.bea.com/products/weblogic/portal/index.shtml
Broadvision	One-To-One Portal	http://www.broadvision.com/OneToOne/SessionMgr/home_page.jsp
CAS Software AG	genesisWorld	http://www.cas.de/Produkte/frameset.asp?nr=1&ex=1
Computer Associates	CleverPath	http://www3.ca.com/Solutions/Solution.asp?ID=303
Convera	RetrievalWare	http://www.convera.com/Products/index.asp
EASY Software	EASY Portal	http://www.easy.de/
IBM	WebSphere Portal	http://www-3.ibm.com/software/info1/websphere/index.jsp?tab=products/portal
Inducad Systemberatungsgesellschaft mbH	induBIS	http://www.inducad.com/indusql/index2.php
Infopark AG	NPS Portal Manager	http://www.infopark.de/de/products/pm/index.html
Knowledge Raven Management GmbH	Knowledge Navigator	http://www.knowledgeraven.com
Microsoft	SharePoint Portal Server	http://www.microsoft.com/sharepoint/
Odissey Development	Isys:Web	http://www.isysdev.com/products/index.shtml
Oracle	Oracle9i AS Portal	http://www.oracle.com/
PeopleSoft, Inc.	Enterprise Portal	http://www.peoplesoft.com/corp/en/products/technology/appcon/portals/features.asp
Pironet AG	Pirobase	http://www.piro.net/servlet/PB/menu/1000170_l1/index.html
Plumtree Software	Corporate Portal	http://www.plumtreesoft.com/products/platform/
Wherewithal	Enterprise Knowledge Portal 3.0	http://www.wherewithal.com/pb/
Xerox	AskOnce / DocuShare	http://www.xerox.com/

Process-Oriented KM

The focus of the knowledge management discipline and practice shifts now from traditional technology-driven to process oriented approaches. Some of the newer products take this into account and offer tools for process oriented KM.

Company	Product	Web
Action Technology	Action Works Coordinator	http://www.actiontech.com/products/
ActionBase	ActionBase	http://www.actionbase.com/
BOC GmbH	Adonis	http://www.boc-eu.com/german/adonis/adonis.shtml
Fraunhofer IPK	Mo^{2}go	http://www.um.ipk.fhg.de/mogo/bprhome.htm
HandySoft Corporation	BizFlow	http://www.handysoft.com/products/products.asp
IDS Scheer	ARIS 6.1 Collaboraive Suite	http://www.ids-scheer.de/english/index.php
Optika	Acorde	http://www.optika.com/solutions/index.cfm
Savvion	BusinessManager	http://www.savvion.com/products/enterprise.htm
Sopho Systems	Tansfer	http://www.sopho-systems.com/
TeamShare	TeamTrack	http://www.teamshare.com/products/index.html
Ultimus	Workflow Suite	http://www.ultimus.com/ultintro.htm
Vitria	BusinessWare	http://www.vitria.com/products/

Search Engines, Categorization Tools, and Intelligent Agents

These tools help to find the knowledge a user is looking for. This involves less user effort if the tool also categorizes knowledge or if knowledge is made individually accessible according to a profile given to an agent.

Company	Product	Web
80-20	Discovery	http://www.80-20.com/products/index.htm
80-20	Retriever	http://www.80-20.com/products/index.htm
Alexa Internet	Alexa	http://www.alexa.com/
AltaVista	AltaVista	http://www.altavista.com/

Company	Product	Web
Ask Jeeves	AskJeeves	http://www.ask.com/index.asp?intl=0
AskMe	AskMe Enterprise	http://www.askmecorp.com/product/default.asp
Ceyoniq Technology GmbH	Content Manager	http://www.ceyoniq.com/ceyoniq/index.cfm
CognIT a.s	Corporum	http://www.cognit.no/home_multi/html/index.asp
Divine	Northern Light	http://northernlight.com/enterprise/ese.html
EasyAsk	Precision Search	http://www.easyask.com/products/index.cfm
Endeca Technologies	ProFind	http://www.endeca.com/products/profind.shtml
Entopia	Knowledge Builder	http://www.entopia.com/products_pg3.10.htm
Entrieva	Semio Map	http://www.entrieva.com/entrieva/products/semiomap.asp?Hdr=semiomap
Exago	QuizXML	http://www.exago.com/quizxml.htm
Google	Google	http://www.google.com/
IBM	Intelligent Miner	http://www-3.ibm.com/software/data/iminer/
Inktomi Corporation	INKTOMI Enterprise Search	http://www.inktomi.com/products/search/enterprise.html
Inmagic	Interactive Web Knowledge Management	http://www.inmagic.com/sol_km.htm
InQuira	InQuira for Search	http://www.inquira.com/products/web_search.html
iPhrase Technologies	iPhrase	http://www.iphrase.com/
Mercado Software	IntuiFind 5	http://www.mercado.com/
Mohomine	Mohomine Classification	http://www.mohomine.com/products/index.asp
Schema GmbH	SchemaText	http://www.schema.de/sitehtml/site-d/wissens0.htm
Smartlogik	Smartlogik Decision Intelligence Platform	http://www.smartlogik.com/apr/technology.asp
Temis Group	Temis	http://www.temis-group.com/

Company	Product	Web
Textology	Categorizer	http://www.textology.com/products.html
Thunderstone	TEXIS CATEGORIZER	http://www.thunderstone.com/texis/site/pages/Products.html
USU AG	Knowledge Miner	http://www.usu-ag.com/english/eProdukte/KnowledgeManagement/products.html

Skill Management

The idea here is to know where the skills of the company's staff lie and how these skills can be managed.

Company	Product	Web
Meta4	Meta4 PeopleNet	http://www.meta4.com/
Orbital Software	Accolade	http://www.orbitalsw.com/
SkillScape	Learning Solutions	http://www.skillscape.com/
SkillView Technologies	SkillView Enterprise	http://www.skillview.com/

Toolkits for Developing Individual Solutions

These are development environments for implementing individual knowledge management solutions. Although every knowledge management tool has to be adapted to individual company needs, the tools in this group need more work to become a knowledge management solution.

Company	Product	Web
Adobe	Adobe Accelio Form Server	http://www.adobe.com/products/server/formserver/main.html
BackWeb	BackWeb eAccelerator	http://www.backweb.com/
CAS Software AG	CAS teamWorks	http://www.cas.de/Produkte/frameset.asp?nr=3&ex=0
Electronic Data Systems	Digital Collaboration	http://www.eds.com/products/plm/dc/
FileNet	FileNet WCM	http://www.filenet.com/English/products/WCM/Index.asp
IBM	Domino	http://www.lotus.com/lotus/products.nsf/fa_prohomepage

Company	Product	Web
IQDoQ GmbH	Hyperdoc	http://www.iqdoq.com/index.html
Lightspeed Software	Lightspeed Knowledge Appliance	http://www.lspeed.com/
Mediasurface	Mediasurface	http://www.mediasurface.com/product/?view=Standard
Microsoft	Exchange	http://www.microsoft.com/exchange/
Novell	GroupWise 6	http://www.novell.com/products/groupwise/
Promatis	INCOME	http://www.promatis.com/products/income_suite/index.htm
Saperion	SAPERION® Server	http://www.saperion.com/start_e.html
SAS	The Power to Know	http://www.sas.com/products/index.html
Savvion	BizCompomemts	http://www.savvion.com/components/
SER Systems	SERbrainware	http://www.ser.com/product_showcase/index.asp
Softarc	First Class	http://www.softarc.com/products/
Stratify	Solution	http://www.stratify.com/products/index.html
Sybase	Enterprise Portal	http://www.sybase.com/products
Tacit Knowledge Systems	KnowledgeMail	http://www.tacit.com/products/knowledgemail/
Tibco	Business Integration Solutions	http://www.tibco.com/solutions/default.jsp
Vignette	Vignette V6 Content Suite	http://www.vignette.com/contentmanagement/0,2097,1-1-1489,00.html
zapwerk AG	ucONE	http://www.zapwerk.com/

Virtual Teams / Collaboration

These tools strongly emphasize the communication aspect of knowledge management. Discussion groups and virtual team rooms are examples here.

Company	Product	Web
AEC communications GmbH	AEC/community	http://www.aeccom.com/

Company	Product	Web
Akiva	ChatSpace Community Server	http://www.akiva.com/products/commserv/index.cfm
Altavier	Knowledge Café	http://62.206.115.2/publicityaltavier/altavier.nsf/index/de_produkte.htm
blaxxun interactive	blaxxun Platform	http://www.blaxxun.com/solutions/applications/teamworkspace/index.shtml
cassiopeia AG	cassiopeia Community	http://www.cassiopeia.de/index.html
CaucusCare	Caucus Team	http://www.caucus.com/prod-services.html
eRoom Technology	Eroom	http://www.eroom.com/DWP/about_dwp.asp
EXITEC AG	Nects	http://www.exitec.de/communities/
ezboards	Gold Community	http://www.ezboard.com/ezcommunity/index.html
Fraunhofer FIT	BSCW 4.1	http://bscw.gmd.de/
Intraspect	Intraspect 5.5 platform	http://www.intraspect.com/products/
Jelsoft Enterprises	vBulletin	http://www.vbulletin.com/
Living Systems	living markets	http://www.living-systems.com/static/lsw_mainTemplate_en.html
Mongoose Technology	RealCommunity CiviServer	http://www.mongoosetech.com/realcommunities/experience.html
Multex	Buzzpower	http://www.buzzpower.com/pd_buzzpower.asp
OpenACS Community	OpenACS	http://openacs.org/
Pendulab Pte Ltd	Chatblazer	http://www.chatblazer.com/features.htm
PeopleLink	PeopleLink Suites	http://www.peoplelink.com/next/products.htm
Prospero	Prospero	http://www.prospero.com/
SiteScape	SiteScape Enterprise Forum 7.0	http://www.sitescape.com/
Someon GmbH	someon community software	http://www.someon.com/paco.php?content=1
	Sun ONE	http://wwws.sun.com/software/pr

Company	Product	Web
Sun		oducts/appsrvr/home_appsrvr.htm l
Thinkfactory.de	Opentopic	http://www.thinkfactory.de/produ kte.html
Webcrossing	Webcrossing 4.1	http://webcrossing.com/WebX/Ho me/products
Webfair	Community Engine	http://www.webfair.com/plugin/te mplate/webfair/1/*?lang=de

Visualization Tools

These tools can be used as a supplement to existing knowledge management solutions or they emphasize the presentation of knowledge.

Company	Product	Web
Autonomy	Visualization	http://www.autonomy.com/
InXight Software	Inxight VizServer	http://www.inxight.com/products/
MindJET	MindManager	http://www.mindman.com/
Moresophy GmbH	L4 Modeller	http://www.moresophy.de
Think Tools AG	Think Tool	http://www.thinktools.com/
TouchGraph	TouchGraph Browser	http://www.touchgraph.com/

7 Intellectual Capital: Measuring Knowledge Management

Robert Schindler, Arne Jaitner

The importance of the management of Intellectual Capital (IC) for a modern company has been widely accepted. Without doubt, this is due to developments of the present economic world in which knowledge has become a decisive competitive factor.[264] In order to be able to cultivate this production factor, it is necessary to measure and evaluate IC. IC, intangibles, intangible assets, knowledge assets, immaterial assets and intangible resources etc. are terms which are commonly used synonymously. In general, these terms designate the non-physical or immaterial assets of a company. Usually, IC refers to the whole of immaterial assets, including patents and copy-rights as well as employees' know how, brand value and corporate image.[265] The following introduction offers a general overview on the development of the issue of intangible assets and suggests a theoretical foundation for the growing importance of IC. The second part of the article discusses various approaches towards measuring IC before giving an outlook on the subject.

The rise of the importance of IC can be interpreted as a fundamental change of the economic strategic paradigm. In this field, market-oriented and resource-orientated approaches compete. The classic theory of market structure explains the strategy of an enterprise and, subsequently, its potential economic success by the structure of the sector in question.[266] In order to succeed on the market an enterprise will have to attain, secure and develop a profitable and sustainable position among its competitors. The intensity and dynamics of the competition for a certain sector as well as its profitability is determined by the structure of the sector for

---

264 Cf. Stewart (1997).

265 There is no generally accepted definition of the term IC. This article will not attempt to clarify this question. Therefore different terms will be used alternatively in order to adjust to the terminology of other authors; however, they designate the same phenomenon.

266 This approach draws on the structure-conduct-performance paradigm determined by Mason/Bain within their theory of Industrial Organization. The following remarks refer in principle to Michael E. Porter's analysis of business structure, which is closely related to Shapiro's approach of strategic conflict. Cf. Mason (1939); Bain (1968); Shapiro (1989), pp. 125. With regard to Porter, cf. the following footnote.

which five fundamental criteria have been identified.[267] The stronger these criteria apply, the further the economic attractiveness of a sector decreases. The strategic behaviour of enterprises therefore has to be adapted to the market structure. From the Fifties to the middle Eighties, the classic market-oriented approach was, with few changes, determinative for the strategic orientation of enterprises. Yet the uncompromising application of market-oriented recommendations, the concentration on comparable generic strategies and instruments led to a relative similarity between market participants. Due to the ever-increasing similarity between the competitors, the attainable competitive advantages dwindled. As a result, other factors relevant for competition were searched after, focusing on the competitive environment and the enterprises themselves. The conceptional critique of market-oriented approaches emphasized that the sector structure as well as the employees′ behaviour was considered to be static and foreseeable.[268] But in the light of increasing competition – e.g. through enterprises acting globally and linked by electronic data processing – the market-oriented perspective proved too narrow. While a leading cost position, unique selling points and market niches can be defined relatively precisely for a national economy, this can be achieved only partly for a multi-national economic space. To name just one example – the worldwide communication networks allow for drastically reduced product development times to be achieved. Yet, because of shortened product cycles, the importance of the classic positioning on the market decreases. The industrial countries are particularly exposed to growing competition and rising costs by newly industrialized countries, which they, in turn, react to by way of concentrating on high technology and innovations.

A one-sided orientation on the market only partly considers the internal structures and processes. The basic topic of resources including the necessary reaction time of internal adaptation to changing frameworks is not considered. Yet the development of staff, the employees‘ competences and innovative abilities are crucial in coming to terms with technological change and increasing competition. As a result of this discussion, the flexibility and the ability of adapting human resources, and with them the intangible assets, has become more and more important since the Nineties. A market-oriented concept is put opposite a resource-oriented approach that defines internal structures and processes as crucial strategic potentials for success.[269] The central concern is no longer the analysis of a company's position within the individual sectors, but rather the analysis of its inherent strengths and

---

267 Namely the rivalry among the established companies in one branch of business, the purchaser's bargaining power, the vendor's bargaining power, the threat arising from substitute products, and competitors new on the market. Cf. Porter (1980), p. 26.

268 Mintzberg (1998), p. 116.

269 Wernerfelt (1984), pp. 171 is understood to be a predecessor of this attitude that has become a commonplace in the 1990s.

weaknesses. Instead of products and markets, internal structural and cultural factors such as IC become the focus of attention.[270]

When the notion of IC gained importance, the first publications on the importance of IC for enterprises came about.[271] A systematic development of an IC concept started in the early Nineties. Methods and tools that should establish documentation and evaluation of intangible resources were developed on both theoretical and practical levels. In the first place, these efforts were made to describe the internal knowledge of an enterprise to external stakeholders. In 1994, the Swedish insurance company *Skandia AFS* published a supplement to its annual report that served as an evaluation of the company's intangible assets. This was the first publication concerned with both IC and a reasonable dealing with it. Subsequently, tools were developed which rather aimed at the internal controlling of knowledge resources than at the external report.[272] In 1992, Kaplan and Norton introduced the concept of the Balanced Scorecard, which was to develop further during the following years.

7.1 Efforts of National and International Organizations

In the 1990s, the importance of intangible assets began to be recognized by national governments and international organizations in Europe and America as a factor contributing to the competitiveness not only of firms, but also of entire economies. In April 1996, for example, the US Securities and Exchange Commission (US SEC) held a symposium on intangible assets. The agenda for the symposium covered topics such as the nature and types of intangible assets, including intellectual property, human capital, software and related items. Authorities sponsoring European economy and research were also interested in the concept of knowledge-based economy from early on. In 1994, *An Industrial competitiveness policy for the European Union* was published, stressing the economic importance of training, research and development, and industrial property.[273] Subsequent studies initiated by the EC concentrated on the economic importance of information-and-communication-technology. The characteristic features of the so-called *New Economy* were to be defined as well as structures and mechanisms of intangible

---

270 Collins and Montgomery (1999) name five typical features of resources: low chance of being copied, persistent relevance for the company, availability within the company, the possibility of being replaced, contribution to the company's competitiveness.

271 Sveiby (1987).

272 In 1993, *Dow Chemical*, an American pharmaceutical company, began to categorize all of its patents and to develop an innovative strategy in order to generate new profits from their intellectual capital. They concentrated on their patent portfolio and established a methodology to evaluate these assets. 1996, *Dow* followed *Skandia's* example and published „Visualizing Intellectual Property in Dow".

273 Commission of the European Community (1994).

assets to be identified. A number of projects were initiated to work out the possibilities of measuring IC.[274] Also in 1998, a work group was established by the Danish Trade and Industry Development Council in order to investigate the way 10 Scandinavian firms manage their IC. Firms such as *Skandia*, *WM-data* and *Celemi* and the savings banks *Nordjylland* (SparNord), *Sverige*, *ABB* and *Telia* took part in the study that aimed at identifying the respective practical dealing with IC in order to create a basis for standardization.[275]

Part of the efforts of both governments and (inter-)national organizations are first attempts at adopting traditional accounting regulations to the new demands. Because of legal regulations, however, only few intangibles have yet been included in balance sheets. The respective German laws, for instance, does usually not allow for the accounting of intangibles, whereas the International Accounting Standard (IAS) indeed acknowledges some intangibles since 1999 (IAS 38), among others patents, copyrights, licenses, customer and supplier relationships.[276] According to IAS 38, the German Accounting Standards Board (GASB) is working on an according instruction (E-DRS 14, immaterial assets). In spite of these efforts, it must be stated that a standardized and internationally acknowledged guideline for the measuring of immaterial assets is still missing.

7.2 Methods for Measuring IC

In the second part of this article, more detailed descriptions and further examples of methods and tools for the evaluation of knowledge shall be introduced. In spite of the growing number of publications, it can be observed that new conceptional and technical developments in IC are commonly created by companies themselves and therefore show a strong practical concern.[277] The methods can basically be distinguished into either monetary evaluation or a system of indicators, trying to cover aspects that are hard to define in terms of money, such as the practical experience of employees, by non-financial or qualitative quantities.

The relation between approaches concentrating on financial/quantitative aspects on one hand, and non-financial/qualitative on the other, can largely be correlated

---

274 For instance, as part of the fourth and fifth R&D Framework, two research projects have been initiated in 1998: MERITUM (Measuring intangibles to understand and improve innovation management); http://www.fek.su.se/home/bic/meritum/, and MAGIC (Measuring and Accounting Intellectual Capital; http://www.profactor.at/magic/index/html/, cited 12-20-2003.

275 The Danish Trade and Industry Development Council: Memorandum (May 1997).

276 Cf. Vallabhajosyula (2001) for a survey of generally accepted accounting principles in the US.

277 A more detailed list, which is impossible to provide within the scope of this paper, can be found in: Sveiby (2001). Those methods not mentioned by Sveiby can be cross-checked in Andriessen/Thiessen (2000).

with the perspectives of company accounting by external reporting or internal controlling. Thus, financially oriented approaches are mostly used in the evaluation of companies themselves – e.g. in company sales – whereas non–monetary criteria are mostly used in internal controlling – e.g. in the development of IC. The approaches introduced in the following paragraphs thus fulfill different, yet sometimes overlapping functions. They are to be represented according to their financial/quantitative or non-financial/qualitative aspects.[278] The sequence in which they are introduced here corresponds to the growing complexity of the individual approaches and nearly reflects the historical sequence of their development as well. Through the bibliography and the recommendations for further reading, the interested reader may find further information about the approaches respectively than we are able to provide within the scope of this paper.

7.2.1 Measuring IC by Financial/Quantitative Indicators

In this paragraph, the approaches are arranged according to their chronology and their complexity as far as this was feasible. The approaches described here provide a survey of the spectrum of those approaches featuring a monetary perspective, but are perhaps just typical for another number of concepts.

7.2.1.1 Market-to-Book Ratio

The basic idea of one group of the given approaches is to calculate the difference between a company's market capitalization and its stockholders' equity as the value of its IC. A relatively simple possibility to evaluate the value of a company's IC, especially from an external perspective, is Market-to-Book ratio: IC can be easily expressed in numbers when one subtracts the book value of a company from its market value. This form of computing IC is simple and easily reproducible, assuming that everything remaining after the subtraction of the book value is intangible. The problem of this method, however, is obvious. The stock market is subject to many macro-economic factors, not least because of speculation. Additionally, this numeric value of IC could only be ascertained in the case of a company's actual quotation, in order to determine a market value without difficulty.

---

278 Different systems of classification have been proposed: Sullivan (1998), p. 20, e.g. opts for a distinction in terms of value creation versus value extraction. Sveiby (2001) wants the classification system to orient on basic forms of *procedere*, such as market capitalization methodologies, return on assets methodologies, scorecard methodologies and direct intellectual capital methodologies. North (1998), in turn, proposes a distinction of deductive-summary and inductive-analytic approaches.

7.2.1.2 Tobin's Q

Tobin's Q is a quotient comparing the market value of an intangible with the cost of its reproduction.[279] If q is less than l, the market value of an asset is less than the cost of its reproduction in consequence. Although Tobin's Q was not used as a procedure for computing IC, it can be used in this context. According to Tobin, q begins to take effect when a company enjoys unusually high profit from something other companies do not have. Therefore, it is an appropriate definition of the potential, which is manifest in IC such as knowledge assets, in whose case the market value exceeds the cost of reproduction. In this case, the company probably has tangibles similar to those of other firms, but it clearly has a competitive advantage in those areas determining IC, such as employees, ICT, or customers. It is possible to apply Tobin's Q to individual assets as well as to an entire company; the market value of the company is then divided by the reproduction costs of the sum of tangible assets it disposes of. It is, of course, a clear advantage that a figure for the economic value of an individual intangible asset can be established. Nevertheless this approach has not been widely accepted what might be due to the difficulty to obtain the relevant data. As well, Tobin's Q does not provide sufficient information when it comes to the operational management of the asset, since it does not consider the correlation between an intervention and the subsequent change with regard to the intangible assets.

7.2.1.3 Calculated Intangible Value (CIV)

Other approaches ascertain a so-called *return on assets* (ROA); so does, e.g., the concept of *calculated intangible value*. Average pre-tax earnings of a company are divided by the average of its tangible assets for a certain period of time. The result, the ROA of the given company, is then compared with its industry average. The difference is multiplied by the company's average tangible assets to calculate an average annual earning from the intangibles. By dividing the above-average earnings by the company's average cost of capital or an interest rate, an estimate of the value of its intangible assets can be derived.

The *calculated intangible value* (CIV) is supposed to represent this special group of methods.[280] IC is calculated in relation to the achievements of the average competitor, i.e. the average of the given branch of business.[281] By establishing CIV, the value of all intangibles together can be derived, regardless of the company's capitalization on the stock market. It has to be provided, nevertheless, that the

---

279 This is a method developed by the Nobel Prize winner and economist James Tobin. For a more detailed description cf. Stewart (1997), p. 225.

280 It was derived by *NCI Research*, Illinois.

281 However, a precondition has to be fulfilled, namely that the data for the entire capital profit in the respective business are available. With regard to this, NCI used numbers from Robert Morris Associates' Annual Statement Studies for firms with the same branch code. Cf. Stewart (1997), p. 222.

company's entire capital profit exceeds the average of the given branch of business. If so, CIV is calculated as follows:

1) Computation of average pre-tax income of the last 3 years.
2) Computation of the average of tangible assets in the last 3 years.
3) The result for tangible assets is calculated by multiplying the average assets by the profit of the branch of business.
4) The difference between the actual pre-tax results and the results for tangible assets equals the results of knowledge capital.
5) Taxes paid for results from knowledge capital indicate the extra cost of immaterial activities.
6) By dividing the extra costs of intangible activities by an appropriate percentage (e.g. capital costs), it is possible to calculate the given CIV, i.e. the up-to-date net value of the intangible activities used.[282]

7.2.1.4 The Value Explorer™

A number of methods can be subsumed under the concept of *value explorer*.[283] All of them work by identifying the components of IC and subsequently evaluating them directly in financial terms, either individually or in the aggregate.

The accounting methodology of value explorer has been proposed by KPMG in order to calculate and assign a certain value to 5 types of intangibles: (1) assets and endowments (brand, networks); (2) skills and tacit knowledge (know-how); (3) collective values and norms (client focus, quality); (4) technology and explicit knowledge (patents); (5) primary and management processes (leadership, communication).

Andriessen/Tiessen have defined a so-called tool-kit, with the help of which intangibles can actually be evaluated. The tool-kit consists of three phases – „examine your core competences", „assess your strengths", „measure your values" –, which can be run through with the aid of several checklists respectively. Andriessen/Tiessen thus choose the relevant intangibles, i.e. the core competences of an organization, and attempt at establishing a monetary value for each of them. Basically, the value of a core competence is defined as net present value of the gross profit that is generated during the period of sustainability of the competence, corrected with a factor for robustness. Problems can be expected mainly with regard to the process of determining the necessary basis of information and the validity of the findings proper.

282 Cf. Edvinsson/Brünig (2000), p. 154.

283 For detailed information cf. Andriessen/Tiessen (2000).

7.2.1.5 Conclusion

The methods offering financial evaluations, such as those discussed above, are useful in situations of merger and acquisition as well as for valuations at the stock market. They are also helpful when it comes to comparisons between companies within the same kind of industry as well as for illustrating the financial value of intangible assets. On the other hand, the examples mentioned above do not provide sufficient information about the character of the various immaterial assets. This can be shown very well by way of methods concerned with the whole of a company's IC. The concrete monetary value of individual assets, however, does still not provide sufficient information to allow for individual measures of managing IC to be derived. A company planning to handle IC from a managerial point of view will find that the mentioned methods for calculation do not suffice. Since financial approaches do not succeed in measuring and evaluating IC and knowledge in a flexible and reliable way, other approaches have been developed in order to assess the IC of a company. They are generally based on more precise information and they predominantly employ non-financial factors supposed to enable a well-aimed development and steering of either IC or the company to happen.[284]

7.2.2 Measuring IC with Non-Financial Indicators

The approaches presented in the following chapter, just as those in the previous one, have emerged from practical work in the majority of cases and are moulded by the specific demands of the context respectively. The main difference, for that matter, is the scheme of structuring IC with regard to selected indicators as well as general aims. Accordingly, we will separately introduce an exemplary scheme for structuring IC at the beginning. Further schemes of classification will be dealt with below, in the context of methods and instruments being closely connected with the respective form of classification.

7.2.2.1 A Structural Model of Intellectual Capital

Sullivan's approach at structuring IC can serve as an example distinguishing several appearances knowledge can have (cf. Fig. 7.1).

---

284 Mavrinac and Siesfeld, in their mentioned study about the evaluation of intangibles at capital markets, come to the following conclusion: „...in a world where ‚knowledge work' and intangible assets have become of profound importance, future financial performance is often better predicted by non-financial indicators than by financial indicators." Cf. Mavrinac/Siesfeld (1998), pp. 49-72.

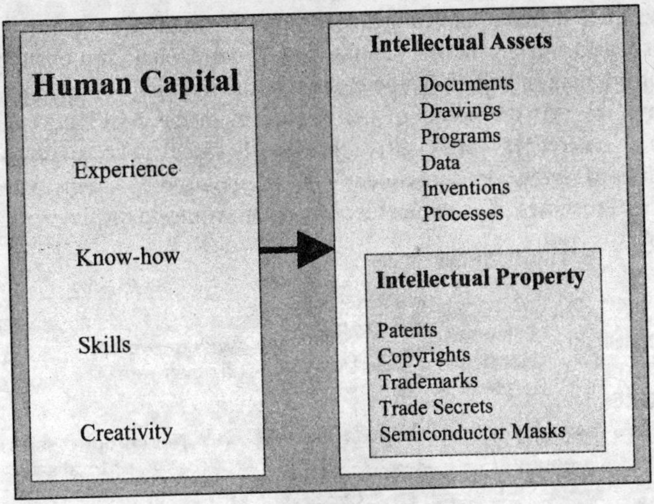

Fig. 7.1: The Intellectual Capital of a Firm According to Sullivan

Sullivan defines the human capital of a company as the capabilities of employees as well as contractors, suppliers and other company people when it comes to the solution of customer problems, i.e. the companies' know-how, collective experience skills and so forth.[285] In Sullivan's understanding this human capital generates the value or wealth of a firm. The human capital can subsequently be transformed into the 2nd component of IC, i.e. intellectual assets, these being the codified, tangible, or physical descriptions of specific knowledge of which the company can claim ownership rights (Intellectual Property). Sullivan claims that this is why managers should aim at transforming human capital into intellectual assets. Structural capital, however, he does not consider a part of IC: „Structural capital is the support or infrastructure that firms provide to their human capital."[286] It includes physical elements such as computers, desks and telephones or even buildings, electricity and so forth. Structural capital also includes intangible elements such as information systems, computer software, work producers, marketing plans, and company know-how.

7.2.2.2 *Intellectual Capital Navigator*

Stewart's model for the determination and visualization of IC (cf. Fig. 7.2) is based on his criticism that an indicator like market-to-book ratio is too one-sided an approach.[287] He therefore proposes a complementation of this indicator by adding further indicators. Stewart uses a radar chart as a sort of ‚navigator' for the IC.

---

285 For more detailed information cf. Sullivan (1998).

286 Cf. Sullivan (1998), p. 23.

287 For more detailed information, cf. Stewart (1997).

It displays three different indicators, each referring to different components of IC; his major categories are human capital, structural capital, and customer capital. Target- or orientation values can be entered at the markings of the individual axes of the navigator. At the points of intersection with the circle, e.g. the company goals can be entered. If representative, previously calculated numbers are assigned to the individual indicators subsequently, an irregular polygon emerges as a result. The inside of it shows the company's property; the outside displays what the company wants to acquire or to attain.

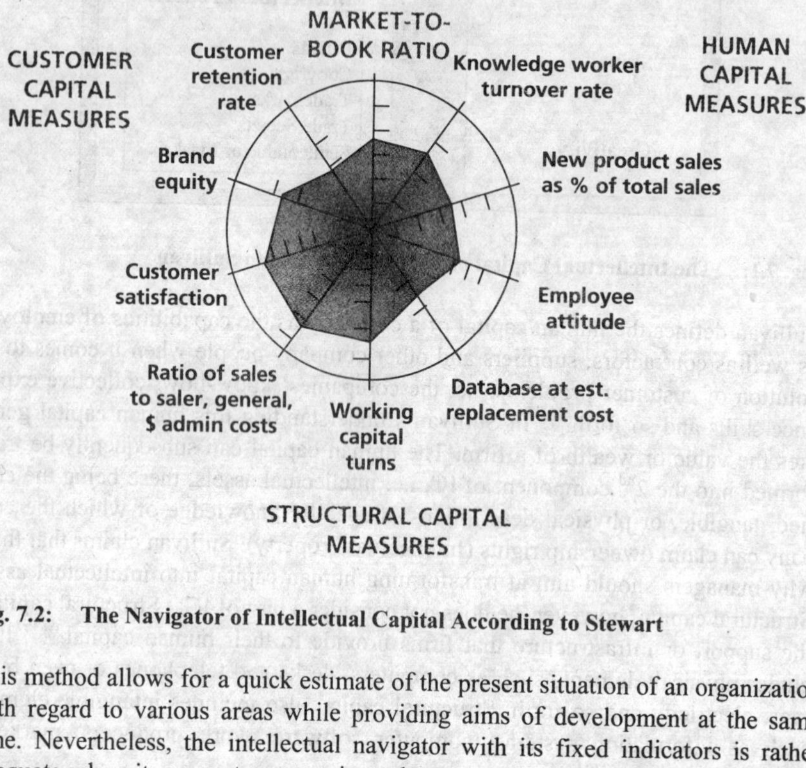

Fig. 7.2: The Navigator of Intellectual Capital According to Stewart

This method allows for a quick estimate of the present situation of an organization with regard to various areas while providing aims of development at the same time. Nevertheless, the intellectual navigator with its fixed indicators is rather adequate when it comes to comparisons between organizations using this instrument. The consideration of the organization as a whole, however, is severely restricting the contributions the instrument can make for the controlling of IC. The indicators employed by Stewart are highly unlikely to allow for a deduction of aims and measures for the development of IC.

7.2.2.3 Intangible Assets Monitor (IAM)

The Intangible Assets Monitor focuses on the employees of a firm, and their know-how.[288] Accordingly, it assigns great value to the competence of employees

---

[288] For more detailed information, cf. Sveiby (1997).

in its classification of IC. Sveiby divides the IC of a firm into the following three categories:

- *Competence of employees* (education, experience),
- *Internal structure* (legal form, management, systems, corporate culture, R&D, software),
- *External structure* (brands, customer relations, supply relations, cf. Fig. 7.3).

The goal is to track how intangible assets are developing by designing indicators correlating with the *growth/renewal* rate of the asset in question, with the *efficiency* the company shows in utilizing it, and with its probable *stability*.

Tangible Assets	Intangible Assets		
Financial Capital	**Customer** (External Structure)	**Organization** (Internal Structure)	**People** (Competence)
Growth/Renewal Net invest ratio	**Growth/Renewa** • Image-enhancing	**Growth/Renewa** • Orga-enhancing	**Growth/Renewa** • Competence-enh.
Efficiency •Profit margin	**Efficiency** •Revenues per custo.	**Efficiency** • Revenues per adm.	**Efficiency** • Value added /expert
Stability •Solidity	**Stability** •Repeat orders	**Stability** • Adm. staff turnover	**Stability** • Expert turnover

Fig. 7.3: Celemi Intangible Assets Monitor

Celemi, a learning design consulting company uses the model of IAM in order to evaluate e.g. the contribution customers make with regard to the three categories of competence, internal structure and external structure. Being reference customers, they add to the respectability of the given company, they improve internal structures by way of introducing innovative methods of working, and they promote the employees abilities. The perspective of tangible assets (which is comparable to the traditional perspective orienting on financial aspects) can be taken additionally. It is an essential feature of this approach that it allows for the assignment of IC elements (e.g. customers) to organizational objectives (e.g. stability etc.), thus producing a relationship of cause and effect. This can be achieved with regard to both individual projects and individual elements of the organization. Yet, there is no explicit connection to financial results at this point. The delicate point is rather that the assignment of indicators to the mentioned perspectives of internal and external structure as well as of competence can be ambiguous. Thus, the relationship of cause and effect between IC and its contribution to organizational success is not depicted with sufficient preciseness.

7.2.2.4 Balanced Scorecard (BSC))

The BSC was first conceptualized in 1992 in the USA, and has been a common instrument for strategic management ever since the mid-Nineties.[289] The BSC can be understood as an integrative approach ascertaining IC-related financial metrics as well as including traditional financial ones. Originally, the instrument was designed to translate an organization's strategic aims into a coherent set of financial metrics. Measures for the achievement of these aims would be defined. Both aims and measures are being ascertained with regard to four so-called perspectives: The financial perspective, the customer perspective, the perspective relating to learning and growth, and the perspective relating to internal business processes (cf. Fig. 7.4).

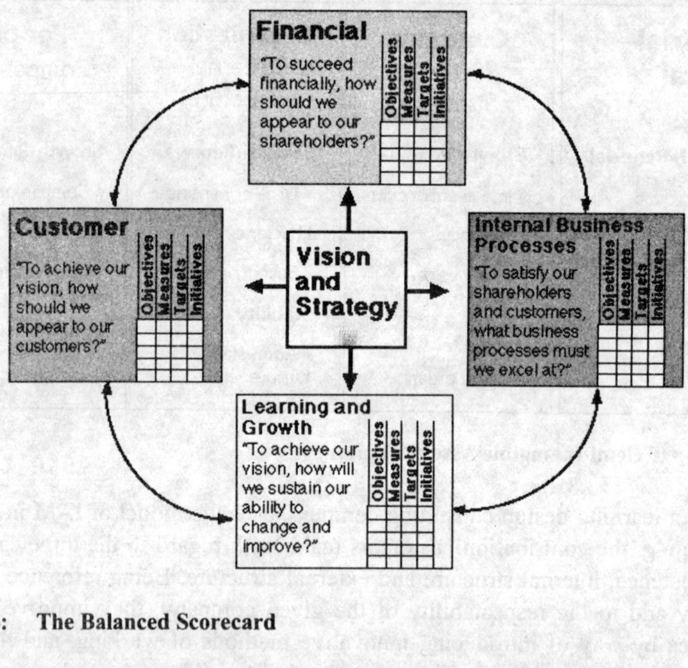

Fig. 7.4: The Balanced Scorecard

The discovery that the financial perspective is insufficient in order to depict a balanced picture of the organization is valid in this case as well. Non-financial measures monitor the building of capacities and intangible assets. The various metrics, either financial or non-financial, are then being related in terms of cause and effect. This relationship allows for the evaluation of measures for their (financial)

289 Cf. Kaplan/Norton (1992).

contribution to an organization's profit. The step from financial to ‚strategic reporting' is thus possible because of BSC. This early vision was then extended towards a new strategic management system, which was supposed to help the given management translate strategic aims into operative ones, thus combining long-term objectives with intermediate and short-term aims.[290] Besides the functions of controlling and reporting, the concept contributes to communicating the corporate strategy internally: a BSC including a whole company, so Kaplan/Norton maintain, would both indicate the relations of cause and effect between individual areas and their contribution to general profit. Four new processes for the use of BSC as an instrument were introduced: *translating the vision* helps to reach and stabilize a consensus with regard to the organization's vision and strategy. *Communicating and linking* enables managers to communicate the strategy throughout the organization and link it to objectives. *Business planning* allows for the integration of various business plans and the change of programmes within the organization. *Feedback and learning*, finally, fosters strategic learning within the organization.

Various approaches have in the meantime drawn on the basic structure of BSC and tried to adjust it to the more specific demands of knowledge management.[291] The idea behind this is to create an instrument for measuring success with regard to the steering and controlling of KM activities. Nohr does so replacing the mentioned traditional perspectives by knowledge-related perspectives, which he deduces from the model of Probst et al., namely those of generating, applying, distributing/sharing, and storing knowledge. He continues, according to the usual procedure of establishing a BSC, with the definition of normative, strategic, and operative objectives and the selection of appropriate units of measurement, while focusing on the supervision of the knowledge-related perspectives. Nohr does mention the importance of an additional financial perspective, yet he does not elaborate on the subject. The fact that this was the first attempt at developing and steering IC/knowledge on a concrete operative level is, however, essential with regard to this derivative of BSC.

7.2.2.5 Skandia Navigator

Skandia, the Swedish financial services company, was the first to develop a set of indicators based on BSC, the so-called *Skandia navigator* that was supposed to add to the measurement of IC. In the 1994 supplement to the annual report, Skandia made statements concerning the company's IC by way of using indicators.[292] Skandia chose the scheme of structuring depicted in the below figure (cf. Fig. 7.5).

---

290 Cf. Kaplan/Norton (1996 a) and Kaplan/Norton (1996).

291 Cf. Kaps/Nohr (2001).

292 Skandia published IC reports in the period from 1994-98. According to official statements, a new format for reports is worked on internally.

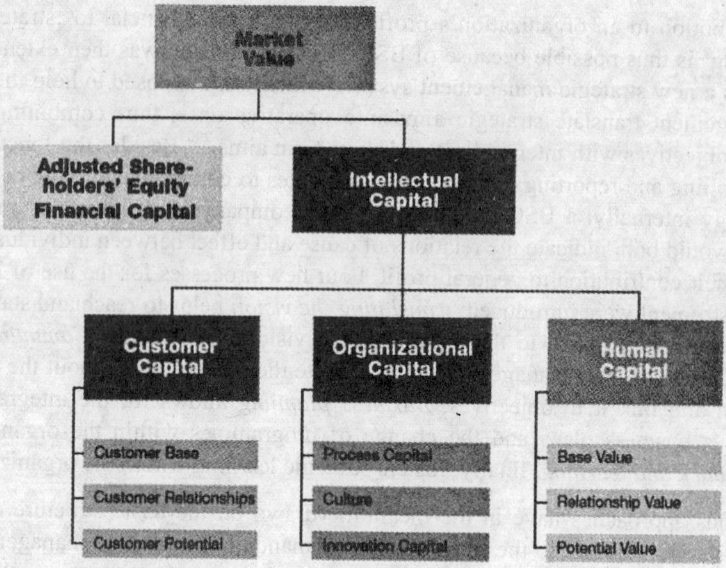

Fig. 7.5: Skandia's IC Structuring in 1998

The perspectives the Skandia navigator concentrated on were customers, employees, business processes, financial aspects and growth/renewal (cf. Fig. 7.6). The supplement was originally designed for external information. By analogy with the development of the BSC instrument, the navigator is rather understood to be a management tool than an instrument for reporting.

The Skandia navigator is tied to Skandia's so called process model. This model determines the critical factors of success, it defines indicators and it allows for the planning of adequate measures. There is a navigator for each organizational site. Some of the navigators have been established on the level of teams or departments. The navigator is used as a basis for staff development and has apparently been tied to an incentive system in recent time. An IT instrument presenting a web-based BSC solution was developed on the basis of the navigator. The selection of financial metrics, however, is more likely to fulfill an external reporting function than to advise in terms of well-aimed development and steering.

	1997	1996	1995	1994*
FINANCIAL FOCUS				
Return on capital employed (%)[1]	21.9	27.1	28.7	12.2
Operating result (MSEK)	1,027	579	355	115
Value added/employee (SEK 000s)	2,616	2,206	1,904	1,666
CUSTOMER FOCUS				
Number of contracts	189,104	133,641	87,836	59,089
Savings/contract (SEK 000s)	499	396	360	333
Surrender ratio (%)[2]	4.4	4.4	4.1	4.2
Points of sale	45,881	33,287	18,012	11,573
HUMAN FOCUS				
Number of employees, full-time	599	418	300	220
Number of managers	88	86	81	62
Of whom, women	50	27	28	13
Training expense/employee (SEK 000s)	2.7	15.4	2.5	9.8
PROCESS FOCUS				
Number of contracts/employee	316	320	293	269
Adm. exp./gross premiums written (%)	3.5	2.9	3.3	2.9
IT expense/admin. expense (%)	8.1	12.5	13.1	8.8
RENEWAL & DEVELOPMENT FOCUS				
Share of gross premiums written from new launches (%)	0.9	23.7	49.2	11.1
Increase in net premiums written (%)	31.9	113.7	29.9	17.8
Development expense/adm. exp. (%)	9.8	9.9	10.1	11.6
Share of staff under 40 years (%)	76	78	81	72

[1] Changed calculation methods for 1996 and 1997.
[2] Surrenders during the year in relation to the average mathematical reserve, net.

Fig. 7.6: Measurement Indicators at American Skandia 1994 – 1997[293]

7.2.2.6 *The IC Index*

Roos et al. consider the indicators of measurement confusing and thus not very important. Working with *Skandia* as an example, Roos et al. found out 24 different indicators in the system supposed for measuring IC.[294] Therefore, they suggest an index for IC (*IC index*) integrating all calculated indicators. The individual indicators for categories of IC and associated indicators thus become one single index,

---

293 Cf. Skandia (1998).

294 Roos/Roos/Dragonetti/Edvinsson (1997), p. 71; Skandia first used the IC index in its 1997 IC Supplement of the annual report: Skandia, Intelligent Enterprising. Intellectual Capital Supplement to Skandia's 6-month interim report 1997.

a relationship capital index which can be calculated considering the following indicators: growth in number of relationships, growth in trust, customer retention, and the quality and productivity of distribution channels. The remaining categories of IC are dealt with in the same way, with the infrastructure capital index and the innovation capital index resulting from it. These indices can subsequently be combined into an overall IC index, an example of which is depicted in the following figure (cf. Fig. 7.7).

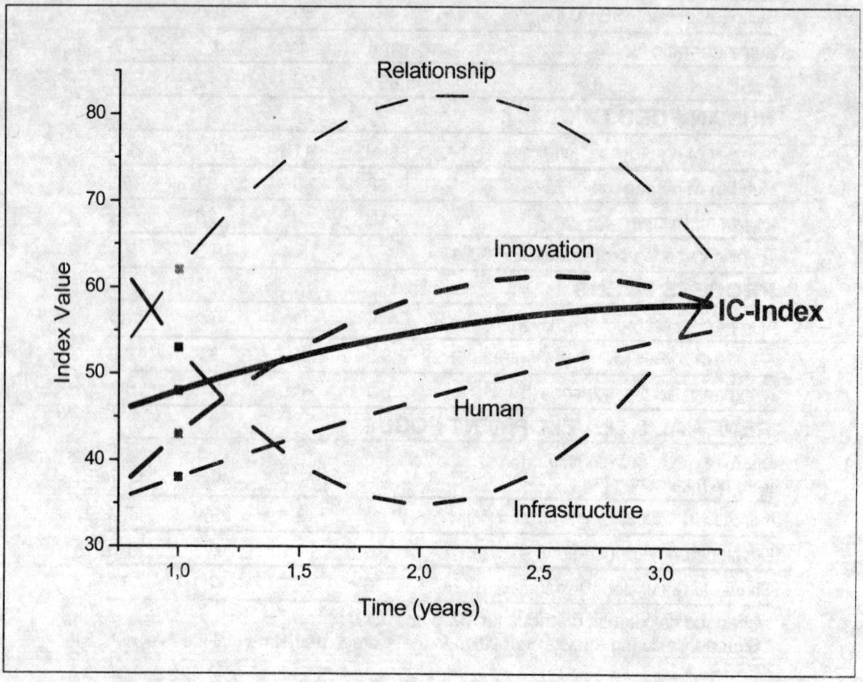

Fig. 7.7: IC-Index

According to Roos et al., the main benefit of such an IC index is that it forces the given management to address the important issues and to come up with solutions. It also allows for comparisons with regard to IC and especially the performance of IC. Thus, benchmarks can be established both in absolute terms and in the percentage of increases. Finally, an IC index provides external shareholders with a visualization of a greater part of the company.

7.2.2.7 *The Intellectual Capital Method*

The IC method was developed as a tool supposed to assist in planning as opposed to a tool for mere reporting.[295] Brooking proposes the following four components of IC:

- market assets (all market-related intangibles, including various brands, customers, distribution channels, repeat business, backlog),
- human centered assets (collective expertise, creative and problem solving capability, also psychometric data and indicators on how individuals may perform in situations such as in a team or under stress),
- intellectual property assets (know-how, trademarks, trade secrets, copyright, patents, design rights, trade and service marks),
- infrastructure assets (all the elements which make up the way the organization works: corporate culture, methodologies for assessing risks, methods of managing a sales force, financial structure, databases of information on the market or customers, communication systems [e-mail]).

The IC method allows for requirements of knowledge to be deduced from specific aims, thus fostering the well-aimed development of knowledge. It refers, for this matter, to the company's objectives, which are being analyzed in terms of knowledge-related demands necessary for the achievement of the given objectives. ‚Dream Ticket' is Brooking's term for the combination of necessary elements of knowledge.

During an audit, the strong as well as the weak points of a given Dream Ticket are being ascertained. The methods employed differ with regard to the asset that is to be measured. An index is created, juxtaposing the current state of the assets with the proposed optimum. This is evaluated with the aid of a ranking from 0 to 5. A so-called target unites all relevant assets. The target additionally excels by a function of developing the asset. Measures are defined and implemented subsequently, aiming at lifting the individual assets towards the desired level of intensity of fulfillment. A second audit is dedicated to the evaluation of the progress, thus making the IC method a management tool. It is not quite clear which criteria are supposed to determine this evaluation of individual assets. Nevertheless, this approach has been a landmark for the perspective on well-aimed development of organizational knowledge.

7.2.2.8 *Intellectual Capital Report*

The ARSC (Austrian Research Center Seibersdorf) has developed a model for IC reports and has published its first IC report in 1999.[296] According to this model, IC

---

295 For detailed information cf. Brooking (1996).

consists of human capital, structural capital and relationship capital. Apart from metrics for the individual components of IC (so-called added value potentials), however, financial metrics referring to the basic ARCS processes of contract research projects and independent research as well as both financial and non-financial results are also being ascertained. The financial metrics are being derived from knowledge objectives. The model in its fundamental design is thus not exclusively based on structure (like, e.g., the Skandia model is), but, given its process-oriented approach, on the European Model for Business Excellence (EFQM), which indeed reveals a similar structure (cf. Fig. 7.8).

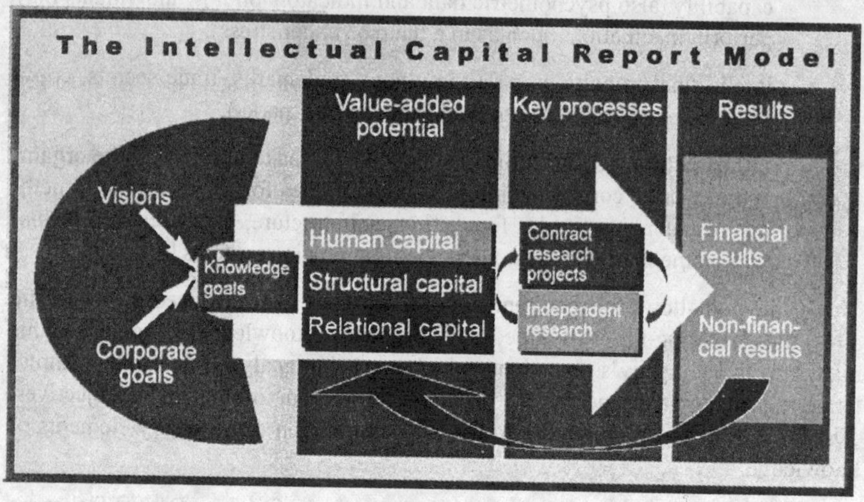

Fig. 7.8: The ARCS Intellectual Capital Report Model

Interestingly, this approach tries to reduce the contingency in the selection of indicators by deducing them from knowledge objectives. By a basic distinction between metrics relating to either added value potentials, processes and results, it furthermore contributes to increased expressiveness as well as to the evaluation of the financial metrics. Hereby, the financial metrics are being established according to the three supposed knowledge-related processes, namely acquisition, application, and utilization. The German Center for aeronautics and cosmonautics (DLR) has largely adopted this model. Thus, the ARCS Intellectual Capital Report of 2001 was able to present benchmarks for the first time. The model was applied by ARCS, and it was innovated and applied by the department for business administration and economics of the university of Graz as well as by Joanneaum, another Austrian Research Center.[297] In the course of higher education reforms in Austria,

296 For detailed information, cf. Leitner/Bornemann/Schneider (2002). The intellectual capital report was published by ARC Seibersdorf (2001).

297 Cf. Joannaeum Research, 2001.

there have recently been plans to obligate the annual establishment of an IC report with references to intellectual properties. The model proposed by ARCS would serve as guideline.

7.2.2.9 *Value Chain Scoreboard™*

In 2001, Baruch Lev has developed a model similar in conception, which also emphasizes strongly on a process-related perspective.[298] The idea behind this was an orientation on the value chain. The model integrates the following aspects: (product) innovation (i.e. discovery and learning), implementation (i.e. the state of development of products), and the transfer of product-related ideas to products and commercialization (i.e. the economic use of products).

Discovery and learning could for instance include financial metrics referring to the following aspects: internal renewal (R&D, work force training etc.), acquired capabilities (technology purchase etc.) and networking (communities of practice, alliances, customer integration etc.) These aspects will vary according to the company respectively.

A value chain sequentially lists the aspects, financial metrics and indicators relevant for the company. It wants to provide information for both internal processes of decision as well as for external stakeholders. Lev specially presses the fact that indicators would have to come up to certain requirements. Apart from quantificability and standardizability, it is decisive that there be a connection between each indicator employed and traditional financial metrics such as share returns and gain, the latter probably being difficult to cope with. One of the major yet unsolved problems, however, is to ensure the validity of financial metrics and indicators for intangible assets. Lev emphasizes the importance of standardization in terms of comparability to systems of indicators individual to each company.

7.2.2.10 *WissensBilanz (Knowledge-Related Balance Sheet) – an Instrument for Knowledge- and Personnel-Related Work*

Within the scope of a project sponsored by the German Ministry of Education and Research, an instrument for practical measurement of knowledge was designed in 2001 and tested shortly after by way of a pilot project at the *Volkswagen* company among others. Apprehension time, i.e. the time of (further) training until a certain qualification is disposable, was chosen as unit of measurement. The qualification required for the operation of a system (Knowledge demand of a System) is expressed in units of apprehension time and juxtaposed to the actual qualification of the employees (Human Knowledge) by way of a so-called *Knowledge-related Balance Sheet*. Discrepancies between the profiles of requirements on one hand and competences on the other (hypo- vs. hypercongruence) can thus be recognized and can become a basis of steering.

---

298 Cf. Lev (2001). Cf. also Lev/Zarowin (1999), p. 353.

Providing a means of quantifying knowledge by apprehension time, this approach has been a landmark. The instrument is, however, restricted to those areas where activities can easily be translated into apprehension time, namely particularly processes featuring a high degree of standardization, such as production etc. Strictly speaking, the Knowledge-related Balance Sheet is an instrument for human resources development apt for the registration, quantification and description of requirements of further training.

7.2.2.11 Conclusion

A standardized IC report for external communication to stakeholders and providers of capital has not yet gained acceptance. Despite the fact that companies seem to show increasing commitment with regard to this matter, *Skandia* has stopped the publication of IC reports at least temporarily. The differences between the approaches restrict the comparability of various companies, which is one impeding factor. Another is that a sound comparison would only be possible between similar companies at the present. A company working in the car industry is likely to have an IC basically different from that of a research center such as DLR or ARCS Seibersdorf. Additionally, the decision might be influenced by political considerations with regard to the disclosure of delicate information.

Apart from external communication, some of the mentioned approaches are supposed to serve as instruments for the steering of IC. The below figure aims at scrutinizing this demand by way of distinguishing the approaches in terms of structural models, structural and process models, and models orienting on the individual elements of IC (content model) (cf. Fig. 7.9). Structural models mostly confine themselves to the classification of IC. They make claims as to the forms in which IC can appear, whether as a patent or as know-how, whether it is tied to human or structural factors and so forth. A structural model is hardly apt for steering. Structural process models aim at connecting IC with material (financial) and/or immaterial results. This concept at least offers a standard that can be used as an informational basis in the course of decisions of steering. It does, however, not offer information about the actual elements of IC, while this does seem to be a necessary prerequisite of steering to display the actual forms of IC in hand. A description of the organization's knowledge basis in terms of contents, however, is lacking in most cases. This description would be actually helpful with regard to decisions of investment and disinvestment.

	Sullivan's Intellectual Capital Model	Intellectual Capital Navigator	Intangible Assets Monitor	Balanced Scorecard	Intellectual Capital Method	Intellectual Capital Report (Wissensbilanz)	Value Chain Scorboard	Skandia Navigator	Wissens Bilanz
structural model	☑								
structural and process model		☑	☑	☑		☑	☑	☑	
Content model					☑				☑

Fig. 7.9: A Distinction of Structural Models, Structural Process Models, and Models Orienting on Actual Elements of IC

A problem of non-financial metrics, which must not be underestimated, is the validity of the metrics employed. North criticizes that „metrics showing a different level of aggregation and meaning are put side by side".[299] It should be individually questioned whether the given non-financial metrics actually make statements about the state of affairs being about to be measured.

	type of metrics used		focus of attention	
	financial metrics	non-financial metrics	IC in its entirety	individual assets
market-to-book ratio	☑		☑	
Tobins's Q	☑		☑	
Calculated Intangible Value	☑		☑	
The Value Explorer	☑		☑	☑
Intellectual Capital Navigator	☑	☑		☑
Intangible Assets Monitor	☑	☑		☑
Balanced Scorecard	☑	☑		☑
Skandia Navigator	☑	☑		☑
IC-Index		☑	☑	
The Intellectual Capital Method				☑
ARCS Wissensbilanz				☑
Value Chain Scoreboard	☑	☑		☑
WissensBilanz		☑		☑

Fig. 7.10: The Different Approaches Can be Categorized According to the Type of Indicator Used and the Aspect They Try to Cover

However, it can still be doubted if the mentioned models can actually provide information relevant for steering. For the controlling of an organizational base of knowledge, statements concerning the efficiency in dealing with knowledge would be important as well. The integration of such statements into an instrument of

299 Cf. North (1998), p. 196.

controlling, however, has not been attempted yet. The following matrix summarizes again the characteristic features of the approaches introduced in this paper (cf. Fig. 7.10).

7.3 Evaluation of KM Measures

The evaluation of KM measures is a special case. The approaches introduced in this paper so far aim at evaluating either IC as a whole or its individual components. They are, however, merely partially apt for the evaluation of an investment in knowledge management in form of capital budgeting. It is largely three factors that are responsible for problems in evaluating knowledge management projects:

Firstly, those effects of KM relevant for profit can hardly be determined. The areas involved and the types of improvements, as well as time-related factors, would have to be identified and defined. Because long-term experiences and studies on KM's potential effect on profit are still lacking, the information available is inadequate and unsatisfying. Secondly, there would have to be a possibility of measuring and evaluating the improvements. It is, however, only with certain reservation that knowledge can be measured and evaluated in terms of quantity. As has been pointed out above there are yet no valid and standardized methods for an evaluation of knowledge in monetary terms. Thus, qualitative data are enlisted, which in turn are not apt for capital budgeting. Additionally, there is the lacking validity of the metrics employed as to whether the aspect in question is actually expressed by the chosen metrics. Finally, the relations of cause and effect have to be ensured especially in big companies with several measures employed at a certain point of time. If changes can be measured, it has to be crosschecked whether these are actually due to KM activities.

All approaches up to now excel by a recursive tendency to emphasize the time saved through KM with regard to special working processes. This is due to the fact that they mainly reflect the efforts of consulting companies forced to justify the usefulness of a certain project towards their customers. Additionally, there are proposals of taking not only financial dimensions into account, but to consider others as well for the purpose of a BSC (internal benefits: efficiency, customer benefits, other financial benefits, growth and innovation benefits).[300] In the course of their knowledge sharing initiative, the German company of SAS proposed a way to estimate a minimum return on investment. SAS, according to this calculation, had an ROI of 5 000 000 $ resulting from an annual investment of 380 000$ in 1999. The benefit was calculated as follows:

---

[300] Cf. Foster (1999), p. 157.

Full Benefit = Number of Accesses/downloads x usage ratio (<1) x consultant cost x benefit (time saved, conservative estimate: 4 hours)

The costs of the initiative proper is set opposite:

Total Cost of Operation = {preparation (time to prepare a contribution: 4 hours each) x contributions (number) x consultant cost} + fixed support cost (to run and support the initiative)

Those numbers are furnished by references emphasizing the conservativeness of these estimations. Some factors less easily measurable, such as the transfer of an entry from one employee to another, the positive effects of an employee network and the reduced cycles of production development are not included.

David Skyrme Associates, a consulting company, argues in turn and rightfully that knowledge management *per se* was a project of infrastructure, and that the benefits would be distributed throughout the whole organization. A tool called KM Benefits Tree tries to link the immediately visible benefits with those lying further down under the surface (cf. Fig. 7.11).

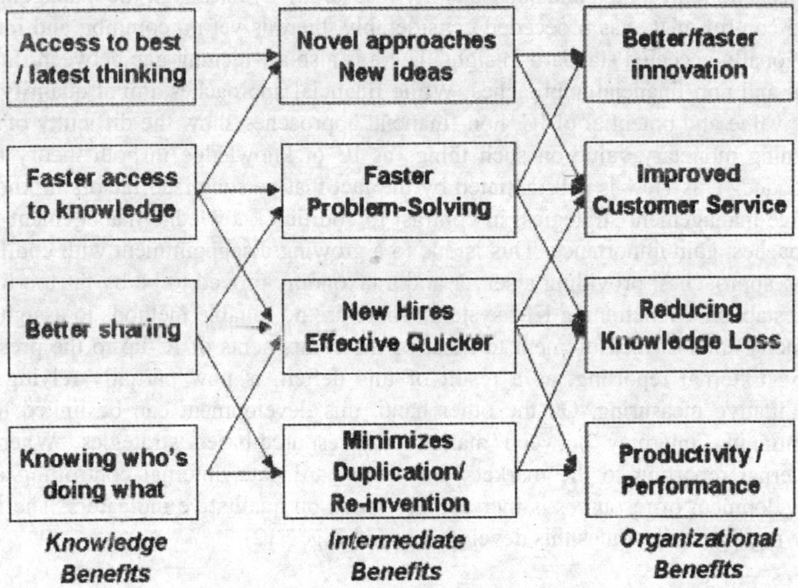

Fig. 7.11: Benefits Tree Introduced by David Skyrme Associates

Ernst & Young Consulting propose to measure the use of KM initiatives on three different levels: (1) hard data concerning the actual use of knowledge data bases, tools and external sources; (2) success stories and other verbal or written feedback is used to draw conclusions about the use of knowledge management; (3) annual global surveys will include a questioning of employees in terms of knowledge tools and resources considering the dimensions of awareness, usage and value.[301]

The approaches introduced here are, in the whole, still unsatisfying. It can be expected, however, that in the near future there will be more balanced procedures for the evaluation of KM projects, for it will be necessary within the concrete field of project acquisition and –management to find adequate arguments in favour of an investment into KM projects. In order to be acknowledged as an independent discipline, knowledge management must be able to prove and quantify the contribution of a KM project to the corporate objectives.

7.4 Outlook

What is included in the knowledge based resources of a firm? Which aspects of this capital are important for company and management strategies? How can these aspects be represented and subsequently measured? Even though the management and control of IC has proceeded considerably, there is yet no common and internationally accepted standard in sight. There is also a widening gap between financial and non-financial approaches. While financial approaches aim at quantifying the value and potential of IC, non-financial approaches show the difficulty of assigning monetary value on such things as IC or knowledge in both theory and practice. This view is substantiated by the fact that personalized factors in knowledge management strategies, in contrast to codified knowledge management approaches, gain importance. This is due to a growing disappointment with codified KM approaches: providing a server and a taxonomy proved to be by far too little to establish a functioning KM system. There are no reliable methods to assign financial units of measurement to most of the components of IC up to the present day. External reporting, as a result of this deficit, is now partially relying on qualitative measuring. On the other hand, this development can be linked to a continuing interplay between market and resource-based strategies. Whereas external reporting to the markets draws on hard data, internal controlling and development of resources is increasingly based on qualitative indicators. The following figure illustrates this development (cf. Fig. 7.12).

301 Cf. Vopel (2001), p. 4.

perspective / approach	internal controlling	external reporting
financial/ quantitative indicators	→ stagnating importance because of strengthening of personalized KM	↑ increasing importance for capital acquisition
non-financial/ qualitative Indicators	↑ increasing importance for HR development	↓ decreasing importance due to a lack of standardized evaluation

Fig. 7.12: Trends for Measuring IC

Internal controlling is increasingly dominated by the requirements of a resource-based company: the development of intellectual potential becomes the central strategic objective. One of the prevailing objectives for external reporting is the acquisition of capital, i.e. the maintenance of the stock market quotation. For this purpose, financial indicators are more popular than non-financial indicators, which have certain effects on the public relations of the enterprise. As a result, the tendency towards measuring IC is divided into either external or internal purpose. As long as there is no generally accepted method for the monetary evaluation of all aspects including IC, the development of such an internationally accepted standard, integrating both financial and non-financial aspects, stays the most pressing challenge for the near future.

Part II

Survey

8 Delphi Study on the Future of Knowledge Management[302] – Overview of the Results

Wolfgang Scholl, Peter Heisig

8.1 Objectives and Methodical Approach of the Delphi Study

Knowledge management has become an integral part of the agenda in business and research. The term knowledge management first appeared in American literature in the seventies (Rickson 1976). Finally, at the beginning of the nineties, knowledge management acquired more management attention because of the structural changes in the economy and society towards the information society and knowledge economy (Drucker 1993, OECD, 1996; Grant, 2000). Since the mid nineties knowledge management could be considered as a new business management approach. Contributions to this new approach are coming from very different disciplines, not only from business administration, computer sciences and engineering, but also from library science, psychology, sociology and social sciences and from ethnology and philosophy as well. This cross-disciplinary attention shows the importance, which has been attached to this emerging business and research field. This is also underlined by the still increasing budgets for knowledge management projects in industry and public as well as the set-up of knowledge management professorships and university courses (Erlach et al. 2000; Ruth et al. 2003). Despres and Chauvel (2000) state that "all parties agree that we are nonetheless witnessing an explosion of interest in the term ‚knowledge management'", whereas Grant (2000) thinks that „among the innovations that have swept through the world of management during the past two decades – total quality management, shareholder value creation, business process reengineering, and competence-based strategy – knowledge management has probably aroused the greatest interest and

302 The First Global Delphi Study on "The Future of Knowledge Management" has been initiated by the Fraunhofer Competence Center Knowledge Management, Berlin and the Institute for Psychology of the Humboldt-University, Berlin. The work has been partly financed with a grant of the Donors' Association for the Promoting of Sciences and Humanities in Germany („Stifterverband für die Deutsche Wissenschaft") and the Fraunhofer Society. The research and statistical work was carried out by Prof. Dr. W. Scholl, Mrs. C. König and Mr. B. Meyer at the Institute for Psychology of the Humboldt-University, Mrs. S. Thiele supported the first Delphi round. Cf. Scholl et al. (2002). The authors thank all colleagues for their contributions, comments and help.

made the biggest impact. (...) Unusual among major management innovations, knowledge management has captured the interest of both practicing managers and business school academics“.[303]

Despite this immense and increasing interest in knowledge management, the field is still complex and there is no consensus about the further development. A multitude of different approaches exists, so it could be said that knowledge management is „one of the most ramified topics in the business lexicon“[304] and „there exists a patchwork of subdomains in and around knowledge management that deal with one set of issues while ignoring others“[305].

Following our European-wide company survey[306] carried out four years ago (Heisig, Vorbeck 2001), we wanted to assess the state of the art of knowledge management from a theoretical and practical point of view as well as looking at the "Future of Knowledge Management" with this First Global Delphi Study[307]:

1. What is the most pressing and challenging *theoretical* research issue for the understanding and advancement of knowledge management?
2. Which theoretical approach and/or scientist is most likely to deal effectively with this *theoretical* research issue?
3. What is the most important recent *theoretical* advancement in knowledge management?
4. What is the most pressing and challenging *practical* problem for the understanding and advancement of knowledge management?
5. Which practical approach and/or organization is most likely to deal effectively with this *practical* problem?
6. What is the most important recent *practical* advancement in knowledge management?

Fig. 8.1: The Delphi Method and Procedure

In order to approach a broad range of researchers from different disciplines and practitioners from various businesses and to leave the evaluation of the problems, achievements and developments in the knowledge management field open to them, we have opted for the Delphi technique. The Delphi study method has been described as a step-wise approach to, first, gather feedback from a selected group

303 Despres and Chauvel (2000), p. 5; Grant (2000), p. 27.

304 Despres and Chauvel (2000), p. 55.

305 Despres and Chauvel (2000), p. 57.

306 The chapter with the detailed survey results has been withdrawn. A summary of the most important results could be found within the Introduction.

307 "Delphi-Study: Future of Knowledge Management"- Announcement and Delphi-Study-Questions by Wolfgang Scholl and Peter Heisig, October 2001.

of respondents, second, re-submit the categorized results for evaluation and ranking, and third, feedback on ranking and evaluation of the results.

So, we aspired to get feedback from knowledge management experts. A KM Expert was defined as "somebody being active in knowledge management research or practice, or in closely related fields, who had published on the subject or presented a case at a KM conference". Around 400 potential panelists have been identified, while 254 actually received the questionnaire. A participation of 45 experts (17,7 %) was achieved in the first round. The second Delphi panel involved 25 experts. The distribution between researchers and practitioners as well as participants with a background in natural sciences versus social science and business administration was relatively equal in both rounds. The Delphi study was carried out in Winter 2001/2002.

After receiving the answers to the six open ended questions in the first round, these answers have been categorized into 61 categories. For the second round, only categories with at least two responses were chosen and presented with the best exemplifying phrases from the first round to all panelists, who were asked to rate these categories on a Likert scale reaching from *-3 = I strongly disagree* to *+3 = I strongly agree*. Instead of a third Delphi round, the results of the first two rounds were presented at a conference on the future of knowledge management to invited knowledge management experts and a Berlin group of researchers, both originating from different disciplines. Their commentaries on the results helped us to come to a more thorough interpretation[308]. The means and standard deviations were calculated for all categories. The categories were listed in the rank order of their means displayed in the tables 1-6. This was done also for each of the contrasting subgroups (practitioners and researchers, social scientists and natural scientists)[309]

8.2 Results on Theoretical Issues and Approaches

This chapter displays the results about theoretical challenges and advancements. The following explanations are based on the evaluations made by the panelists in the second Delphi round.

---

308 The conference entitled *First international conference on the future of knowledge management* took place on March 8th – 10th 2002 near Berlin and was financed by the Stifterverband der Deutschen Wissenschaft (Donors' foundation of German Sciences). We thank all participants for their helpful comments.

309 The detailed Delphi results have been submitted for publication in a peer-reviewed international journal, cf.: Scholl, König, Meyer, Heisig (forthcoming).

8.2.1 Most Challenging Theoretical Research Issues for KM

Results for question 1: What is the most pressing and challenging theoretical research issue for the understanding and advancement of knowledge management?

Category
1. Integration into business processes, e.g. integrating knowledge management into the common business processes
2. Knowledge sharing, e.g. identifying the knowledge bearers within an organization, convincing and motivating people to share their knowledge
3. Organizational learning, e.g. forming and developing organizational competence, its connection with business success
4. KM framework: integrating human resource management, organizational management and information management
5. Knowledge assessment, e.g. valuing contributions to a knowledge pool, identifying invalid knowledge as well as measuring valuable knowledge and intellectual capital in unambiguous terms
6. Learning, e.g. the differences between implicit and deliberate learning and between non-formal and formal learning, as well as the social aspects of learning and their connection with business success
7. Motivation, e.g. motivating people to participate in KM
8. Terminology, e.g. definitions, taxonomies, classifications and ontologies
9. Knowledge enabling; enabling knowledge management e.g. by using KM infrastructure
10. Implicit knowledge
11. Knowledge-orientated data-bases, e.g. structuring and integrating text documents and data-bases into knowledge bases

Fig. 8.2: Results for Question 1

The panelists considered the *Integration of knowledge management into business processes, e.g. integrating knowledge management into the common business processes* as the most important theoretical challenge. *Knowledge sharing* ranked second followed by *Organizational learning* and *Knowledge management framework.*

The high importance attached to the *Integration of KM into business processes* reflects three aspects: First, the lack of time is one of the most mentioned barriers to KM (Bullinger et al. 1997). Second, additional departments for KM are increasing costs and do not address the main issue. Third, knowledge handling takes place within the tasks performed in business processes and should not be something apart (Davenport, Prusak 1998, Heisig 2000). Several KM approaches have started to address this challenge and provide some practical methods.[310] The second rank for *Knowledge sharing* is not surprising due to the fact that several KM approaches[311] have placed their main emphasis on the logistical aspect "the right knowledge at the right time at the right place".

Somewhat surprising is the low rank attached to IT-oriented research issues, like *Knowledge-oriented data bases*, which got the lowest rank. This validates the experiences made by KM pioneers, that "80 per cent is culture and twenty per cent is technology"[312], and the results from our former benchmarking company survey, where IT (27,9 %) was ranked only on the fifth place.[313] The low importance of IT-aspects might reflect the sometimes disappointing results of costly knowledge databases (Prusak, 2002; Snowden, 2002).[314] The expectation that the new IT-possibilities for information processing can be easily translated into huge progress in knowledge management is rejected. Nevertheless, IT-support like relational data bases, e-mail and discussion forums, one-to-one chat, etc. is important, but, of course, never a sufficient element of a KM system.

8.2.2 Most Promising Theoretical Approach for KM

Results for question 2: Which theoretical approaches are most likely to deal effectively with these theoretical research issues?

Category
1. Interdisciplinary and multi-disciplinary approaches, combinations of respective methods and techniques
2. Empirical research designs, e.g. action research, case studies, survey studies, qualitative studies, statistical studies
3. Social network analysis
4. Aspects of knowledge sharing, e.g. transactive memory, common knowledge, connecting people
5. Organizational learning
6. Social science approaches
7. Instruments and practices
8. Psychological approaches
9. Knowledge enabling

310 Cf. overview of process-oriented KM approaches in Abecker et al. (2002) and Remus (2002), Heisig (2001); other European process-oriented KM approaches are presented by Weggeman (1998), Grundstein (2000), El Sawy, Josefek (2003).

311 Davenport, Prusak 1998; Probst et al. 1997; O'Dell, Grayson (1998), Weggeman (1998).

312 Best Practice Transfer in Chevron (1997).

313 Cf. Introduction Fig. 1.5.

314 Cf. chapter 16.

10.	A different conception of management
11.	Economic approaches
12.	Complexity theory
13.	Philosophy of knowledge deliberations

Fig. 8.3: Results for Question 2

The most promising theoretical approach to knowledge management was seen in *Inter- and multidisciplinary approaches, combinations of respective methods and techniques*, followed by *Empirical research designs*. These answers suggest that the many disciplinary perspectives and publications on KM are not completing each other and that there is a dearth of sound empirical studies. The "patchwork" metaphor, cited above as a characterization of the state of knowledge management, points into this direction but may be still a flattery picture. Single disciplines received lower ranks in the second round. Despite the fact that practitioners and participants with a background in natural science are equally represented in the sample, disciplines like computer science or engineering are not between the most promising approaches.

8.2.3 Most Important Theoretical Advancements in KM

Results for question 3. What is the most important recent theoretical advancement in knowledge management?

Category	
1.	Priority on human factors: shift from an IT-perspective to a behavioural science perspective
2.	Social network analysis
3.	Matching social and technical aspects
4.	Organizational learning, collective learning models
5.	Systems theory, e.g. autopoiesis, systemic thinking
6.	Nonaka & Takeuchi: spiral of knowledge creation, Ba
7.	Knowledge enabling
8.	Implicit knowledge: the distinction between implicit and explicit knowledge
9.	IT-systems such as web tools and portals, IT capabilities
10.	Boisot: I-Space, Knowledge Production
11.	None (there is no recent theoretical advancement)

Fig. 8.4: Results for Question 3

The statement that *There is no recent theoretical advancement* received the lowest ranking and was therefore rejected by the panelists. The results show a clear emphasis on human factors: *Priority on human factors, shift from an IT-perspective to a behavioural science perspective*, followed by *Social network analysis* and *Matching social and technical aspects*. IT-aspects are again ranked low. This validates the results to the former questions and the organizational practice, where after starting with IT-driven KM solutions the human factors are acquiring more and more importance. These results remind us of the evolution of Computer Integrated Manufacturing (CIM) concepts. These first technology-driven CIM concepts failed and had been enhanced by human factors aspects mainly because of overlooking human expertise.[315]

8.3 Results on Practical Issues and Approaches

Let's have a short look at the results on the most pressing practical problems and the most promising practical solutions for knowledge management. Are the responses to the practical questions in line with the answers to the theoretical ones?

8.3.1 Most Challenging Practical Problem of KM

Results for question 4. What is the most pressing and challenging practical problem for the understanding and advancement of knowledge management?

Category
1. Barriers: organizational, technical and emotional barriers; breaking the dominance of Taylorist thinking
2. Organizational culture promoting KM
3. Knowledge assessment: measuring and validating knowledge, inventorying knowledge; distinguishing between data, information and knowledge; quality measures
4. Knowledge sharing, e.g. identifying the knowledge bearers within an organization, convincing and motivating people to share their knowledge
5. Knowledge organization: transforming an organization into a knowledge organization, reducing the KM overhead
6. Matching social and technical aspects
7. A different conception of management, new mindset for management
8. Time for knowledge management
9. Knowledge creation, knowledge selection and use of knowledge
10. Sensibilization, awareness raising for knowledge management
11. Motivation for knowledge management
12. Instruments and practices, e.g. activity reports and measures to increase usability

315 Mertins et al. (1991), Mertins et al. (1993).

13.	Implicit knowledge: externalization of implicit knowledge, distinction between implicit and explicit knowledge
14.	Standards: standardization of knowledge management vocabulary and knowledge management processes
15.	Knowledge-orientated data-bases, e.g. knowledge formats, extracting knowledge from documents
16.	IT-systems: intranet, internet and groupware

Fig. 8.5: Results for Question 4

In practice, the most challenging problems are *Organizational, technical and emotional barriers*, followed by an *Organizational culture promoting KM* and *knowledge assessment*. This result reflects the experiences with first KM projects, which are struggling with several barriers, and culture is still seen as one of the most critical factors[316]. *Knowledge assessment* comes third and points to the need for indicators, measuring and control systems to facilitate a more strategic management of the development and usage of knowledge.[317] Similar to the theoretical point of view, also from the practical point of view, *knowledge-oriented databases* and IT-*systems (intranet, internet and groupware*) received the lowest ranks. Taken together, the most challenging practical problems are stated in a somewhat more concrete way than the parallel theoretical issues.

8.3.2 Most Promising Practical Approaches for KM

Results for question 5. Which practical approaches are most likely to deal effectively with these practical problems?

Category	
1.	Integrating KM into business processes
2.	Communities of practice
3.	Knowledge assessment, e.g. evaluation systems, verification of knowledge, follow-up analysis and project success measures
4.	Focus on organizational culture; cultural change away from command and control, fit between the business culture, KM potentials and information systems
5.	Sensibilization: raising the awareness for the importance of KM
6.	HR management, e.g. capabilities development, management by knowledge objectives, altering assumptions about people and human nature, measuring employees on KM
7.	Knowledge roles, e.g. internal auditors, knowledge workers
8.	Learning, e.g. internally by encouraging user interaction or externally by capturing

316 Cf. Introduction on Critical Success Factors.

317 Cf. Sullivan (1998); cf. chapter 7.

experience from consulting firms
9. Incentives: providing (im-)material rewards for sharing knowledge
10. IT-systems: intranet, internet, groupware; computer-based information-systems, web tools, networking, chat rooms
11. Knowledge trading, e.g. via knowledge market places or E-commerce
12. Artificial intelligence tools, e.g. human language technologies

Fig. 8.6: Results for Question 5

While the most pressing theoretical research issue was the *Integration of knowledge management into business processes*, this approach was considered as the most likely practical approach to deal with the current problems in knowledge management and achieved the highest corresponding evaluation of all weightings. The increasing importance related to human factors is reflected in the second rank with *Communities of practice*. *Knowledge assessments, e.g. evaluation systems, verification of knowledge, follow-up analysis and project success measures* are ranked third, which implies that easy to use measurement systems are more acceptable than sophisticated, difficult to communicate ones. While *IT-systems* and *Artificial intelligence tools* are not surprisingly on the lowest ranks, *"Knowledge trading, e.g. via knowledge market places or E-commerce"* is very low rated too, despite the fact that it is forcibly proposed for internal knowledge trade (North 1998, Skyrme 2003).

8.3.3 Most Important Practical Advancement in KM

Results for question 6. Which practical advancements are the most important recent practical advancements in knowledge management?

Category
1. Priority on human factors, e.g. the non-technological reflection of KM, emphasizing social aspects
2. Communities of practice
3. Emphasizing human approaches, considering human values, trust etc
4. Knowledge assessment, e.g. knowledge value added (KVA) technique, knowledge as long term investment goal
5. Transfer techniques, e.g. storytelling
6. Sensibilization: awareness raising for the importance of knowledge management
7. HR management, e.g. supporting grass-root KM-initiatives, group intervention techniques
8. Knowledge roles, e.g. knowledge management officers (KMOs)

9.	IT-systems: intranet, internet, groupware; web conferencing, filing systems, instant messaging, collaborative knowledge creation tools, portals and e-mail
10.	Artificial intelligence tools, e.g. sophisticated information extraction and document management systems

Fig. 8.7: Results for Question 6

The trend of low rankings for *IT-systems* continues with the second lowest rank in the second round. The importance of *Priority on human factors* is underlined with the highest rank and *Emphasizing human approaches, considering human values, trust etc, ranking third.* Communities of practice is the one concrete KM-measure which ranks consistently high as a recent advancement as well as a promising approach for the future (preceding question). Knowledge assessment, ranking only a bit lower on both questions, is a broader category encompassing many possible concrete measures. Thus, it is still an important practical problem (Q 4) as well as an open theoretical issue (Q 1).

8.4 Conclusions and Recommendations

The Delphi study on "The Future of Knowledge Management" gives a differentiated picture and the following conclusions could be drawn from this first analysis.[318]

Corresponding agreement could be observed regarding the advancements from the theoretical (Q3) and practical (Q6) perspective. Both responses point to a *Priority on human factors*, understood as a *shift from an IT-perspective to a behavioural science perspective* (Q3: 1st. rank) and *e.g. the non-technological reflection of knowledge management, emphasizing social aspects* (Q6: 1st. rank).

Despite these advancements, *Organizational, technical and emotional barriers* are ranked as the most challenging practical problem of knowledge management (Q4: 1st. rank). As IT-related answers have been always received the lowest rankings (Q1, Q3, Q4, Q5, Q6), further research on human and organizational factors influencing knowledge management is definitively required for the understanding of these barriers.[319]

It is clear from the Delphi study that research on knowledge management should be *Interdisciplinary and multi-disciplinary approaches, combinations of respective research methods and techniques* (Q2: 1st. rank) and *Empirical research de-*

318 The detailed Delphi results with the frequencies of the first round and the standard deviations of the second round are reported in Scholl, König, Meyer, Heisig (forthcoming).

319 Cf. chapters 3 and 4.

signs, e.g. action research, case studies, survey studies, qualitative studies, statistical studies ($\overline{x}$ = 2nd. Rank). This approach has been followed not only by the Fraunhofer Competence Center Knowledge Management presented in this volume but also by Probst et al. (1997) and Bach et al. (1999, 2000) and Eppler et al. (1999) in Switzerland.

The *Integration of knowledge management into business processes* (Q5: 1st. rank) has been assessed in this study as the most promising practical approach for KM. The opinion that knowledge management can be nothing apart from the daily work processes and the (knowledge) workers is seemingly acquiring more and more acceptance (Wiig 1995, Davenport, Prusak 1998, Weggeman 1998, Bach et al. 1999, Heisig 2000, Abecker 2002, Firestone, McElroy 2002, El Sawy, Josefek 2003). Although practical methods to achieve this integration are underway[320], more practical tests with empirical validated design criteria are required.

A surprising outcome of this study is, that the popular ideas of Nonaka and Takeuchi (1995) regarding the *Distinction between implicit and explicit knowledge* was most often mentioned as a theoretical advancement in the first Delphi round, but received low weights in the second round (Q1: 10th rank; Q3: 6th & 8th rank, Q4: 13th rank). Furthermore this endlessly discussed distinction has even not been mentioned at all among the promising theoretical and practical approaches (Q2, Q5). One may speculate whether this distinction is too difficult to handle or whether it seems not fruitful for solving the real KM problems.

Another not expected outcome is the low importance attached to *Motivation for knowledge management* and *Incentives, providing (im-)material rewards for sharing knowledge*. Although this motivational issue is one of the most frequently asked question in discussions about knowledge management, it is neither considered as one of the most pressing practical problems (Q4: 11th rank) nor are incentives seen as a very promising practical approach (Q5: 9th rank). Maybe, a *Focus on organizational culture; cultural change away from command and control* (Q5: 4th rank) is a better answer to the motivation problem, fostering intrinsic motivation fueled by self-control and responsibility for challenging tasks themselves.

Reflecting the results of this First Global Delphi Study on "The Future of Knowledge Management" with the background of our applied research and practical projects, the following recommendations could be made:

A knowledge management initiative has to start with an evaluation of *Organizational, technical and emotional barriers* with an *Interdisciplinary approach* and *Empirical research design*. All relevant stakeholders have to be involved in order to *Raise awareness about the importance of knowledge management* and to gather the user requirements for relevant IT-support and process changes. *Knowledge management* tools and techniques have to be *integrated into the* daily *business*

---

320 Cf. chapter 2.

processes. *IT-systems* have to be *better matched with the human factors*, e.g. usability testing of systems and programs.

The sciences should work on a common *Knowledge management framework, integrating human resource management, organizational* (process) *management and information management* with some degree of *Standardization and a common terminology*.

First efforts to establish an European KM Framework are currently underway within the CEN/ISSS Workshop on Knowledge Management. The aims of this workshop[321] are "to investigate those soft areas related to KM which can be the subject of common approaches, good practice identification or standardization initiatives, and to situate and describe these in the wider organizational context. The overall intention is to provide meaningful and useful guidelines to companies, and notably SMEs, as to how they might align their organizations culturally and socially to take advantage of the opportunities of knowledge sharing within and beyond their organizational boundaries." (CEN/ISSS 2002)

The main progress in knowledge management will be made by joint work and research of scientists and courageous practitioners developing new knowledge management solutions in organizational practice and experimenting with them. This should be done with more emphasis on human factors, a thorough understanding of the manifold barriers and the cultural environment, such that both organizations and sciences will yield more of the potentials of the knowledge-enabling organization.

---

[321] For more information cf.: http://www.cenorm.be/isss/Workshop/km/Default.htm or first KM standards are: BSI 2001, Standards Australia HB 275 – 2001: cf. also Kemp et al. 2002, Weber et al. (forthcoming).

9 A Survey on Good Practices in Knowledge Management in European Companies

Rob van der Spek, Geoff Carter

9.1 Introduction

In 1997 the European Foundation for Quality Management (EFQM), in partnership with CIBIT and the American Productivity & Quality Center (APQC), undertook a Benchmarking Study Project searching for Good Practice in the area of Knowledge Management. At the time this was probably the first Benchmarking Project on Knowledge Management driven from a Pan-European perspective.

In the subsequent five years many more organizations have come to recognize the importance of knowledge management as a key enabler for their future success, viewing it as a legitimate management discipline rather than being the latest, here today, gone tomorrow, "flavor of the month" fad.

EFQM therefore judged that the time was right to undertake another search for good practice in this challenging area and so, in late 2001, working in partnership once again with CIBIT Consultants| Educators, the EFQM undertook further research. This article is based on the key findings to emerge from that research. Readers should note that because the benchmarking activity was carried out based on the European Benchmarking Code of Conduct, the article does not name specific organizations, thus respecting the confidentiality that was promised at the time of the research

The research was based on a Screening Survey created by CIBIT & EFQM, which was then sent to a number of organizations across Europe in the period between November 2001 and January 2002. It was sent with two purposes in mind:

- To identify the current state of knowledge management practice across Europe.
- To identify a short list of potential good practice organizations in the Knowledge Management field.

The individuals who received and responded to the Screening Survey were all actively involved in working with KM in their organization, so our responses were from the heart of the KM engine room.

Once the analysis of the Screening Survey responses had been completed each respondent received a complimentary copy of the final report, thus CIBIT and EFQM:

- Delivered what was promised to Screening Survey respondents, enabling them to both learn how other organizations manage their knowledge management activities and also to compare their own position in relation to these same organizations.
- Furthered their respective missions to spread information on good practice across Europe.

9.2 What Do We Mean by Knowledge Management?

Common sense says that learning from successes and failures, sharing knowledge with others and smart application of lessons learned from the past will lead to better results now and in the future. Excellent companies know that by heart and apply this common sense everyday, preferably in a natural way. However for many reasons these learning processes might not function properly anymore and need attention and support. Competing instead of collaborating divisions, differences in culture, pressure of the daily challenges, lack of communication tools and places to meet, stubbornness of people or lousy discipline within the company might get in the way. Dealing with these barriers and organizing practical tools to improve the required learning processes is considered to be the scope of knowledge management.

In the context of this study, we described 'knowledge management' as follows:

All the necessary activities to orchestrate an environment in which people are invited and facilitated to apply, develop, share, combine and consolidate relevant knowledge in order to achieve their individual and collective ambitions.

Based upon the experiences in our first best practice study and the client work of CIBIT knowledge management is seen as one aspect of the total focus of management in organizations and not as a goal in itself. The ultimate goal of knowledge management is to improve an organization's effectiveness by leveraging three core-learning processes in a smart and lasting way:

- Learning from success and failures, on individual, team or company level.
- Learning from each other, both from co-located colleagues as well as colleagues which might be located at a further distance (geographical as well as disciplinary-wise).
- Learning from 'outside-in', from partners, suppliers, customers and even competitors.

Knowledge management should empower employees to integrate these learning processes within their ways of working and to apply the outcomes of these processes on a daily basis. But learning should take place in a focused way because learning is not a goal in itself but a tool to improve the company's capabilities. In order to be able to learn in a focused way, a company should understand how these learning processes could fit in the business strategy and which knowledge areas are most important to focus on. Depending on the type of business and strategy, important knowledge can range from the professional knowledge that underlies products and services, knowledge about customers, markets and common practices and knowledge about processes and the internal organization.

The benefits of knowledge management should ultimately be demonstrated in the Key Performance Indicators of the company or organization, where else? Any ROI of knowledge management initiatives should be connected with the model that is used to measure corporate performance. In this context we strongly believe there should be a strong connection with models for business excellence including indicators for areas such as customer satisfaction, employee satisfaction, financial results and impact on society.

9.3 What were the Issues Addressed in the Study?

Based upon our core principles of knowledge management as explained in the previous paragraph we structured the screening survey. In addition to requesting the normal general information on the respondent's organization, i.e. industry sector, number of employees and turnover, the screening survey sought answers to a mixture of qualitative and quantitative questions within each of the issue areas:

- Knowledge management in organizations; what is the general approach?
- Strategies for knowledge; how is knowledge management aligned with the business strategy and which objectives are pursued?
- Enablers for knowledge management; which tools are applied?
- Cultural and motivational Issues; how is employee and management morale addressed?
- Leadership and Involvement of management; are senior and middle managers involved?
- Competency building in knowledge management; are employees trained in knowledge management principles, processes and tools?
- Communication about knowledge management; how do companies communicate about their activities?
- Results and metrics; how do companies measure the outcomes of their initiatives?

- Best Efforts; what are there most successful practices?

We were overwhelmed by the output of the screening survey and the lessons learned which the respondents were so kind to provide us with. For the purposes of this article we have decided to give just a flavour of the overall analysis and focus primarily on the key learning points provided by the respondents.

9.4 General Information on Respondents

27 companies responded to the screening survey, originating from a wide cross section of industry and with a relatively strong contribution from telecommunications, software / IT-services, fast moving consumer goods, consulting and automotive sectors.

Most companies deployed their KM activities either on a global level (63 %) or within Western Europe (26 %) and respondents were a good mixture of large through middle-scale and on to the smaller companies with 22 % employing more then 40.000 staff, 19 % employing between 10,000 & 40,000 and 33% employing less then 10.000 staff.

9.5 Knowledge Management in Respondent Organizations

This section searched for data, information and knowledge on the general KM environment and maturity level of the organizations.

A majority of the respondents had a formal KM-initiative (77 %) while 55.5% of the companies reported that they had a Chief Knowledge Officer (CKO) and 53.3% of these reported that this CKO-role had been in place between 2-4 years.

Reasons for not having a CKO were listed as:

- The board thinks that there is no added value,
- The company is too small for such a function,
- KM-activities are performed at the local level.

63% of respondents reported that knowledge management was still hot on the strategic agenda and 15 % of companies' reported that it had now became a natural way of doing business. 15 % stated that they were still building pilots. Only 7 % of organizations reported that KM never took off (see also Fig. 9.1).

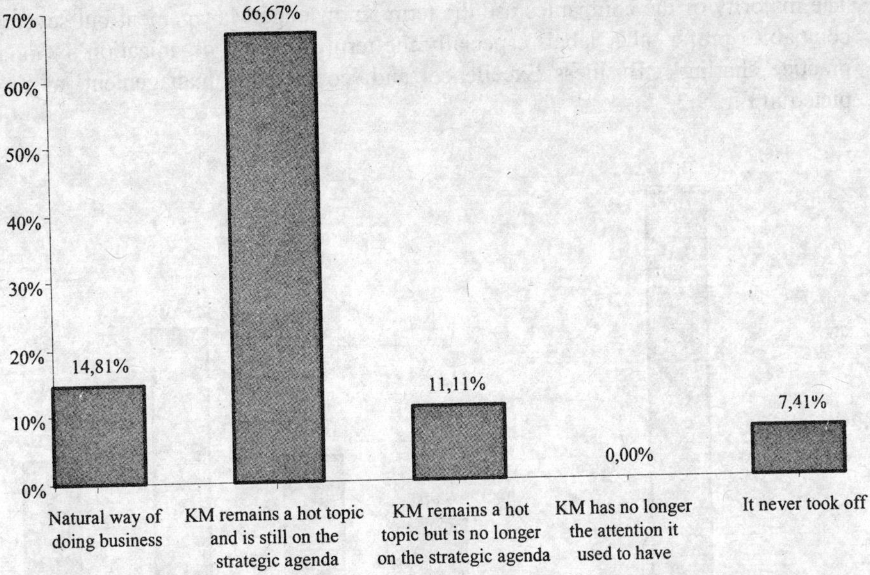

Fig. 9.1: In Your Organization, How Hot as a Topic has KM Become in Recent Years?

Respondents reported a relative high level of maturity according to the APQC-framework[322] for knowledge management initiatives. The phases in the APQC framework are: (1) getting started, (2) develop a strategy, (3) design and launch a KM initiative, (4) expand and support, (5) institutionalize.

55 % of companies score themselves beyond phase 3 and of these 44 % are involved in expanding their KM-activities. Only 3 companies claimed that they were at phase 5.

Most KM initiatives were started by a small team of pioneers acting as catalysts, creating awareness, starting pilots and convincing management. Successful pilots were used to communicate the "business case' and potential benefits. Companies who are now expanding their KM activities work with core teams that are multi-disciplinary, networked within the company and strongly business oriented. They facilitate local champions, share their insights within the network of 'knowledge managers', and act as internal consultant and change agents. Most KM-teams are guided by a steering committee in which representatives of the business are placed. Reporting lines vary from directly to the CEO, the CIO or executives who are responsible for business development, R & D, internal communications, TQM and IT.

322 For more information cf.: www.apqc.org.

The majority of the companies use the term Knowledge Management but several companies prefer other labels especially the term 'learning organization', 'Good practice sharing', 'Business Excellence' and 'competency management' as depicted in Fig. 9.2.

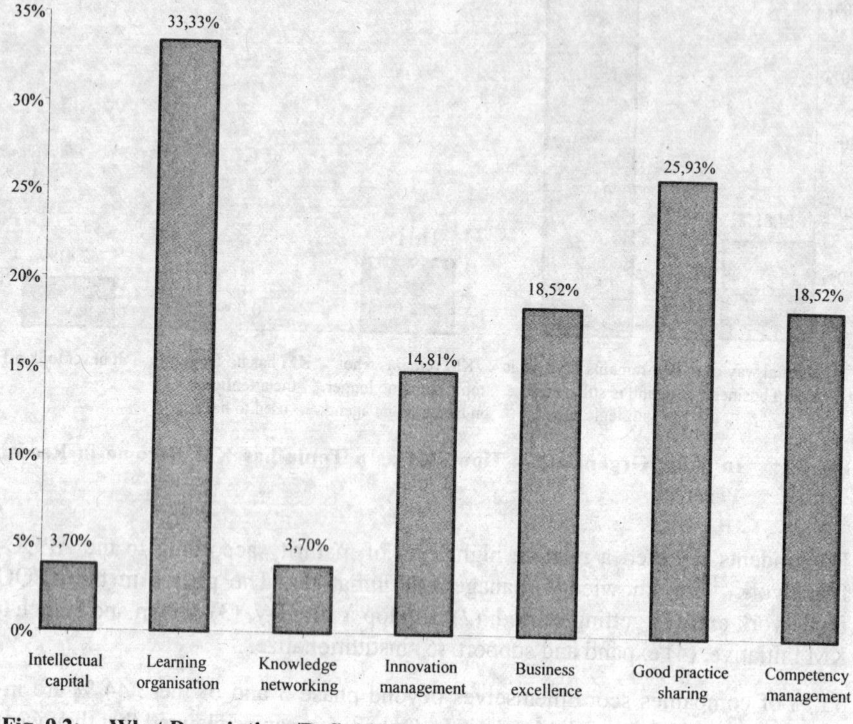

Fig. 9.2: What Descriptions/Definitions are Used?

Common lessons learned by respondents about knowledge management initiatives

- Leadership from the top is a pre-requisite.
- All KM activities should fit in the strategy and the value impact should be clear from the start.
- Starting with pilots and building on success works better then the "big bang' approach.
- Learn by doing.
- Be aware of overselling knowledge management and deliver first.
- Communication is an important factor. Keep the message simple and consistent.

- Address the 'soft factors'; they are real!
- It is especially difficult to organize knowledge management in a decentralized company though the potential benefits of sharing knowledge across operations may be the highest in these companies.
- KM-activities should be focused top-down but also emerge bottom-up. To find a balance between these two strategies is a major key to success.
- There is no blueprint and one-size-fits-all within companies. Allow for flexibility in your corporation.
- Communities are key to success.
- Knowledge management is mainly change management and people oriented. Tools are very important but not decisive. At the same time a majority of the best efforts are based upon ICT-tools!
- Create 'Quick wins' from the beginning but do not forget to create a sustainable environment.
- In the end knowledge management should be integrated in daily operations and processes.

9.6 Strategies for Knowledge

This section addresses the way companies formulate focus for their knowledge management initiatives and align them with their business strategy.

The survey showed that a vast majority (88%) identified crucial knowledge areas, strengths and gaps and formulated actions to deal with these as part of the business strategy (see also van der Spek, Kingma and Hofer-Alfeis 2002). Leaders involve several stakeholders (including senior executives) during workshops in order to create knowledge portfolios and knowledge management roadmaps. Integration of these approaches within existing business planning and strategy formulation processes is considered to be key to sustainable success.

The most important objectives for KM-activities were identified as:

- Facilitation of re-use and consolidation of knowledge across operations.
- Development of new knowledge to innovate products and services in the future.
- Learning from customers to innovate products and services.

Companies reported clearly that they operate both codification (focused on creating tangible information resources based upon knowledge in the heads of employees) and communication strategies (focused on exchange of experiences via collaboration and communication between employees). Which strategy they use depends on the situation.

Common lessons learned by respondents about Knowledge Management strategies

- One needs a corporate strategy to align to and it is necessary to have explicit goals! When the business strategy is unclear or implicit, knowledge management initiatives will suffer because there is no clear focus possible.
- Different knowledge strategies, focused on codification and / or communication, can live next to each other in companies.
- Communities are the major platforms for knowledge exchange and transfer and also for the codification of experiences and re-use of this information. They emphasize the fact that knowledge must be shared before it can be managed! Networks of people should ensure that the relevant experiences are codified in shared databases, web sites or any other tool, which they think, is useful. Information without ownership is useless. Re-use of information relies heavily on teamwork, trust in others and shared passion.
- Codification costs money, energy and time. Before undertaking it an organization must think about the added value and decide whether it is worth it.

9.7 Enablers for Knowledge Management

In this section we addressed the enablers companies to use within their knowledge management initiatives. In order to give direction to the survey we investigated whether the most common used instruments within the KM context were considered to be of high impact and implemented at a satisfying level. We will highlight two core issues in the context of enablers: IT-support and communities of practice.

IT as an Enabler:

Though many respondents argued that IT is not the most important enabler for KM, all but one respondent reported that they use IT as an enabler. When respondents "best efforts" were taken into account it showed that the majority of these best efforts made strong usage of IT. This seems to be one of the most interesting paradoxes in the knowledge management community. An often formulated statement was: 'IT is not the most important enabler but at the same time we cannot live without it'. We assume that most pioneers have learned the hard way that good IT-tools are no less then expected but not enough. Commitment and culture determine whether good IT-tools will provide optimal results, but it is much easier to deliver good results with passionate and committed people and sub-optimal IT-tools then the other way around.

According to the survey most used IT-tools are:

- Intranet with email & discussion forms
- Intranet with document management functionalities
- Intranet with people finding functionalities
- Video- conferencing
- Tools for communities and project teams
- E- learning

Least used IT-tools in the context of KM are:

- Gaming and simulations
- Decision support tools
- Data or text mining tools
- Customer relationship management (CRM)

Common lessons learned with regard to IT-enablers are:

- People must be involved and motivated to use IT-tools. It must make life easier for them.
- IT-enablers should be integrated in daily operations. When the gap between work and IT-tools is too large, people will not spend the required additional efforts.
- People should be trained to use the IT-tools. You cannot expect that all employees can use then in the best way.
- Yellow pages or personal home pages only work when people provide content, keep them updated and when they have clear benefits form profiling themselves via these media.

Communities of Practice as an Enabler

It was clear that *Communities of Practice* are a major tool for companies to create knowledge sharing platforms between people. 74 % of all respondents use Communities of Practice in order to:

- Sharing of knowledge and good practices (all)
- Solve problems (75% of those using CoPs)
- Create knowledge (65% of those using CoPs)

It seems that communities are less used for the development of tools, methods and frameworks (45% of those using CoPs).

Communities of Practice emerge mainly both from individual needs as well as from corporate needs (70% of those using CoPs).

The benefits of COP's are measured by most respondents (75% of those using CoPs), support to COP's is provided by several companies (70% of those using CoPs) and supporting technology is offered (70% of those using CoPs).

Common lessons learned by respondents about Communities of Practice as an Enabler

- It was important to demonstrate the power of communities both to the members as well as involved managers.
- The role of the community facilitator / content editor is key to success.
- Thriving communities will take their own responsibilities for their knowledge processes and the proficiency of its members.
- Communities require a marketplace that enables its members to collaborate, co-learn and to share their resources.
- Communities should be empowered to create their "own identity", shared language and products.
- You need social behaviors in order to design good virtual spaces. It is necessary to lead the intranet by people not only from IT-departments. It is a cultural change working with intranets. "We are at the beginning of this new way of behavior and we need to develop new spaces for the people working at the shop floor."

Specific learning tools

Respondents were asked to score the impact and satisfaction of the use of several typical KM-instruments such as after-action reviews, peer assists, story telling. It became clear that they were most positive about the use and outcomes of after-action reviews, peer assists, and coaching. They were least satisfied with the way their company had organized yellow pages, master-apprenticeships and story telling

9.8 Cultural and Motivational Issues

In this section we addressed the cultural and motivational issues that companies encounter related to knowledge management initiatives.

According to a majority of the respondents' common aspects of the organizational culture, which supported knowledge management, initiatives were:

- Interpersonal exchanges, respect and trust
- Lack of hierarchy
- Perceived urgency to change because of market environment
- Focus on quality and customers

The main cultural and motivational *barriers* were:

- Not enough time to think about the future and pressure on billability
- Decentralized and federative structure which emphasizes focus on local results instead of company-wide benefits
- KM perceived as additional to the workload of employees
- Turn-over of employees which generates knowledge erosion
- Not-invented-here syndrome
- Internal competition
- "Knowledge is power"

Respondents reported several tactics to deal with these barriers including the demonstration of practical results, good tools and clear added value, repeated communication what the company would like to achieve and involvement of employees.

63% of respondents reported that the culture has changed because of KM-initiatives in favor of a better understanding of how business operations can be improved by explicit attention for knowledge related issues.

50 % of the respondents had KM connected to their recognition and rewarding systems while 40% rewarded employees in addition to their salary for their contribution to KM-activities. Most companies rewarded employees who contributed to KM-activities on a personal 1-1 basis. Some companies have introduced awards for innovation, Best-practice sharing or for excellent teamwork. One company with extensive experience in KM (> 4 years) introduced an incentive system where users of a knowledge-sharing intranet earn points for their contributions and for feedback from other users. These points can be transformed into gifts in the form of additional educational budget, computers, PDAs or budget for conferences.

Common lessons learned by respondents with regard to motivational and cultural issues

- Change takes time; it is necessary to be patient and to invest in creation of the required pre-conditions.
- Leaders are necessary to initiate, support and guide.
- It is not possible to force people to share knowledge. It can only happen voluntarily.
- Money is not the only way to reward people for their contribution. Recognition and providing more challenges were seen as important ways to stimulate these employees.

9.9 Leadership and Involvement of Management

This section focuses on the role, involvement and support of senior and middle managers.

Respondents reported that leaders were involved in KM-initiatives in several ways. Most common were:

- They fund and support KM-activities
- They recognize and appreciate people efforts and achievements in the area of KM
- They are active and personally involved in KM-activities

Leaders were less involved with regard to the following issues:

- Acting as role models when it came to optimal creation, sharing and use of knowledge
- Formulating a knowledge strategy for their activities
- Defining priorities for KM-activities

The support of middle management was surprisingly high with 51.9% of respondents claiming that middle management in general supported KM-activities within their company nowadays. However in a significant minority of organizations, respondents reported that middle management considered KM a buzzword and doubted whether it would add value to their daily business.

Common lessons learned by respondents with regard to leadership and middle management

- Managers seem to be resistant to the term knowledge management. Language is important here. Managers do not like jargon and creating a shared language about KM is an important step towards understanding and thus support.

- It is necessary to have support at all levels of the organization.
- In order to convince managers, a business case is crucial. However, evidence that KM activities deliver better results is very difficult to provide because many factors can influence corporate performance.
- KM should be aligned with the strategy and demonstrate how it fits.
- Do not sell cheap! Real KM activities, which create sustainable benefits, require investments that should not be hidden from management upfront. Be clear on what you would like to achieve, how it fits in the strategy and what it will cost.
- Start from strategy: 1) must be directly and visibly linked to the delivery of the strategy; 2) look for hooks; 3) specify clear purpose, vision and mission for KM in terms of strategy.
- Balance sustainability versus quick wins: 1) If quick win achieved, then "so this is KM", if not achieved "so this was KM"; 2) Think big, act small; 3) Show a tangible and sustainable impact on business performance; 4) Be comprehensive in terms of approach and business coverage; 5) Clarify that this doesn't come for free!

9.10 Competency Building in Knowledge Management

This section focuses on the way companies build competencies within their employees focused on knowledge management principles.

50 % of the respondents organized stand-alone programmes about KM. Most KM teams/task-forces ran short training programmes to create awareness and to educate in the usage of tools and instruments. They considered these training programmes as crucial in their campaigns. These programmes focused on project- or line managers, senior executives and specialists. Reasons for not organizing specific training programs were:

- Lack of time, staff and budget
- No interest or support from management

37% of respondents address KM-issues in general training programmes for managers, executives and employees.

63% of respondents use a Community of Practice to connect people who are involved in KM-initiatives. These communities are perceived as a powerful platform to share experiences and to multiply KM-activities across companies.

9.11 Communication about Knowledge Management

This section focuses on the way companies communicate about their KM –initiatives.

The survey shows that companies use a broad set of communication channels to communicate about KM in their organization(s). Repetitive and consistent communications via different channels was perceived as crucial for the success of KM-activities. In this communication, theories about KM should be limited; focus should be on the success stories, anecdotes and personal impressions of managers and employees.

Some companies communicate on purpose via external media because they believe it adds value to their corporate image.

Main lessons learned by respondents with regard to Communications about KM

- Keep the messages simple and consistent over time. Use few but clear models to explain concepts. Always link to real issues that the audience can relate to. Think carefully about developing awareness programmes. Persist! Build a marketing plan and implementation. Don't oversell. Undersell but outperform.
- Be concrete; give examples rather than theories. Show people where they and their business can benefit. It is a good opportunity to get some people "into the limelight".
- Know your audience and know their interests.
- The more you talk about it, the more people will understand it and want to get involved.
- Use the complete spectrum; communication is a major part of change management!
- Communication is the main element of success. You must be willing to repeat, repeat and repeat.
- We would rather communicate after there is a deliverable rather than making promises and setting up expectations that might not be fulfilled.

9.12 Results and Metrics

One of the main issues to emerge during the 1997 benchmarking project was the difficulty in measuring success that could be attributed to a specific KM initiative. During this research project, although progress would appear to have been made, it remains one of the biggest challenges.

A minority of respondents measured the impact of their KM-initiatives. Of those who did not measure, 50 % were planning to do so in the future. However a vast majority reviewed their KM-initiatives in terms of progress and output. These reviews involved internal clients, stakeholders, steering committee members and relevant disciplines.

Those companies who did measure the impact of their KM-initiatives reported outcomes in the following result areas or enablers (see also Fig. 9.3):

- Practice and process improvements
- Enhanced employee capabilities and satisfaction
- Increased customer satisfaction

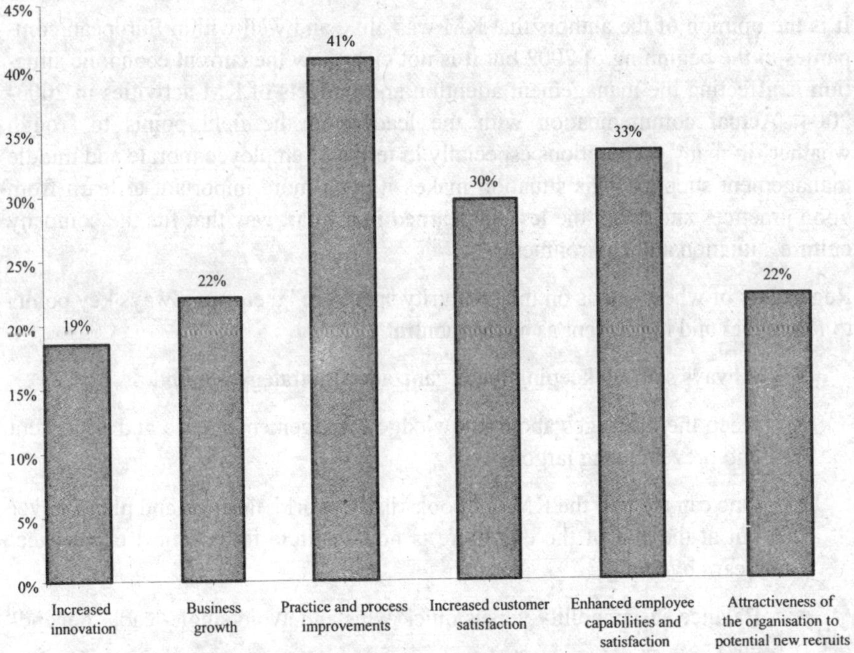

Fig. 9.3: Which of the Possible Outcomes are Most Relevant?

40.7% of respondents connected their KM-initiatives to their performance measurement system. A relationship between KM-activities and the Balanced Scorecard is reported most often.

Only 7.4% of respondents' measured their intellectual capital and only a limited group of companies plan to do that in the future.

Main lessons learned by respondents with regard to Results and Metrics

- Be careful with trying to prove that KM has a significant impact on business performance. It might bring more costs in terms of measurement than it will bring benefits.
- Be aware of too simplistic and one-dimensional metrics.
- Start from result areas that are well known in the company such as market share, employee satisfaction, financial results and customer satisfaction. That's why the Balanced Scorecard and the EFQM Excellence Model form a good starting point.

9.13 Conclusions

It is the opinion of the authors that KM was alive and well within European companies in the beginning of 2002 but it is not clear how the current economic situation is affecting the management attention and budgets of KM-activities in 2003-2004. Actual communication with the leaders in the field points to 'rough weather' in many corporations especially in terms of employee morale and middle management support. This situation makes it even more important to learn from good practices and adopt the lessons learned in a smart way that fits the company culture, situation and environment.

Regardless of where one is on the "maturity spectrum" there are always key points to remember and implement as a KM mantra.

- Always start by keeping the organization's strategy in mind.
- Keep the messages about knowledge management simple and consistent and prevent using jargon.
- One can read all the KM textbooks in the world, theories and plan forever but at the end of the day there is no substitute for practical experience. Learn by doing.
- Balance sustainability versus quick wins and avoid single-leader dependency.
- Technology is a significant and necessary enabler but it is not sufficient. The "people dimension" is crucial.

10 How German Companies Employ Knowledge Management. An OECD Survey on Usage, Motivations and Effects.

Jakob Edler[323]

10.1 Introduction: Context and Scope

In the year 2000, the OECD, more concretely the Center for Educational Research and Innovation (CERI) within OECD,[324] launched an initiative to conduct comparative studies in member countries in order to learn about the existence, dynamics and effectiveness of knowledge management (KM) practices in the private sector, and in order to systemize these practices and to work on comparable indicators and measurement guidelines. An international working group was established that designed a questionnaire to be used in a broad survey within participating countries. This paper summarizes first major, mainly descriptive findings of the German survey[325] that was conducted at the Fraunhofer Institute for Systems and Innovation Research in Karlsruhe in summer and autumn 2002.[326]

The impulse for this task within the OECD stemmed from the notion that there is growing information on knowledge management within OECD countries, however, this information lacks a common reference framework needed for international comparison. Much of the information is based on case studies or used to formulate managerial guidelines rather than analyze KM in depth, and, above all, all the various endeavours have not led to comparable datasets yet.

---

323 With the indispensable support of Rainer Frietsch for the statistical analysis. I am also indebted to Michael Bordt from StatCan and to Dominique Foray from OECD/CERI for their extremely valuable comments on a first draft. Any mistakes and inconsistencies remain of course within the full responsibility of the author.

324 The OECD set up an international study group to prepare the common reference framework and a core questionnaire. Responsible project managers at the OECD are Dominique Foray and Kurt Larsen.

325 More detailed analysis will be published at a later stage, first in the context of a forthcoming OECD publication – to be edited by D. Foray/ F. Gault- and second in a more comprehensive report.

326 The study on which this article is based was made possible by the Donors' Association for the Promotion of Sciences and Humanities in Germany who fully funded it, and by the willingness of the German Federal Ministry to officially support it. We are very grateful to both institutions.

It is the aim of the OECD practice to contribute to a comparable understanding of knowledge management practices and their usage within the OECD.

The idea pursued by the OECD is that one should start off by employing a broad understanding of KM. According to the OECD working definition, *Knowledge management (KM) involves any activity related to the capture, use and sharing of knowledge by the organization*. This definition encompasses activities that can be grouped under four headings: communication (databases etc.), human resource (training, mentoring etc.), policies and strategies (using partners, defining KM strategy etc.) and finally capture and acquisition of knowledge from outside the company.[327] Above all, this understanding is meant to be as broad as possible and serves an exploratory purpose.

The core questionnaire is supposed to be used for the country studies. It contains four research dimensions.[328]

(1) **Usage:** What practices are used, and how is the usage developing?

(2) **Motives:** What are the driving forces to employ KM practices?

(3) **Effects:** What effects are attributed to the usage of KM practices?

(4) **Institutionalization:** How is KM institutionalized within the companies organizationally and/or financially?

10.2 The Sample

Almost 3.500 questionnaires were sent out within seven industrial sectors, 497 responded positively (response rate above 14%), 410 participated in a non-responsive analysis (cf. annex for details). The sectoral and the size distribution are shown in Fig. 10.1 and Fig. 10.2. To choose seven distinct sectors reflected the hypothesis that there are sectoral differences for KM that can only be analyzed with sufficiently large sector samples.[329]

327 Most importantly, this definition sees ICT-based KM practices as one major dimension among a range of others, and not as the key to KM, at least *ex ante*.

328 The German country survey added a fifth dimension: *KM, external interfaces and Innovation* in order to analyze the meaning of KM for innovation, especially the capture of knowledge from the environment needed for the innovation process. However, the analysis presented here focuses on the core of the OECD survey.

329 The size of the initial sector samples reflected the experiences of a preceding pilot study and the analysis of response behaviour.

Sector	*N*	*sample*	*share (%)*
Chemical Industry (except pharma and biotech)	409	**48**	9.66
Pharmaceuticals	344	**31**	6.24
Biotech	612	**76**	15.29
Mechanical Engineering	395	**51**	10.26
Vehicles (including transport equipment)	394	**36**	7.24
Electrical Engineering/ Electronics (ICT)	614	**61**	12.27
Business-related services	727	**160**	32.19
No sector/company name given in response		**34**	6.84
Total	*3495*	*497*	*100*

Fig. 10.1: Company Sample – Sectoral Distribution

To include the service sector is based on the hypothesis that this sector has a very special need for and usage of KM.[330] For the understanding of sectoral patterns it is important to note the size distribution of the seven sectors: the service and especially the biotechnology sector are dominated by companies with less than 250 employees, while the pharmaceutical sector is dominated by companies above 250, but below 2000 employees. The remaining four sectors are dominated by companies with more than 250 employees.

Size	*1-49*	*50-249*	*250-1999*	*2000+*	*no size given*	*total*
Number	121	150	139	64	23	*497*
Share (%)	24.35	***30.18***	27.97	12.88	4.67	*100*

Fig. 10.2: Company Sample – Size Distribution

10.3 Usage of KM Practice

According to our broad understanding of KM, we asked the companies about their usage of 19 KM practices. To trace the dynamic with which the practice has been taken up lately or will be taken up in the near future, we differentiated "in use" into "in use before 1999" and "in use since 1999", and we added the possibility to tick "plan to use" if a practice was not in use yet. Fig. 10.3 shows the result for the whole sample, in the order of decreasing rate of usage.

330 As service sectors are of high relevance to the question of KM resp. innovation management, we have selected market/opinion research, strategic and PR company consulting, architecture and engineering services, technical, physical and chemical expertise, and consultation.

Cat.[a]	Practice. Your company...	in use total	in use before 99	in use since 99	plan
C/A	uses knowledge obtained from other industry sources	97	89	8	0
HR	offers off-site training to workers to keep skills current	95	84	11	2
C/A	uses the Internet to obtain external knowledge	95	57	38	2
HR	encourages experienced workers to transfer their knowledge to new or less experienced workers	93	78	15	3
HR	encourages workers to continue their education by reimbursing tuition fees for successfully completed work-related courses	90	79	11	2
C/A	uses knowledge obtained from public research institutions	88	78	10	2
Com.	prepares written documentation such as lessons learned, training manuals, good work practice etc. (organisat. Memory…)	85	69	16	6
C/A	encourages workers to participate in project teams with external experts	81	65	16	4
C/A	dedicates resources to obtaining external knowledge	70	56	14	5
P/S	uses partnerships or strategic alliances to acquire knowledge	68	50	18	6
P/S	has KM within responsibility of top management	61	44	17	11
Co	facilitates collaborative work by projects teams that are physically separated	59	33	26	12
Co	regularly updates databases of good work practices, lessons learned or listings of experts	57	36	21	18
P/S	has a values system or culture promoting knowledge sharing	45	30	15	18
HR	uses formal mentoring practices, including apprenticeships	39	26	13	7
HR	provides informal training related to KM	34	21	13	12
P/S	offers monetary or non-monetary incentives	30	21	9	12
P/S	has a written KM policy or strategy	23	10	13	18
HR	provides formal training related to KM practices	16	11	5	9
Com.	uses the Internet to obtain external knowledge	95	57	38	2

a: Category of practice: C/A = Capture and Acquisition, Co = Communication, HR = Human Resources, P/S = Policies and Strategies.

Source: Fraunhofer ISI Survey 2002

Fig. 10.3: Percentage of Companies Using Selected KM Practices – Total Sample – in Order of Decreasing Importance

Clearly the practices most widely used are those to capture knowledge from outside the company. Of the eight individual practices used by more than 80 % of the companies, four are related to knowledge capture, three to training, only one to communication and none to KM strategies. The two most popular practices are the use of knowledge obtained from other industrial sources and the use of the Internet (capture), followed by off-site training, inter-personal knowledge transfer and work-related formation (training), using knowledge from public research (capture), written documentation (communication) and encouraging collaboration with external experts (capture). At the low end of usage, out of the six practices used by less than one third of the companies, three are related to policies and strategies (appropriate value system, incentives and written KM strategy)[331], three stem from the training category. It is clear that, first; KM is not dealt with in strategic terms very often. Second, in contrast to general formation practices KM practices geared towards the build-up of *KM capabilities* are not broadly established, in fact only 16% of the companies have a formal KM training – which is the lowest rate of use.

In any case, this distribution shows that a broad approach for knowledge management going beyond ICT-based practices is fully justified. In fact, the limitation to ICT is justified only if specific aspects of KM are analyzed. The employment of these practices differs according to the size and the sector of companies. As to size, the findings are rather straightforward: the larger the company, the higher the average number of KM practices used (Fig. 10.4, left box), the larger a company, the bigger the need for broad KM.

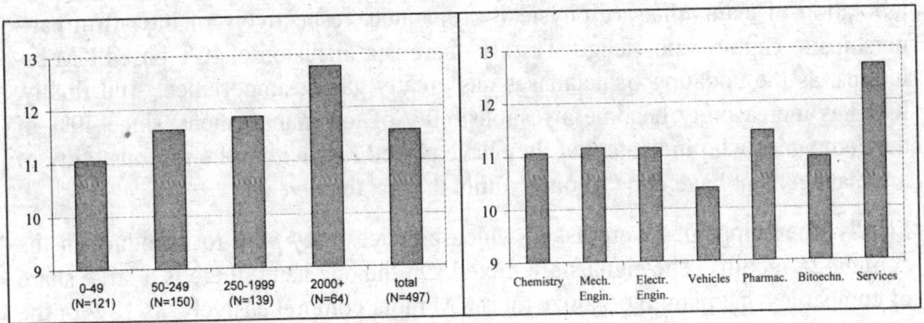

Source: Fraunhofer ISI Survey 2002

Fig. 10.4: Average Number of Practices Used – by Size and Sector

The pattern for the sectors is less clear-cut (see Fig. 10.4, right box) as only two sectors stand out while the rest show a very similar average number of KM prac-

---

331 This is true, although 60% of the companies indicate that KM lies within the responsibility of top management (as was asked additionally in the German questionnaire). Apparently this high institutionalization has not yet led to formalized KM policies.

tices. Obviously, the companies of the service sector use significantly more KM practices, at least those from the service sub-sectors we have selected, which are business-related and knowledge-intensive[332]. These service companies on average employ almost 15 out of the 19 practices we asked about, although the sector sample consists mainly of SMEs. Exactly the opposite pattern is true for the vehicle sector, here the sample is characterized by large companies, and still the average number of KM practices is lowest.

Interestingly, however, the order of practices used is strikingly similar between sectors and size groups. Especially at the low end of practices, there are almost no differences; the distribution of policies and strategies is low for all sectors and size groups. The sectors deviating most from this general pattern are mechanical engineering with a special focus on human resource practices in use, and electrical engineering, a sector that is allegedly prepared to undertake comprehensive, strategic KM in the near future. Finally, the stronger usage of KM by large companies in general is also characterized by a different pattern, as very large companies lay much more emphasis on the acquisition of knowledge from outside the company (especially from research institutes (95%)), with 88% of them dedicating resources to do so.

In which areas have companies complementing their KM in recent years (column "since 99" in Fig. 10.3). Not surprisingly, the usage of the Internet has diffused most in German industry lately (38 % out of the 95 % using it now have introduced this only recently). Secondly, there is a growing need to integrate knowledge across organizational borders and distances, be it from inside or outside the own company, as indicated by the increased importance of attempts to ease collaboration of teams that are physically separated, respectively of inter-firm partnerships to capture knowledge. Thirdly, there is a diffusion of ICT-based KM solutions, as the updating of databases has greatly gained importance. And finally, KM has increasingly become a responsibility of top management, since 40% of the companies who indicate that they have placed KM within the responsibility of top management have done so only within the last three years.

Finally, the companies were asked which practices they *plan* to introduce in the coming 24 months. The signals are mixed. On the one hand, there is a large share of companies that plan to organize their KM more comprehensively, as 18% of the companies indicate to foster an appropriate value system or culture and another 18% plan to formulate a written KM strategy. In addition, the rather low share of companies that have an informal KM training will grow by 12%. At the same time, the tendency to employ ICT-based databases and to ease collaboration across distances remains. On the other hand, however, this development should not be overrated, especially as for policies and strategies and as for KM-related human

332 Cf. footnote 330.

resource practices there seems to be a stable and big portion of German companies that will continue to do without.

10.4 Motivation Patterns

What are the most important reasons for German companies to use KM? Can we see a pattern of motivation? In line with the broad understanding of KM, the motivations to use KM practices are manifold. The companies were asked to rate the importance of 19 different motivations on a scale from 1 (extremely important) to 6 (not important at all). Fig. 10.5 indicates the motivations in the order of decreasing importance for the whole sample. There are eight most important motives for which more than 75% of the companies have attributed an importance of 1 or 2 (top two boxes).[333] The single most important driving force to employ KM practices is apparently the sharing and *integration* of knowledge *among the workforce within* the company, represented by the two most important single motivations (transfer to new workers, integration of knowledge) plus the support for intra-company collaboration across distances. The second most important driving force, made up of three out of the top eight variables (motive 3, 6, 8), is rather defensive. Many companies rate the importance of *stock taking of knowledge* and *its protection* as highly important. This reflects the increasing fluctuation of the workforce as well as the growing importance of knowledge as a strategic asset. Finally, the companies grasp the opportunity provided by KM tools for the upgrading of their workforce internally, as KM is a major tool for *human resource development* (motive 4, 5). In short, German industry is employing KM driven by three major purposes: *internal integration and internal transfer of knowledge, taking stock and protection from loss of knowledge* and *the improvement of the workforce.*

333 Top two category reflects the percentage of companies who indicated a value lower than 3 on the scale from 1 (extremely important) to 6 (not important at all).

Motive	top two	mean
To accelerate and improve the transfer of knowledge to new workers	91	1,64
To help integrate knowledge within your firm or organization	86	1,75
To protect your firm or organization from loss of knowledge due to workers' departure	82	1,77
To encourage managers to share knowledge as a tool for professional promotion of their subordinates	80	1,91
To train workers to develop their human resources	77	2,00
To identify and/or protect strategic knowledge present in your firm or organization	76	1,95
To ease collaborative work of projects or teams that are physically separated	75	2,03
To capture workers' undocumented knowledge (know-how)	75	2,06
To ensure that knowledge resident in all international work sites is accessible to the entire firm or organization	69	2,29
To train workers to meet strategic objectives of your firm or organization	68	2,21
To help managers to focus their attention to key information	67	2,28
To improve the capture and use of knowledge from sources outside your firm or organization	67	2,22
To increase worker acceptance of innovations	65	2,30
To avoid information overload problems within your organization	59	2,45
Following merger or acquisition to help integrate knowledge within your new firm or organ.	47	2,92
To promote sharing and transfer of knowledge with suppliers	47	2,75
To improve sharing or transferring of knowledge with partners in strategic alliances, joint ventures or consortia	37	3,05
To promote sharing and transfer of knowledge with customers	36	3,07
To update your firm or organization on KM tools or practices used by competitors	31	3,23

Source: Fraunhofer ISI Survey 2002

Fig. 10.5: Motivation to Employ KM – Decreasing Order of Importance[334]

The motives at the low end confirm the result that for the German companies KM is still very much connected with internal knowledge stock and flow, including the integration of knowledge – or information – obtained from external sources. The *sharing* of knowledge with actors external to the organization is of lower relevance; three out of the last four least important motives are about the sharing of knowledge with customers, suppliers and co-operation partners. This marks an important characteristic of the relation with the outside world if it comes to KM. While the practices used to obtain knowledge from outside are rather prominent

---

334 Within a scale ranging from 1 (extremely important) to 6 (not at all important). „Top two" indicates the percentage of companies that rated the motive 1 or 2.

and important (cf. Fig. 10.3 above) and while using the environment as a knowledge source gets at least a medium mean value and is an important motive for two thirds of companies (motive 12), the inclination to actually *integrate* the internal circulation of knowledge with the relevant environment is weak.

There is not enough space here to discuss in detail the sector and size differences. However, the bottom line is that – again – the differences are relatively minor. As for size, the only obvious – and logical – deviation from the general pattern is that for large companies the motives that relate to the transfer of knowledge across borders within the company (but across distances) respectively with close partners are not important. Regarding the different sectors, the *service* companies are – in relation to other motives – driven rather weakly by the need for sharing of knowledge with customers and suppliers. This is somewhat counter-intuitive, as service companies, especially the knowledge-intensive ones that we included in the service sector, are dependent upon the exchange of knowledge. This might indicate that service companies are not driven by the need to exchange knowledge with their environment that much, but rather capture the necessary information needed to deliver their specific service. The little relevance of sharing knowledge with the environment is also true for the *chemical* companies, which – in addition – indicate least importance of transfer of knowledge across intra-company interfaces or with close partners. According to our survey data, the chemical companies seem to be least open to letting their knowledge circulation come in touch with outside actors. The opposite is true for the *pharmaceutical* and, to a lesser extent, for *electronic* companies, for which the sharing of knowledge with external partners, especially vertically (market) is significantly more important.

10.5 Reported Effects of KM

How effective are the companies in employing KM? As yet, indicators for KM are still to be defined and the empirical findings on the actual practices of companies to measure and even report on their intangibles and the related management practices are poor.[335] The majority of companies, although often reporting on their activities as part of the knowledge economy, do not have measuring practices and reporting systems, and those who do are rather reluctant to disclose them. Secondly, our knowledge of the relative impact of KM on certain business indicators we might have is still very limited. Therefore, up to now effects of KM cannot– beyond the level of case studies – be measured systematically. The simple solution chosen in our study was to ask those responsible what they think about how effec-

---

335 One recent example of measuring effectiveness of KM is given by Kremp/ Mairesse in a study on French industry to be published by the OECD in 2003. They show that there is statistically significant correlation between usage of KM and labour productivity. Their basis is the linkage of questions on KM practices and data stemming from the regular French industry survey panel.

tive the ensemble of their KM – not single practices – are. The German survey aske. for nine possible effects on a range between 1 (extremely effective) and 6 (not effective at all). The effects are displayed in Fig. 11.6 in decreasing order of magnitude.

First of all, there is a strong correlation between the number of practices used and the effects reported. The more practices are employed the higher the score for effects.[336] *Secondly*, the score for KM effects is highest when it comes to the improvement of human resources and the direct market effects. Fig. 11.6 indicates that two of the three top rated effects are human resource effects (skills, productivity of workforce). This is interesting, since the improvement of human resources is not the most important driving factor for KM (cf. above). *Thirdly*, the single biggest effect (adaptation in the market), as well as number 4 and 5 (Fig. 10.6, are directly linked to the market success of companies. Again, we have seen that the companies at the same time rate the motivation for external transfer or sharing of knowledge with clients very low. In other words, the companies either see no necessity to share and transfer knowledge with their clients in order to meet their needs properly, or they are reluctant to do so – for whatever reason. The fact that they still rate the market effect as high rather points to the general effects obtained through the efficiency gains of internal KM practices. *Fourthly*, the direct KM effects (capture of knowledge and the improvement of the organizational memory) are rated low. It would be interesting to find out, through more qualitative research, why these direct KM effects are rated lower than the functional effects (human resources, market). One explanation – as indicated above – might be that the companies simply have no measurement, maybe not even a feeling for their KM abilities, and thus are not able to assess the effects in the first place. Furthermore, the limitations of the direct KM effects (capture, memory) might point towards the slow reaction of the companies to a KM culture that needs to be institutionalized in order to be effective. The functional KM effects (Human Resource, market), on the other hand, are traditional dimensions that might very well have improved through KM, however, KM on that level is only one explanatory variable among many other managerial tasks, and effects hard to attribute.

---

336 We conducted a Chi-Square test, that compares the *expected* distribution in a matrix "effects – number of instruments" if there was no systematic relationship between number of instruments and effects at all and the *real* distribution in that matrix given by the data set. The higher the deviance between the two distributions, the stronger the relationship. To do so, an *index of overall usage* was constructed and the sample was grouped into those companies that employ not more than 50% of the practices (N=128) and those who employ more than half the practices. The *total effect* was calculated building the overall mean value on the scale from 1 (extremely effective) to 6 (not effective at all). Three groups reporting high (mean below 1.5), medium (mean between 1.5 and 3) respectively low effectiveness (mean above 3) were built. The resulting cross table was tested, correlation showed high significance on the 1% level.

Cat. [a]	Effect	top two [b]	mean [c]
Market	Increased our adaptation of products or services to client requirements	73	2,07
Human Res.	Improved skills and knowledge of workers	73	2,08
Human Res.	Improved worker efficiency and productivity	69	2,12
Market	Helped us add new products and services	61	2,34
Market	Improved the relation to customers and/or clients	59	2,38
Organ. Mem.	Improved the memory of our organization	57	2,47
Organ. Mem.	Helped avoid duplicating R&D activities	53	2,55
Capture	Increased our ability to capture knowledge from other businesses	51	2,56
Capture	Increased our ability to capture knowledge from public research instit.	38	2,87

a: clustering of effects: Organ. Mem. = Organizational Memory
b: top two indicates the percentage of companies who have rated one or two on the scale from 1 (extremely effective) to 6 (not effective at all).
c: scale ranging from one (extremely effective) to 6 (not effective at all).
Source: Fraunhofer ISI Survey 2002

Fig. 10.6: Effects of KM – Whole Sample

In order to compare the reported effects for sectors and size, we have grouped the effects into four families (cf. Fig. 10.6: market effect, effect on the workforce, organizational memory and finally capture of knowledge) and calculated average values for these four categories. While size does not matter much, the sector makes a difference as for two dimensions. First, there are three sectors that severely lag behind in their ability to capture knowledge from outside the company (chemicals, mechanical engineering and vehicles), while the knowledge-intensive biotechnology sector is situated best. Second, the sectors differ considerably in their ability to build up and improve organizational memory. Here the pharmaceutical sector is by far the most effective, while the vehicle sector – again – lags behind most.

10.6 The Institutional Commitment for KM

To trace the organizational design of KM in our broad understanding would not be possible in a survey, given the multitude of practices and their complex interplay. What can be done, however, is to identify the institutional commitment for KM. One first indication of a rather low formal commitment is the fact that only 16% of the companies provided formal training related to KM (cf. above, Fig. 10.3). In addition – three further proxies for institutionalization – or dedication – as regards KM have been asked about: (1) *dedicated budget for KM,* (2) *organizational unit or a specific manager mainly responsible for KM* and – as additional question in the German questionnaire – (3) the *responsibility for KM at the top management* level.

For the whole sample, top management responsibility is by far most important, indicated by more than 60% of the companies, while a quarter of the companies have a dedicated budget and slightly less a functional unit or responsible manager for KM (Fig. 10.7, left box). Regarding the institutional commitment, it is obviously the size that matters rather than the sector (Fig. 10.7). There is a negative correlation between size and top management responsibility and a positive correlation for specific KM functional units and size on the other hand. This is of course partly structurally determined to a large degree, as the functional differentiation, especially for a relatively horizontal task like KM, is more difficult – or less necessary – for small companies. Therefore, it is hard to assess the explanatory share of the dedication for KM as compared to the minor necessity for small companies to create functional units for each specific task. The sectoral patterns (Fig. 10.7) therefore reflect the size distribution of the sectors rather than showing a clear sectoral pattern – with the notable exception of the pharmaceutical sector, which contains a very large share of companies employing between 250 and 2000 employees, but still shows a very high level of top management responsibility.

Given the size bias for top management responsibility and functional units, the dedicated budget might be better proxy for the institutionalization of KM. The connection between KM budgets and size is not as clear-cut, as the two middle categories show rather similar values. The sector distribution shows, first, the overall importance of KM for the service companies and, second, the importance of budgets for – next to the service sector – the knowledge-intensive pharmaceutical and biotechnological sectors in which it is apparently necessary to invest strongly in intellectual capital.

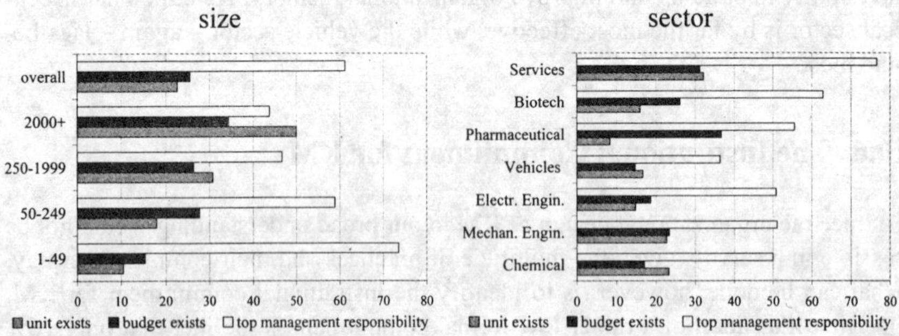

Source: Fraunhofer ISI Survey 2002

Fig. 10.7: Institutionalization of KM – Percentage of Companies

Do the three different forms of organizational dedication towards KM make a difference to the effects of KM? To find out, for all three different forms just discussed we conducted a comparison of mean values for the nine effect variables with the help of T-Tests.[337] *Top management* commitment makes the biggest difference, the mean values for *all* effects are higher with a high statistical significance (below 1%).[338] Companies with a *dedicated budget* for KM also report higher values for all effects, however, only five out of nine effects are significant at the level of at least 10%. What is striking is the fact that if *functional units* or key managers are mainly responsible for KM on the operational level, while most effects are higher the reported effects for the management of knowledge interfaces with the environment (capture of knowledge from public research institutes, relation to customers and suppliers) are *lower* with a statistical significance of 10%. The centralization of KM through organizational units in fact may hamper the openness to the outside world, as the interface function itself is reduced to – or can be delegated to – a core KM group rather than placed within the responsibility of the whole workforce. While this might improve the central overview on external effects – and support the control function – it reduces the number of possibilities for exchange.

10.7 Conclusions

The starting point of this survey was a broad understanding of knowledge management as an ensemble of very different types of practices, driven by very diverse motives and being effective on different levels. The data shows that a conceptualization of KM along these broad lines is fully justified. Not even the non-response analysis (cf. annex below) showed a strong diffusion of the idea that KM needs to be focused around ICT applications. Rather, knowledge management practices are very diverse, the most important category employed is practices to capture and acquire knowledge from external sources. At the same time, however, it is still a rather uncoordinated, spontaneous endeavour rather than a systematically organized and strategically guided management task, even if the policy dimension has been indicated as the most dynamic for the future.

The number of KM practices is systematically related to size while the sector difference, with the notable exception of the by far most active service companies, is rather limited. Even more striking, the diffusion pattern of different practices is very similar for different sectors and sizes, with only the very large companies being more strategic and systematic than the rest. Apparently, KM is not only a

337 A T-Test compares the mean value of two sub-samples, e.g. one can compare the mean value as regards effects for a sub-sample of companies that have a budget and a sub-sample that do not have a budget for KM.

338 The only exception being the avoidance of duplicate efforts in R&D, where the significance is below 5%.

horizontal task within companies, but the challenges of KM are similar across the whole range of industries.

This is confirmed by the motivation patterns, which again are very similar for all groups analyzed. The motivations to use KM are broad, but three sets of practices stand out: internal integration of knowledge, human resource development and capture and control. Interestingly, the data shows that the importance of various motives is not mirrored by the related reported effects. The functional effects on human resources and market (respectively customer relations) are rated higher than the effects related to knowledge management effects in a more narrow sense (organizational memory and knowledge capture). Furthermore, the dedication for KM is strongly dependent on company size. However, the survey revealed that a higher degree of organizational institutionalization, i.e. in centralized KM functions, might also have negative effects on the capability to exchange knowledge with the environment.

What these major findings make clear, above all, is the necessity to go on analyzing KM in industry. The relations among the many variables for which data has been collected must be worked on more intensively. Furthermore, aggregated data must always be checked with qualitative findings on the basis of existing case studies. In addition, we must go on comparing countries and sectors. A prime line of future work, however, will certainly be the conceptualization of a framework that enables us to measure the effects of KM much more accurate and broadly than we can do it up to now.

Annex: Non-Response

The non-response analysis served the purpose to test the relevance of the whole topic and to ask if companies had a totally different understanding of KM. 410 companies sent back the non-response, meaning that altogether 907 companies responded to the survey. The Figure below gives the possible answers that were formulated (multiple responses possible) and the counts as well as percentage of responses and cases.

One can see that the broad understanding of KM was no major problem for the companies asked; only very few indicated that they followed a narrow, ICT-focused KM approach. Furthermore, there are only very few companies that do not have KM at all, but plan to introduce it. That means that KM is already started, or is not considered at all. The most important reasons for not participating – next to the practical ones time and objections to surveys on principle – is that many companies have KM in place, but it is distributed, loosely connected and not systematically managed. 86 companies, out of more than 900 companies who answered to the survey, indicated that KM plays no role whatsoever and is not on the agenda either. While it is clear that most of those non-users of KM might not have answered in the first place, the percentage below 10% indicates that KM – one way or the other – is an important topic in German industry.

	count	percentage of	
		responses	cases
Reasons related to KM			
KM is a horizontal task within the responsibility of every manager, therefore systematic statements for KM as such are hard to make.	99	17,4	24,1
KM plays no major role and there are no plans to build up systematic KM.	86	15,1	21
KM is a major task of our ICT management (databases, information systems) and not as road as in the definition given in the questionnaire	37	6,5	9
KM plays no major role, but a build up of systematic KM is planned	14	2,5	3,4
General reasons, not KM related			
Answering takes too much time	124	21,8	30,2
No participation for principle reasons	93	16,3	22,7
Other reasons (company dissolved etc.)	116	20,4	28,3

Source: Fraunhofer ISI Survey 2002

Fig. 10.8: Non-Response Analysis, N=410, Multiple Responses Possible

[illegible]

[illegible]

[illegible]	[illegible]
[illegible]	[illegible]

[illegible]

Part III

Case Studies

11 Knowledge Management – Results of a Benchmarking Study

Cornelia Baumbach, Anja Schulze

11.1 Project Overview

Starting in August 2000, the TECTEM Benchmarking Center (www.tectem.ch) conducted the Benchmarking Project "Knowledge Management" on behalf of 10 companies (consortium). TECTEM belongs to the Institute for Technology Management of the University of St. Gallen in Switzerland. The head of the Compe-tence Center Knowledge Management of the Fraunhofer IPK was invited as subject matter expert. The consortium identified the following subjects to be the main subjects of this Benchmarking Project:

- Strategic positioning of knowledge management (KM) and integration into organization,
- KM processes and tools,
- Motivation and acceptance of KM.

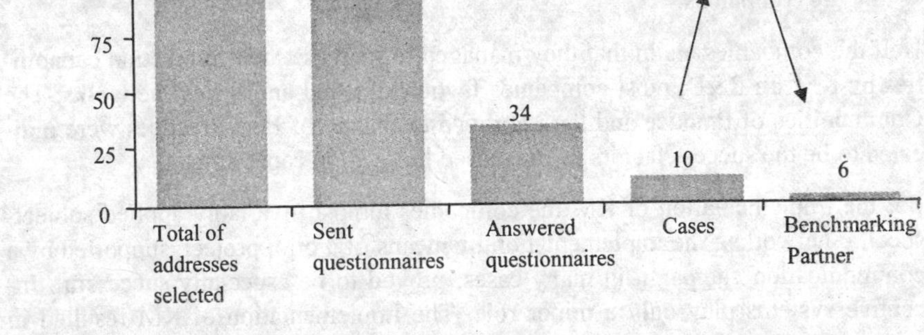

Fig. 11.1: "Selecting the Best"

These subjects of main emphasis formed the basis for an extensive screening phase, a worldwide search for companies, which have successfully put them into practice. On the basis of electronic database queries, expert knowledge and suggestions made by the consortium, over 150 potential Benchmarking Partners were selected from among several hundred indicated firms. The selected firms all feature potential Best Practices in the areas determined by the consortium. The firms were contacted and requested to fill out a questionnaire. 34 out of 90 questionnaires were returned, 10 telephone interviews took place and 10 anonymized case studies were written.

The present report dates from March 2001 and is based on the 34 questionnaires, 10 telephone interviews and 6 company site visits.

11.1.1 Demographic Background and Organization

The firms, which were surveyed during the screening phase, include 65% large companies (>50000 employees), 12% companies with 1000-50000 employees, 14% companies with 1000-10000 employees and 9% companies with fewer than 1000 employees. Almost a third of the surveyed companies operate globally, regardless of their size. The companies were intentionally selected from very different business areas.

11.1.2 Strategic Positioning of Knowledge Management and Integration of Knowledge Management into the Organization

45% of these companies have a hybrid KM organization, 35% have a decentralized organization, and 20% have a central organization of KM.

Most companies have integrated KM into their business processes, although the level of integration into the business processes still varies. The only recognizable constant is said to be the integration into Information Management. Most of the companies are aware of weak points of their KM approach. As far as the integration of KM into the strategy is concerned, there is still quite some uncertainty within the companies.

Half the companies claim that they managed to increase their innovation capabilities by KM. In R&D, most companies favour informal and formal networks. The Communities of Practice and the improved exchange of Best Practices were indicated to be the success factors for increased innovation capabilities.

For the implementation of KM the companies almost invariably applied subject-specific networks. The implementation by means of a pilot project, supported by a communication campaign in many cases, proved to be especially successful. Incentive systems play only a minor role. The implementation of KM resulted in some changes in the companies. As far as the employees were concerned, the following conclusion could be drawn: work becomes easier for the individual if the culture is appropriate! From the point of view of the companies the following changes could be named:

- Company culture: awareness of KM – from skepticism to curiosity!
- Learning process: going through new forms of learning and exchanging tips and tricks!
- Co-operation by simplifying co-ordination creates common values!
- Innovation by linking up existing knowledge!
- Cost and time saving by increased transparency!

11.1.3 Knowledge Management Processes

So far, only two thirds of the surveyed companies have a systematic process for capturing knowledge. In companies, which store knowledge centrally, the employees mostly have access to that knowledge.

Almost all companies carry out quantitative and/or qualitative measurements of the utilization of KM. There seems to be room for improvement in the area of controlling. Two thirds of the companies indicate that the employees are aware of the knowledge requirements of their colleagues so they can prepare knowledge and pass it on specifically.

To ensure the quality of knowledge contents, the company uses heteronymous measures (for example feedback and project reviews) as well as autonomous measures (MbO and success stories). Both processes look promising.

The companies state that in the past they mainly made learning experiences relating to the employees and the culture. Future learning potential is expected mainly in the area of organization.

11.1.4 Knowledge Management Tools

There is a strong agreement on the tools used. Specialist conferences, regular knowledge exchange meetings and the utilization of groupware programs were named by almost all the companies. As regards the methods and tools used, there is a balanced mixture of electronic as well as non-electronic tools.

11.1.5 Motivation and Acceptance of Knowledge Management

Almost all the surveyed companies give their employees time for knowledge leveraging. The time, which the employees take for this is normally below 5% of their working time. Incentive systems based on purely extrinsic motivators are not used. All companies emphasize that the success directly depends on the employees' acceptance of KM. Lasting acceptance, in turn, is achieved by the fact that KM gives the employees direct benefits in their daily routine.

11.1.6 Conclusion

In order to pursue KM successfully, it is first of all necessary to create cultural prerequisites. A starting point is to facilitate teamwork, with KM becoming a natural part of all employees' daily routine. Whether or not activities are initiated by a special KM department or a person responsible depends on the individual company situation.

11.2 On the Road...

Signposts alone are not enough – the right preparations for the journey are crucial for success!

All Benchmarking Partners prepared well for the "Knowledge Management Project". It was pointed out, however, that preparations alone were not enough. During the introduction – as well as later on – regular activities were necessary to revitalize KM.

The majority of the Benchmarking Partners have embedded KM in their visions, and all of them have formulated a KM goal. Uniform terminology defined for the entire organization as well as the appropriate status of KM was also established by most of the companies. A close interlinking of KM with the business strategy and management commitment has led to the fact that KM is taken seriously.

11.2.1 Is there a Common Understanding?

Basic Understanding

To the Benchmarking Partners, a solid base was the prerequisite for successful KM. This base included a fundamental understanding of what the company meant by KM. The point was to define the relevant terminology, but also to determine the activities. However, the Benchmarking Partners agree that the definition alone does not make up KM!

In order to avoid each employee having a different understanding of KM, the first step for Volkswagen was to create a common understanding of the KM activities with its *ww.deck*. It stands for worldwide development and exchange of corporate knowledge. With its characteristic understanding, VW has defined the general term of KM on a company basis.

Terminology

As a first step, Deutsche Telekom made sure that there was a basic understanding of KM and a definition of terms in the company. The company created a corporate understanding through a common language. Furthermore, it ensured from the outset that strategic and operational measures far beyond the IT component were also taken into consideration.

Apart from defining the scope of KM, Arthur Andersen made sure that a terminology was agreed upon and that KM was valued appropriately. They defined KM as $K=(P+I)^s$, i.e. Knowledge equals People plus Information (= knowledge which is available) to be shared. For Arthur Andersen, the consistency began already in the approach, i.e. names of business units or definitions of certain terms were the same worldwide.

Status

Deutsche Telekom attached great importance to the overall availability of KM according to the requirements that KM had the appropriate status.

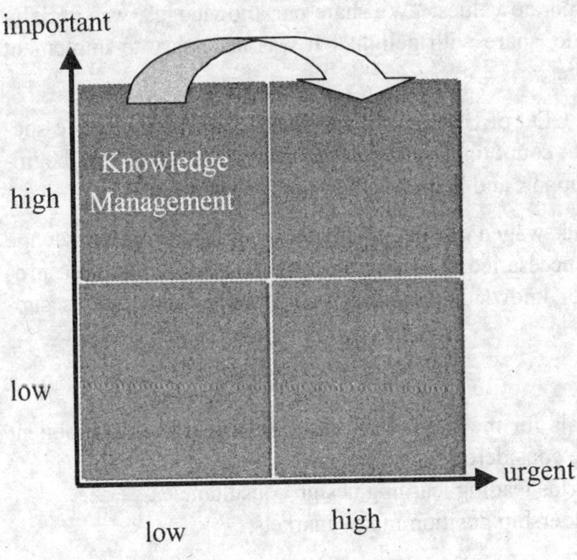

Fig. 11.2: Emphasis of KM at Arthur Andersen

To reach the appropriate status, KM must be defined and positioned in a way that it is rated important and urgent. To this end, Arthur Andersen employed staff to support the KM activities. Furthermore, Andersen integrated KM directly into the existing processes. Project members, for example, are now responsible for documenting project results, thus making them available for future projects.

11.2.2 Where to? – Vision, Strategy, Goal, and Measurement

Vision

What all Benchmarking Partners have in common is a clear idea of the goal to be achieved with KM. In most cases, it is even a part of the vision. Deutsche Telekom, for example, formulated a "Corporate Knowledge Management Vision" separate from KM goals. As the employees' knowledge is the most important resource, Deutsche Telekom wants to make it possible for each employee to share relevant knowledge at his place of work, and take an active part in building up knowledge within the company.

Strategic Establishment

Most Benchmarking Partners came to the conclusion that KM needs to be integrated into strategy. KM at Celemi, for example, is not only part of the business processes, but also part of the business strategy.

ABB Switzerland also recognizes KM as an element of the business strategy, and in addition, their top management formulated a separate cultural value complementing the existing 6 corporate values: "We share our knowledge", was initially known as "Copy with pride, share with delight". It was necessary to implement this required cultural change.

With his commitment, the CEO of BP made a significant contribution to the success of KM. BP was able to count on strong management support from the beginning, including financial support and help in setting up a project team.

The managing board at Volkswagen decided to create a central staff of knowledge managers. Its continuing success led to a long-term financial backing and a progressing cultural change of knowledge sharing across brands and regions supported by the management.

Goals

Celemi formulated its "goals for the future" of which KM is a fixed component rather than an element to be considered separately…
…to become one of the worlds leading learning design consultancies,
…to attain a recognized leadership position in key markets,
…to grow significantly,
…and to maintain the business Celemi has.

ABB Switzerland's goal is to achieve better competitiveness with KM by using existing knowledge in products, in the organization and in business.

At Volkswagen, KM stands for:

- spreading innovative solutions throughout the VW Group,
- making knowledge available to all regions and brands,
- enhancing the networking activities within job families,
- learning from experiences all over the world,
- having the relevant know-how at the right place,
- retaining knowledge from employees leaving the company.

With regards to KM, Arthur Andersen pursues the goal of not having to re-invent the wheel. It consists of an active, worldwide network of partners, not only via databases but also through personal contact.

Deutsche Telekom wants to optimize processes, increase productivity and improve responsiveness and quality. For 2001, Deutsche Telekom set the following strategic goal: "The basic knowledge of the corporation is to be generally available according to the requirements." Accordingly, several sets of measures have been worked out.

Keeping the Goal in Mind – Measurement

Depending on the nature of their business, it seems that the Benchmarking Partners worked with different parameters. Industrial businesses tended to work with qualitative measuring methods, while service companies preferred to apply quantitative methods.

VW is convinced that the benefit of KM is best demonstrated with success stories. Nevertheless, the benefit of KM still has to be proved by qualitative and quantitative measures. Effects of interventions are measured from the perspective of employees and work units. In the long term, expenditure for KM can only be justified if it enjoys a profitable yield.

Besides qualitative reviews, Arthur Andersen also implemented a quantitative approach, carrying out measurements on two levels, i.e. the corporate level and, through annual reviews, on an employee level. The motivation to pursue KM at work should principally result from a "give and take" attitude towards work and learning.

Celemi follows the development of KM within their company with the use of its "Monitor". The real value of this is that it provides a complete picture of the customers, of the employee's competencies and of the internal organization. They are also able to identify the changes, which KM brings about. Various indicators are used for this purpose, and although Celemi uses the monitor, the company emphasizes that "less control creates more innovation".

BP uses a mixture. The company makes sure that the principle "What gets measured, gets managed... Be practical!!" is maintained. However, individual performance assessments have not yet been linked with KM aspects. In some units there are assessments of knowledge gained from others and knowledge shared with other business units. BP has been very successful asking the simple question "Who have you learned from?" Despite their simplicity, these processes used by BP are very effective.[339]

11.3 Not Re-Inventing the Wheel

Why roam so far afield...?

The Benchmarking Partners have realized that tying KM to existing instruments and methods increases the acceptance among the employees, creating "Quick Wins" at the same time. Furthermore, the companies, which are successful in implementing KM instruments, follow the motto "the simpler the better".

11.3.1 Building on the Existing

Successful companies integrate KM processes into existing business processes. ABB Switzerland, for example, first identified the value-creating business processes, in order to subsequently create its knowledge processes accordingly.

When KM became an issue for VW, the Board raised the question whether KM should be tackled as an independent issue. As a first step the existing processes, methods and instruments from a KM perspective were analyzed. Invariably, they found activities, which may have had different terms, but today are classified as KM activities. BP, for example, held peer assists before the term KM was introduced into the company. The second step was to develop the existing processes, methods and instruments further in view of KM.

From Experience Groups to Global Networking

At VW the agreement reached was to specifically pursue KM in areas where existing knowledge offered the best starting point. This thus required an analysis of the existing conditions. Internal and local contacts and experience networks as well as a systematic and scheduled experience exchange with other companies already existed. That is why networking at VW has always been started with groups, which already exchanged experiences (mostly on a local level).

---

339 Practices and Knowledge Management Tools are discussed in detail in "Learning to fly - Practical lessons from one of the world's leading knowledge companies" by Collison, C. and Parcell, G., 2001. In 2001, BP has been recognized as Europe's Most Admired Knowledge Enterprise (MAKE Award).

At ABB Switzerland, Communities of Practice (CoP) have been in existence for years in different forms. They were used as starting points for the development of international networking with regard to KM. Moreover, the work of these CoP's was supported by so-called team rooms, which were based on Lotus Notes. Today, they also facilitate communication and knowledge exchange among task forces and project teams in different locations.

From Employee Lists to Yellow Pages

In a pre-project, ABB Switzerland took over information on the contents of the Yellow Pages from existing lists (based on Lotus Notes) containing information about the employees. Since 1997, there had already been preliminary versions of these lists, for example in the area of research centers. To ensure the relevance of the information, all employees were asked whether they had updated their data. After the updating, the data were transferred into the new system, from where it was extended.

From Suggestion Policies to Idea Management

High Voltage Technology, an operational unit of ABB Switzerland Ltd., reshaped its suggestion policy at the end of 1995. Since then it has been called idea management. Already in the first year, 1000 employees handed in over 400 suggestions of which 130 were actually put into practice. 20 to 30 of the ideas and suggestions put into practice are awarded at the end of each year.

In summary, one can say that building up KM on existing conditions contributes to quick wins and to the acceptance of KM in the organization, and likewise the motto "keep it small & simple!".

11.3.2 KISS – Keep it Small and Simple!

The simpler the better! The KM instruments used by successful companies are pragmatic, structured, simple, understandable – and therefore very effective.

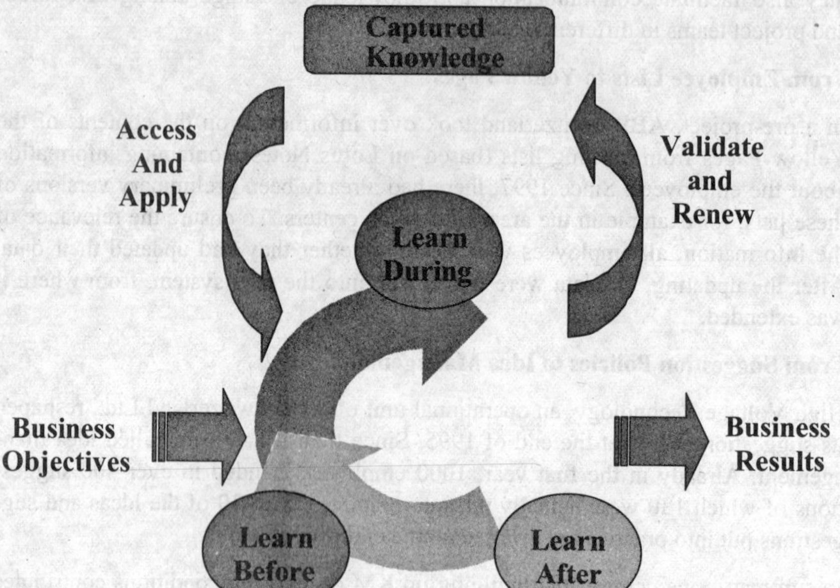

Fig. 11.3: KM Framework Model at BP

The structure of all KM activities at BP is based on the existing framework model, which consists of only a few elements.

BP has continued to work with the analogy of a "knowledge bank" of captured knowledge, which can be "withdrawn" at the start of a project (learning before) and learning experiences can be "paid in" as knowledge at the end of a project (learning after). The KM framework model is commonly understood in the organization.

11.4 The Recipe

An introductory concept creates the necessary sensitization

The successful introduction of KM requires very good planning of its implementation. This includes creating a concept. The step-by-step introduction of KM is an element, which the visited Benchmarking Partners have in common too. Evidently it is worthwhile to plan enough time for an accurate introduction.

11.4.1 Sound Decision-Making Preparation

It was the HR Management of VW, which initiated the creation of a KM project group including the board. Very careful and structured preparations were made, and team and workshop experiences formed the foundation for the internal project of KM. A first overview was obtained by visiting conferences and seminars as well as initiating contact with universities. With an external analysis, the scope of KM was determined, whilst also determining which gaps could be closed through the use of KM. It was only after this analysis that projects were started.

Arthur Andersen believes that the skill mix required for KM is a mix of strategy, processes, technology and employee culture. They managed to introduce KM with a focus on the employees' requirements and the list of top 10 "lessons learned" gives insight into what should be paid attention to.

1. Importance of Leadership - Vision, Values, Behaviours
2. Focus on Strategy: "Knowledge to do What?"
3 Focus on Users Needs: "How will this improve my day?"
4. Need to Create a Network of Content Experts
5. Need to Create a Network of Knowledge Managers
6. Need a Common Organizing Framework (e.g. Content Classification Scheme)
7. Need a Common Content authoring and publishing template
8. Need Human Support (e.g. Call Hotline)
9. Need flexible technology platform and protocol
10. Focus on what works, not perfection

Fig. 11.4: The List of Top 10 Lessons Learned at Arthur Andersen

11.4.2 Step by Step!

ABB Switzerland emphasizes that a clear mission and an initial top management commitment are very important success factors. In addition, they created a common understanding and a KM concept, with the goal being achieved in a three-step process. In the first step the value-adding business process was evaluated. This process focused on the employees, and KM would only be appropriate if it could help them to work towards better customer satisfaction. In the second step the knowledge process was defined. This needed to be integrated into the value-adding process in a way so that it would feed knowledge into the business processes. However, new knowledge should also be gained in this process. In the third step, the company examined the knowledge enablers. It was important to use them in a way that they would support the knowledge process as well as implementing and actually living KM.

Arthur Andersen also attaches great importance to a realistic estimation of employees and their capabilities. Optimum training of the employees is ensured by co-ordinated measures. Yet the firm was wrong in the assumption that all employees would be able to handle the systems. For example, after user tests within the development of a specific KM solution, 75% of the system functions had to be inhibited to increase the acceptance of the system. At regular intervals, the functions were then gradually connected again.

11.4.3 "Implementing Knowledge Management – The Long Road"

In retrospect, BP calls the implementation of KM a "long road". The individual steps of the introduction were very well documented; so even for outsiders it is easy to follow the course of events.

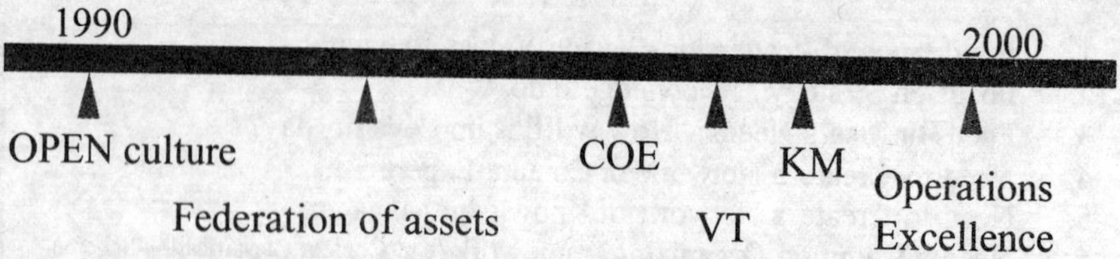

Fig. 11.5: BP's Long Road

The first step towards KM was the introduction of an open culture, which took place by relaxing the hierarchical structure. At the same time, the employees were prepared for the future through programs to encourage empowerment. With the business units working on a more empowered basis including budget responsibilities, their independence began to grow. One of BP's KM tasks was to fulfill individual business units' expectations with regard to knowledge sharing. This was the starting signal for the first Communities of Practice. After putting the different IT systems on one common platform in 1996, nobody had an excuse to claim that it was impossible to exchange information with others, meaning that important barriers had been removed. At the same time, virtual teamwork was introduced, increasing the employees' ability to work together across geographies.

When the project of KM was initiated, the company first searched for a common KM model or process. A team of 12 people was set up for this purpose. The team, headed by a senior manager, had top management attention. BP's KM was started "where the energy was". The fact that the company managed to create a common language was very beneficial to the acceptance of KM. Once the business areas had been "infected", the central KM team was deliberately reduced in size to minimize dependency on a central resource and to increase local ownership. Today it consists of only two people.

11.5 Branding + Marketing = Positive Attraction

Market KM successfully

The Benchmarking Partners spared no efforts in marketing KM internally. Many approaches (from logo to video) were tried. All Benchmarking Partners agree that it was worth the effort, however, in all cases the selected solutions suited the corporate culture and philosophy. The internal marketing efforts had a positive effect on the acceptance among the employees in all the companies.

11.5.1 Branding – Beating the Big Drum

Good branding smoothed the way for creating acceptance of KM among the employees. Some Benchmarking Partners have used highly imaginative processes. In order to introduce a common understanding of KM to the organization, VW formulated *ww.deck*. Arthur Andersen managed to spread a logo in the organization, which reduced the KM activities to a common denominator.

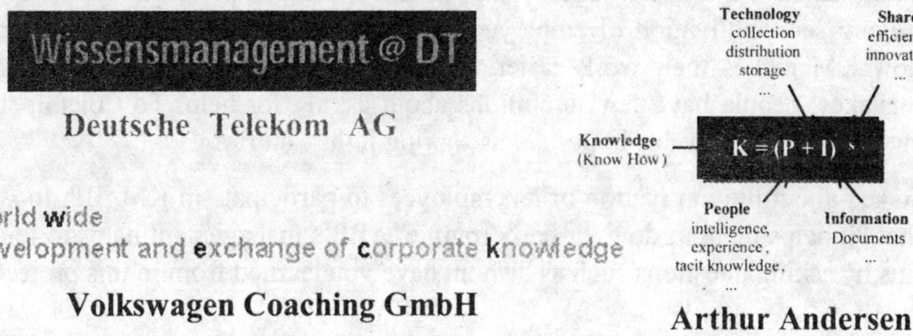

Fig. 11.6: Logos

BP supplemented the term "Knowledge Management" with more practical expressions, like "shared learning", "work smarter" or "performing through learning".

11.5.2 Marketing

All Benchmarking Partners do active KM marketing. Depending on the different phases, the use of marketing instruments can be more or less intensive. KM should be seen as an internal product with its own life cycle, just like a normal product. VW, for example, has produced a video clip and recorded a song called "king of knowledge". The KM team hosts a regular corporate conference where knowledge managers at VW and from other companies exchange their experiences with projects. Corporate newspapers as well as informal meetings are used to inform about opportunities in KM.

11.5.3 Attractive (Knowledge) Management Ensures Utilization

Companies who have a good branding and marketing of KM hardly need to worry about the acceptance among their employees.

An interest in deploying KM at VW was first and foremost created by the fact that KM noticeably reduced the workload. Motivation within expert networks resulted from the fact that employees saved time by exchanging experiences and that their daily routine was made easier. People belonging to a network quickly learned to appreciate this advantage. The goal was to make employees realize which personal benefits they get from KM. Therefore; VW made an effort to spread success stories, as this also creates motivation. In addition to that, participation in KM became an element of the employee assessment and even had an influence on the variable pay. Among other aspects, contributions to KM activities measured with a point system are relevant to salaries.

At Celemi, motivation starts already with furnishing and equipping offices. Celemi's offices make it possible for flexible groups to work together in an uncomplicated way. Open doors also enhance contact between the employees. Celemi sees motivation of employees in the fact that they experience every day how KM makes their work easier. Thanks to the tolerant way of dealing with mistakes, people have few inhibitions about asking for help. To Celemi, sharing success stories is just as important as sharing failure stories.

Asked about the motivation of its employees to participate in KM, BP, too, states that "it helps them to do their daily routine". BP's management helps to reinforce this by asking questions such as "whom have you learned from in this project?".

11.6 Who Does Knowledge Management Belong to?

A change of perspective – and therefore a change of therapy

The introduction of KM entails the necessity of new roles and functions. Which roles and functions are optimal for a company depends on the organizational structure and culture. In any case, the introduction of KM requires people who take specific and full-time responsibility for it. Once KM has been integrated into the employees' daily routine, this function is no longer required, or only to a small extent.

11.6.1 Who is Suited for What?

The introduction of KM involves the necessity of new roles and functions. Which ones are optimally suitable for each company depends on the corporate structure and culture.

A Bottle Can only be Filled from the Top!

It is common to all Benchmarking Partners that senior management actively supported KM! Especially at the beginning of introducing KM, strong senior management support is required!

One Person Responsible on a Strategic Level

At ABB Switzerland, the introduction of KM brought about only one new role, that of a KM project manager. At first belonging to the consulting area, this position was recently assigned to the area of Human Resources. Apart from developing an integrated KM approach, the Knowledge project manager's main task is to carry out different KM activities on the level of Switzerland. Moreover, he is responsible for the co-ordination and communication of Best Practices in the individual organizations of ABB Switzerland. Co-ordination and communication are necessary, as people often do not know about on-going activities of others.

Divided Responsibilities on a Strategic Level

The organizational responsibility for the KM of Deutsche Telekom is shared by Human Resources, IT and the area of Organizational Structuring/Corporate Strategy. At the same time, the people in these areas responsible for KM have other tasks, i.e. they do not exclusively work for KM. The responsibilities have been clearly defined. It may, for example be stipulated in an employee's goals that 40% of his responsibilities are related to KM.

Many Responsibilities on an Operational Level

One of the main tasks of a knowledge manager at Arthur Andersen is 'to know and to facilitate the community'. Knowledge managers are employees who support the communities and the process of information preparation. They act as facilitators. A knowledge manager must be very communicative and ideally master different languages. Besides employees who have been specially recruited for a KM position, there are also interim knowledge managers coming from the 'field' who take on this position for a defined period of time, mostly 1 to 2 years. Afterwards, most of them return to their original functions.

Apart from the knowledge manager function, Arthur Andersen also has content experts responsible for the correctness of the contents. In addition, there is a central organization called knowledge enterprise, which supports the publishing process (e.g. classification of content, editing) of knowledge assets on the global corporate intranet (KnowledgeSpace).

The KM Team

At VW, there is a central team for KM. This team belongs to VW Coaching and currently has 12 full-time employees. Furthermore, a certain number of external people are involved. The KM team has recruited external and internal people and they have a very interdisciplinary background, as far as competencies are concerned. Apart from creating awareness of KM among the employees, they analyze the external market of KM providers for innovative solutions for developing and implementing KM instruments.

The team uses its central position to co-ordinate the continuing exchange of KM solutions between different brands (like Audi, Seat, Skoda, and VW) and different regions. By co-operating with local project partners, a Best Practice exchange for KM solutions has been created that is used to increase the quality of KM. The team has a consulting role. Whereas pilot projects are carried out without financial contributions from the business units, subsequent projects are settled according to the team's work.

All Employees are Responsible

At Celemi, there are no specific roles and functions for KM. KM has been integrated into daily routine. This is based on a culture of trust among the employees. Only the design center has an assisting function, supporting the development of illustrations. Since Celemi wants to visualize as much as possible, this function is needed for a direct and easy exchange of knowledge.

11.6.2 A Change of Direction Requires Caretakers!

All Benchmarking Partners were noticed to have no permanent – and therefore static – KM functions. They are dynamic and change over time. The introduction of KM requires people who are specifically responsible and work on it full-time. These caretakers are people who show the company and its employees a new direction. KM is gradually integrated into the daily routine of the employees. Once it has been integrated, it is everybody's responsibility to manage knowledge, and caretakers are not needed anymore, or only to a small extent. So the number of full-time KM people will decrease again.

The Dynamics of the KM Functions

When, in 1997, the term of KM was increasingly used in BP and senior management began to talk about it, the question was whether KM was just a trend or whether there was more to it. Was KM relevant to the company and which concrete tasks it would entail? Initially, KM was only planned as a two-year initiative. A small focus group of 4 people was formed to take a good look at KM. Later on, a team of 12 people was created. It got strong management support and was headed by a senior manager. As far as the starting points for KM were concerned, BP decided on those areas where the respective "energy" was present. The greatest challenge was to create connections between the individual business units. Identifying and supporting the people who were enhancing the technology and the culture to facilitate knowledge exchange were recognized to be the key factors. In this context, the hierarchical structure of BP had recently been flattened out. After a common language and KM tools had been created and BP's employees had been introduced to KM principles, it was not regarded as a separate issue anymore. In the year 2000, BP's 150 business units started to pursue KM independently. It was becoming more and more apparent that a large central resource was a barrier to local ownership, so the very successful KM team was reduced to just 2 employees. Most of the former team members left the organization or returned to their original functions.

11.7 A Toolbox for Knowledge Management

...including instructions

The Benchmarking Partners specifically use tools to manage the knowledge of their organizations. These tools are mostly pilot-tested before their company-wide implementation. In order to achieve a high efficiency of the tools while ensuring their user-friendliness, the Benchmarking Partners keep working on further development and improvement of tools.

11.7.1 Building Gateways to Knowledge

Finding Those Who Have Knowledge

Successful companies do not aim at documenting their entire knowledge. Documenting would, as far as the major part of knowledge available in the organization is concerned, involve a disproportionately great effort. So the knowledge should remain in the "heads" of the employees. In order to make this knowledge usable, companies use "knowledge guides": VW has the "expertfinder", and BP has a system called "Connect". With these systems, the knowledge, respectively the people who have it, are supposed to be found.

In order to fill these knowledge guides as much as possible with contents, the companies rely on their employees. Respective systems, invariably IT-based today, need to be made as user-friendly as possible for the employees with their different ways of working. Some employees work in a very structured way, others do not. This is why each of BP's Connect page has a structured part as well as an unstructured part.

The Connect page of one person contains…

… the link with the team and the business unit he/she belongs to,

… internal and external contacts (connected with links),

… formal and informal networks he/she is involved in,

… an open text field for each person to freely enter information,

… the curriculum vitae and photos, if needed.

Nobody is mandated to update the Connect page. However, there is a certain mutual, social pressure among the employees because it shows on the Connect page when it was last updated.

Arthur Andersen's "Knowledge-Space" allows looking for employees by competencies. The information in the system is provided by the people themselves, so everybody can decide on what information they want to make public and what not. So there is no problem with data protection. Each capability is graded (basic – intermediate– advanced). The quality of the self-assessment time has shown that the ratings are quite realistic.

Contacting Knowledgeable People

This, however, is not sufficient. The point is to motivate people in the next step to speak to the respective person. The more information one has about a person the easier it is to speak to this person. This is why BP encourages its employees to also display hobbies, family photos etc. on their Connect page.

Obtaining Knowledge from Knowledgeable People

Furthermore, the knowledgeable person needs to be willing to share his/her knowledge and to spend time on this. At BP, this is ensured by keeping it voluntary. Out of its 80000 employees, 20000 registered in Connect on their own accord. BP consciously keeps this on a voluntary basis, as it is better to have 20000 valuable pages than 80000 pages with poor contents. Also, people who register in Connect are aware of the possibility of being contacted. So they are prepared to take 10 minutes or half an hour per day to answer questions.

The information in Arthur Andersen's intranet-based scheduling system is used to staff projects with employees. Since people are interested in exciting and demanding projects, they will avoid underestimating their capabilities. Overestimating could also have a serious consequence – a person may be appointed as an expert having to use his supposed knowledge…

Network if You Can...

Finding knowledge, exchanging it and thereafter developing it further is also a goal of networking. Networks offer an enormous potential for the utilization of knowledge. The Benchmarking Partners have realized this. However, they are aware that the success resulting from these networks is not measurable for sure.

Who Networks with Whom?

The Benchmarking Partners have networks with a different degree of formalization and a different geographical range. It takes structuring and good management of the contents to not only initiate global networks but to sustain them. They require a certain number of formalities. Local networks, on the other hand, are more informal. It was also noticed that formal networks can be initiated by the company, whilst informal networks cannot be forced on people since they require the respective culture and the right conditions. BP believes that all networks have the same value. This value, however, can only be measured in the formal networks, as there are clear goals.

	Local	Regional	Global
Formal (Commitment)			Expert Networks
Practical (Practice)		KSS	
Informal (Interest)	Friday Night Club		

Fig. 11.7: Networks

Local and Informal – Celemi's "Friday Night Club" is an informal gathering on a local level; the employees meet up once per week in a restaurant in town.

Regional and Practical – A good example of practicality on a regional level is Arthur Andersen's KSS (knowledge sharing session). The KSS is held once per month from 12:00h – 13:30h. Over lunch, anybody or any team can give a report on any subject and initiate a discussion. The lunch packages are paid for by the company for this purpose. The topic may or may not be related to the business. An office secretary is responsible for the organization of the KSS. One subject is dealt with per session. An important criterion for practical networks is that they are voluntary, contrary to project groups. In addition, there are strategically initiated formal networks. They are financially supported and they are expected to produce results. If such a network does not perform, it will be stopped very soon.

Global and Formal – VW's expert networks are a good example of formal networks on a global level. The admission of employees to an expert network is based on identifying their competencies. Such a network starts with a face-to-face meeting, which is repeated after one and a half years. The moderator is elected by the group and is also a member of it. It is important that the topics are structured in advance to give people an incentive to use the system, mainly to make contributions.

11.7.2 Opening up Sources of Knowledge

The Best of the Past

The "processors" – The Benchmarking Partners make a great effort to process their experiences in order to make further use of the gained knowledge. They all have people in the organization who are responsible for processing the experienced knowledge. They can be Knowledge Managers who have been coaching a project from the start and draw up a summary of the most important experiences, for example at Arthur Andersen. Deutsche Telekom, too, has Knowledge Managers for this task. However, at Deutsche Telekom each of them carries out a debriefing in form of a structured interview when a project has been finalized. Celemi only has one person who collects knowledge from different areas, works through, documents, systematically prepares and standardizes respective solutions and experiences for further use. BP uses facilitators who moderate the capturing of the most important experiences from a project.

One standard for all – A comprised and, at the same time, standardized structure of the experienced knowledge is a success factor for further utilization of knowledge. It must be easy to comprehend for all employees who potentially use it. Deutsche Telekom, for example, took great care of the right formulation. The individual knowledge components must be integrated into a standardized structure. They must be formulated in short sentences, which are easy and quick to understand.

Where should this knowledge go? – Successful companies categorize knowledge in a way that every user in the company has safe and quick access to the knowledge appropriate to his role. Deutsche Telekom has its ECIP (Enterprise Collaborative Information Portal). This personalized access portal is an IT tool, which meets the requirements mentioned. VW has organized its knowledge databases according to job families. Each knowledge database is supported by an IT tool, whose specific adaptation to the respective job family is very important. Celemi, on the other hand, feeds its experienced knowledge into a manual of methods, which every employee has at his disposal.

Being up to Date

Knowledge is not developed by newly combining the existing, but also by adding new knowledge and information. This is why the Benchmarking Partners have news services. To avoid a lack of information in the abundance of data, these news services must be adaptable to the respective information requirement of every employee. So, every employee can compile the news topics he/she would like to receive information on in a personal profile.

VW's Newsboard has been integrated into its B2E (business-to-employee) portal. The employee defines certain sources (A) as well as certain topics (B). Based on the combination of A and B, a daily background search takes place after the browser has been opened, resulting in a certain hit rate. In most companies, it is also possible to manually search the extensive list of articles and magazines.

Nearly all instruments described so far are based on information technology only. Furthermore, the Benchmarking Partners also have employees who process news for further use in the organization. Deutsche Telekom, for example, has RACE, an information portal of its Research & Analysis Center. By bundling highly qualified information together, it provides access to current, well structured information about markets and competition, meeting the company's need for knowledge about current developments in those areas, in today's global times markets.

11.7.3 Learning, Learning and Learning...

For the most part, knowledge is embedded in the employee's heads. So, one way of building up knowledge is to have learning processes in place. There are many ways to specifically build up knowledge through learning programs. Although this is an HR task, successful companies have devoted some of their KM activities to learning. The reason is that it would take too long to integrate new knowledge into seminars and further education through HR alone. In today's knowledge society, knowledge needs to be integrated faster. This is why the Benchmarking Partners have decided on a co-operation of KM and HR Management. The organizational responsibility for Deutsche Telekom's KM, for example, is shared by Human Resources, IT and Organizational Structuring.

In most cases, learning programs rest on two pillars, namely conveying knowledge via the Intranet and via further education and seminars. Through innovative intranet application tools, a close interlinking with seminar and training programs, the transfer of knowledge in the organization is pushed and optimized.

Another successful way is the combined use of virtual and face-to-face learning. Successful companies have both! Whereas virtual learning is highly flexible in terms of time, knowledge level, place etc., face-to-face learning has the advantage of being more intensive and of enhancing the development of personal networks.

Virtual Learning

For virtual learning, companies use the possibilities of information technology. Virtual learning offers great flexibility because it does not involve fixed times. Competencies can be built up and developed further online (virtual learning network at Arthur Andersen, cockpit for leaders at Deutsche Telekom). In concrete terms, this means that, apart from conveying educational contents online, courses may also contain test questions, feedback on the respective answers etc. These web-based training courses are open to all employees and are free of charge. The only "cost" involved is the working time of the employees.

Face-to-Face Learning

This way of learning, adapted to personal requirements, appeals to different senses and is therefore more intensive. Face-to-face learning is particularly important to new employees.

Get the new people on board – Benchmarking Partners give their new employees a first knowledge basis in introductory programs (Arthur Andersen, Celemi) or through mentoring systems (Celemi). All graduates starting with Andersen are given a two-week introduction in their office by colleagues based in the same office. In addition, depending on their focus area, they follow that up with a three-week course at the Corporate University in St. Charles, Illinois where Arthur Andersen has a training center to introduce new employees to the consulting processes and inform them of consulting methods. Celemi has a mentoring system to integrate new people into the company culture. Juniors always work together with seniors on a project. Furthermore, new employees get training on the corporate philosophy as well as on KM. This training takes place in form of a "business game" and takes a day or two.

When teaching fails – try learning! – Celemi's simulations are an example of active learning, where employees communicate directly with each other. They are a kind of business game. There may, for example, be a simulation on project management, where knowledge concerning problems, which could arise in the course of a project and possible solution approaches is shared. Celemi sees simulation as a means of understanding and acquiring principles, creating "story telling" and exchanging knowledge by intensive utilization of visual effects. Besides, visualization is the key to the enormously efficient transfer of knowledge via work mats. Simulations as well as work mats were originally developed for Celemi's customers. Today both instruments are frequently used internally.

Learning after doing – BP's retrospect meeting is moderated by a person, who does not belong to the project team. Each participant writes down answers to the following questions:

- What was the objective of the project?
- What did we achieve?
- What were the successes? Why? How can we repeat the success?
- What were the disappointments? Why? How can we avoid repeating them?

At first, the moderator asks the team members for their answers to these questions. This way, everybody's opinion can be taken into consideration, meaning that the experiences and the knowledge of all participants are captured. Besides, the moderator applies the 5-why method. The results of this meeting are not communicated to the outside before the team has given its approval.

Learning during doing – After action reviews are held at the end of each (mostly short) activity. At BP this takes place very often. These meetings are focused on the discussion of the following 4 questions:

- What was supposed to happen?
- What actually happened?
- What was the difference between expectations and what really happened?
- What could we learn from it?

Learning before doing – At BP this is called peer assist. Employees get together to exchange knowledge. While one group is just starting a project, the other group has already run a similar project in the past. Both groups share their understanding and experience within their own contexts. In most cases the knowledge of both groups overlaps. Based on this, what can be achieved in the project and which steps should be taken is discussed. Peer assists are structured according to the graphic below. The advantage of this visual display is that it is very clear and understandable and thus easy to remember.

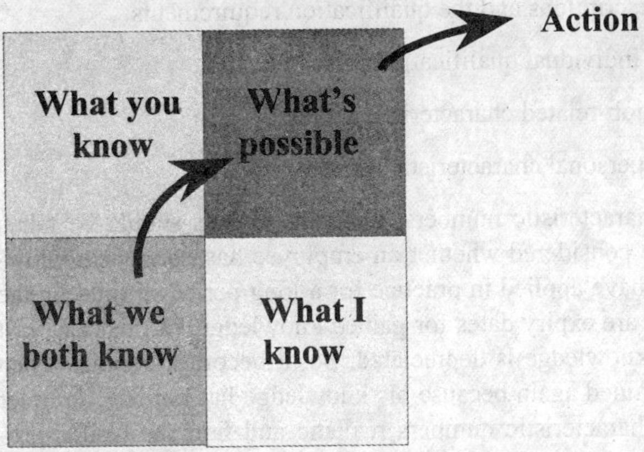

Fig. 11.8: Visualization of BP's Peer Assist Steps

Based on this, there is a discussion of what can be achieved in the project and which steps should be taken. It is critical for the success of the peer assist to invite the right people with experienced knowledge.

11.7.4 Raising the Competency Level

The River Diagram[340] – At first, each business unit of BP compares its practices to a set of key practices, which involves analyzing and identifying the current status. The next step is to determine which competency level is to be achieved. Finally, the business unit works out how the goals are supposed to be achieved. The results of the analysis are visualized in a diagram showing the status quo of the business unit and its desired status. The diagram also contains the average value of each competency for all business units as well as the range of existing levels at BP.

The Stairs Diagram – When a business unit at BP has clarified which competencies it wants to improve, it can determine with a Stairs Diagram[2] which other business units feature a high level of the respective competencies. These business units can then be contacted for an exchange of knowledge.

Knowledge Balance – Raising the Employees' Competency Level

VW's knowledge balance is aimed at capturing, assessing and directing the development of expert knowledge. The company wants to depict the competency level of its employees and then raise it. In a kick-off meeting, the employees are informed about what a knowledge balance is. Further steps towards establishing a knowledge balance could, as in the case of VW's area of production, be:

---

340 For a detailed description and visualization of the River and the Stairs Diagram at BP cf. Collison et al (2001).

- Clarifying job descriptions and the qualification requirements
- Determining the individual qualification steps
- Establishing the job-related characteristic number
- Establishing the personal characteristic number

When establishing the characteristic numbers, different aspects should be taken into account. It should be considered whether an employee has theoretical knowledge, which he may not have applied in practice for a long period of time. In the knowledge balance, there are expiry dates for gained knowledge. In this case, just like in normal balances, knowledge is depreciated. So, it becomes obvious when an employee should be trained again because his knowledge has expired. In order to keep the established characteristic numbers realistic and fair, the results provided by the different foremen are compared. If they differ from each other, which seldom happens, the deviations are either very minor and do not influence the total result, or an agreement between the experts involved can be easily reached.

Concrete results of knowledge balancing at VW's car body works are:

- Transparency in terms of the existing qualification structure
- Identification of qualification deficits
- Training requirements
- Integral planning and assessment of the efficiency of qualification measures
- More precise budget planning (training budget)
- More effective use of employees
- Electronic data management support
- Consequences resulting from changes (for example from transfers)

The results have a direct effect on further education, setting up training budgets and assigning people to respective jobs.

11.7.5 Preserving Treasures

Successful companies are aware that they lose a lot of valuable knowledge when employees leave the company. That's why they use instruments to actively preserve this knowledge.

Making Knowledge Explicit

One way to preserve the knowledge of employees who leave the company is to put it down in documents; another is to establish personal contacts. ABB Switzerland, for example, developed an "exduction program" called "capturing knowledge".

This program is a reversal of the introductory program. It includes exit lectures, exit conversations, farewell dinner, oral debriefing etc. The oral debriefing is the main element of the exduction program. A form describing the procedure in a table serves as a guideline for the interview. The protocol of the interview at the same time serves as an instruction for further steps.

Transferring Implicit Knowledge

Other approaches focus on the transfer of implicit knowledge. VW puts this approach into place with its "Wissensstafette" (knowledge relay) to capture the expert knowledge from key employees and managers who change their position or retire. A combination of structured interviews, moderated transition workshops, and documentation in databases is used to retain and pass on the experts' implicit knowledge and specific experiences about past, present, and future projects, best practices, lessons learned, and the organizational culture as well as knowledge about important social relations to the successor. Another method to prevent the loss of knowledge is to establish consulting companies made up of employees who have retired. At ABB Switzerland, all top management employees automatically transfer to the ABB Consulting Company when they have reached the age of 60.

There are other methods to maintain access to the knowledge of people who have left the company. ABB Switzerland has formed a Community of Practice of employees who have left.

- These former employees are sent information by ABB Switzerland on a regular basis.
- They are invited to a workshop once or twice per year.
- They can pass on their knowledge by giving consulting services.
- The company tries to retrieve certain former employees (There is a list of these people and the company keeps in contact trying to maintain a network with them, e.g. by sending them the company report, Christmas cards etc.).

BP allows people who have left to keep their Connect page in the system, which 25% of them have done. One employee is responsible for updating this page with contact addresses, e-mail, telephone numbers etc. So the valuable experienced knowledge of the ex-employees is still accessible.

11.8 Summary

Whilst the majority of the Benchmarking Partners have integrated KM into their visions, all of them have formulated a KM goal. A uniform terminology defined for the entire organization as well as the appropriate status of KM also seems to have been established by most of the companies.

In a first step towards the introduction of KM, the companies carefully and systematically analyzed existing processes, methods and instruments from a KM perspective. Afterwards, these KM approaches were taken over and developed further according to current requirements and possibilities, building on the existing facilitated Quick Wins and contributing to the acceptance of KM in the organization. This is how the Benchmarking Partners created the conditions for a successful integration of KM processes into the business processes. Moreover, Benchmarking Partners use KM instruments, which are pragmatic, simple and easy to understand. At the same time these instruments are very effective. As a result, they are used daily by many employees thus spreading over the organization.

Common to all visited firms was the fact that they all managed to create a lasting sensitization to KM in their organizations. Sound decision preparation, "Just do it! ", "Focus, Focus, Focus", "Step by Step! ", "Create the big picture" or "The long road" – it is important that their way suits the corporate culture and strategy!

The Benchmarking Partners spared no efforts to market KM internally. Everything – from logo to video – was tried. All Benchmarking Partners agree that it was worth the effort. In all companies that were visited, the employees have accepted KM, appreciating its advantages in their daily routine.

Introducing KM entails the necessity of new roles and functions. Which roles and functions are optimal for a company depends on the organizational structure and culture. In any case, the introduction of KM requires people who take specific and full-time responsibility for KM. Once KM has been integrated into the employees' daily routine, this function is no longer required, or only to a small extent.

Companies, which successfully implemented KM specifically, use tools to manage the knowledge of their organizations. These tools are mostly pilot-tested before their company-wide implementation. In order to achieve a high efficiency of the tools while ensuring their user-friendliness, the Benchmarking Partners keep working on further development and improvement of tools.

12 Knowledge Management: The Holistic Approach of Arthur D. Little, Inc.

Peter Heisig, Frank Spellerberg, Patricia Spallek

In 1886, Arthur Dehon Little, a chemistry professor at the Massachusetts Institute of Technology, founded the company with the same name. His goal was to advise companies on how to develop more efficient production processes and organizational forms. Today, the company is a world-class consulting firm, with 3000 staff members based in 51 offices and laboratories around the globe. In 2000, Arthur D. Little Inc. had an income of $ 590 million.

Arthur D Little

http:// www.adl.com

Industry:	**Consulting**
Business Process:	**Development and Delivering of Services**
Employees:	**3.000**
Sales:	**$ 590 Mio. (2000)**
KM Best Practice:	**Case Debriefing** **Holistic KM Approach**

Fig. 12.1: Company Overview – Arthur D. Little

In Germany there are offices in Wiesbaden, Berlin, Düsseldorf, and Munich with approximately 200 consultants in various "practices." These "practices" are organized according to the industry they serve: chemistry and pharmaceuticals, consumer goods, automobile manufacturers and suppliers, mechanical and plant engineering, electronics, information technology and telecommunications, banks and insurance companies, energy, resources and utilities, and public administration and services. The "practices" are further divided according to the services offered: safety and risk, strategy, design and development, EHS management, operations management, program management, technology and innovation management, information management, organization, and environmental science and technology. Product and process development is handled by staff members of Cambridge Consultants Ltd., a subsidiary of Arthur D. Little. ADL helps organizations create innovation in the full spectrum of their activities, from creating strategies and shaping organizational culture to developing cutting-edge products and technologies.

The responsibilities of global practice leaders at ADL include combining information with knowledge about its practices.

12.1 Distributed Teams as the Starting Point

Arthur D. Little has offices in many parts of the world. This worldwide distribution and the focus of individual staff members on certain industries led in the early nineties to the question: "Are we one company? What are the common features of telecommunications and administrative experts?" A pilot project with Lotus Notes in 1995/96 quickly led to the realization that the computer system was but one element in a broad initiative to maximize the potential of ADL's knowledge resources. In addition to the hardware/software, ADL had to concern itself with issues of content, context, culture, and process. In fact, technology provided only about 20 % of the overall solution.

As a result of these early results, ADL set up two task forces with directors of different practices and providers of internal information. The goal was two-fold. First, ADL wanted to analyze current developments in management technology, particularly in consulting in order to come to a common understanding of knowledge management. Second, it wanted to test and determine the strategic importance and direction of knowledge management for its own development. The members of the task forces had to have very good knowledge of the corporate philosophy and the strategic goals of the firm. The first important result of the discussion – and the first important step in the introduction of knowledge management – was the regular involvement of top management in talks and joint meetings. The goal was to make sure that managers supported the initiative in the long-term. Staff members were acquainted with management's views and visions on knowledge management through presentations in an internal communications program and through articles in the company journal.

According to Arthur D. Little, the vision of knowledge management is combined with the idea of a "one company platform":

- From individual to institutional knowledge – simply, quickly, and globally.
- All staff members make an effort to contribute to the knowledge base for the benefit of the company and the customers.
- To be able to respond to the needs of customers faster and better than our competitors.

12.2 Content, Context and Culture are in the Permanent Process of Being Assessed

According to the approach of ADL, managing knowledge is a multidimensional process. It requires the effective concurrent management of five design fields: content, context, culture, process, and infrastructure.

"Content" includes the task of identifying all elements of knowledge and of understanding the relative importance of each element to individuals, groups and corporate objectives. It is also important to analyze the „Context" of different elements of knowledge and their actual or possible applications. These estimates are very important for setting priorities and creating plans. Arthur D. Little has recognized the importance of a cultural dimension for successful knowledge management.

The traditional culture of "industries" in the consulting business that focuses on generating billed days of consulting work dominates the cultural orientation of each consulting firm. For ADL, this meant that the consultants are driven by what is called "business impact" toward selling its services and keeping its people billable. In the end, that focus does not leave much time for knowledge management.

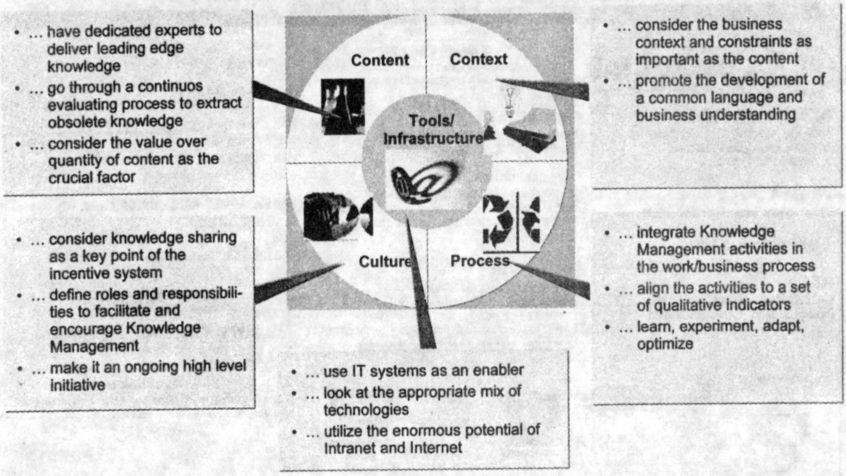

Fig. 12.2: The Five Key Dimensions of Knowledge Management at Arthur D. Little

The general corporate culture has been termed "network-oriented". This is illustrated by the fact that 90-95% of all co-workers quite willingly offer support and information during telephone calls, even though they have never met the caller in person. The philosophy of the German branch that was developed by the founder of ADL Germany, Tom Sommerlatte, is in accordance with the central motivation factors of the staff: team spirit, levels of freedom, great interest in jobs and responsibilities (handling challenging problems), high innovation, and a lower degree of career-focused ambitions. In this context, "you need knowledge management to avoid being too innovative: `Oh, a lot has been done already in this area`" is often an enlightening experience.

ADL attempts to indirectly influence cultural realities that function as barriers by taking steps in the areas of organizing knowledge management (process), rating performance, allocating earnings, and project budgeting.

12.3 ADL-Link: A Straightforward Intranet Approach

ADL describes its activities to develop and make accessible the hardware and software basis of knowledge management as the design field "infrastructure". In this respect, ADL does not differ from other global companies that operate an Intranet through which they provide search possibilities and information. ADL Link is based on a set of Lotus Notes and other applications hosted by a series of Domino and other Intranet servers and available to staff members globally via Netscape browsers.

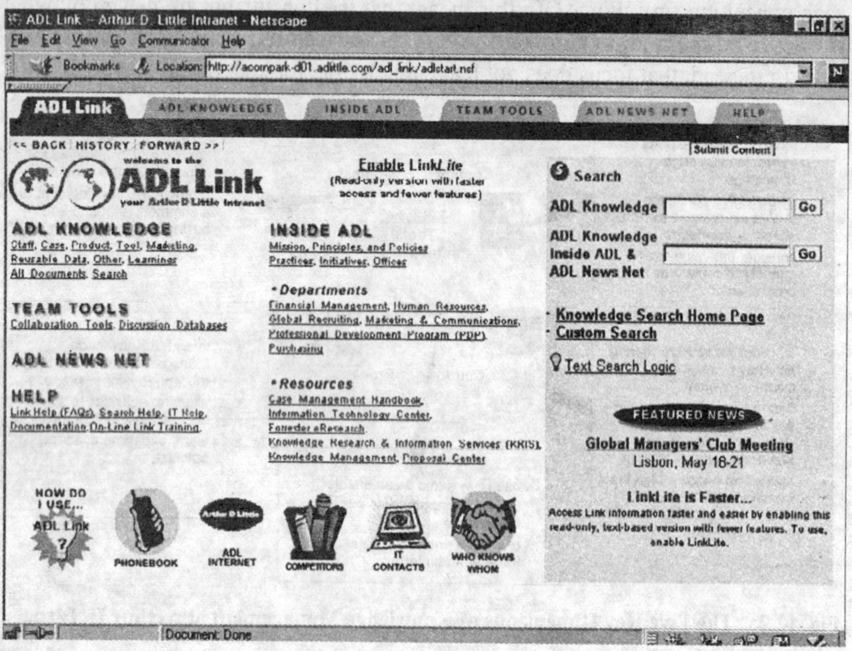

Fig. 12.3: Homepage of ADL Intranet

The ADL Link serves as a portal to knowledge bases and repositories, information bulletins, collaboration tools and resource guides. The main aspects of knowledge accessible to all staff members can be divided into the following main categories: staff members, projects, tools and products, i.e., the firm's functional expertise.

The category "staff members" includes a global telephone directory with basic information on each person (title, department, etc.), detailed staff profiles with information on casework and additional qualifications, and staff résumés as text documents.

Casework is described in different categories. A "case summary" includes general management information, customer data, and brief job descriptions. It also contains information on finances, participating staff members, and case abstracts as a case history. The case abstracts include detailed information on work and keywords reflecting the business context and applicability. In addition, there is information on the location of case documents and their accessibility, and on staff persons to contact for further information. Finally, users can download reference descriptions and regulations for their use.

In the products and tools section, users have access to the description of methods, training materials, analyses, reports, benchmarking data, and customer presentations. The user is also provided with templates, tools, and case proposals that are internally understood as a best practice.

This infrastructure ensures fast and simple global access to elements of knowledge. Frequent hyperlinks allow to quickly access relevant information in related areas. The usage of the ADL Intranet highly depends on the quality of its content. Staff members have to enter knowledge in a pre-defined process. This brings us to another important design field of knowledge management at ADL, the "process."

12.4 Elaborated Knowledge Roles Complement the Knowledge Management Process

The core concept of knowledge management at ADL comprises two essential domains: (1) The definition of the individual steps necessary to develop and maintain the knowledge base. (2) The description of roles required to be filled by the participants of knowledge management.

12.4.1 The Process

The process steps defined by ADL describe the general management of different knowledge elements that have been identified as being relevant (consulting cases, offers, etc.), summaries of staff members, customer information and presentation documents. Subprocesses, such as updating biographies or writing project abstracts, are considered routine processes.

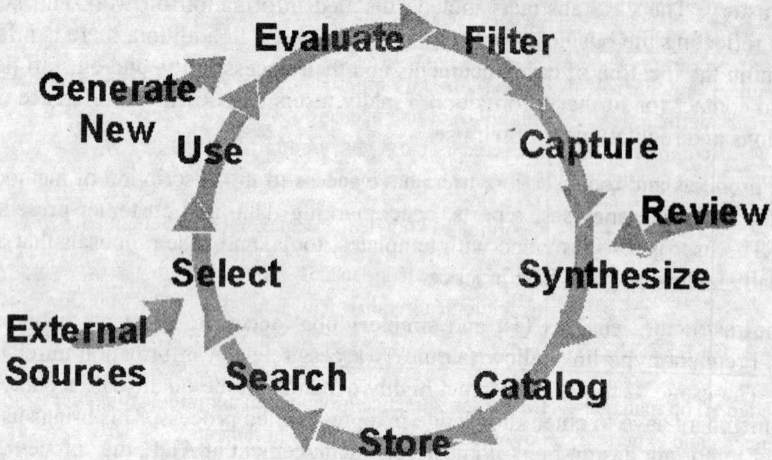

Fig. 12.4: Process Steps of Knowledge Management at Arthur D. Little

12.4.2 The Roles

To apply knowledge management, ADL defined the roles of "Knowledge Stewards", "Knowledge Advocates", and the "KM Board". The central tasks of these roles are to ensure the lasting support of knowledge management by all members of the organization, and to implement knowledge management in each practice/group. The original idea to fill the role of the „Knowledge Steward" without additional personnel had to be given up. However, due to more efficient processes in the long run, ADL succeeded to manage with fewer consultants. Each practice/group finances the expenses for its "Knowledge Steward(s)" and, therefore, autonomously manages the tasks. A "Knowledge Steward" has to have professional know-how (industry, methodology), communicative capabilities (interview skills), and capabilities to prepare and condense data and information. He has to be curious and persistent ("There must be something interesting and new."). Working exclusively as a "Knowledge Steward" is hard to reconcile with full dedication to the daily consulting business. This had been identified as a major problem, since the so far prevailing corporate culture and career criteria had not promoted the fulfillment of this role by consultants. ADL has devised a new profile for knowledge professionals combining KM tasks with specified high quality research tasks and only minor, occasional consulting employment.

Role	Description	Task
KM Board	Strategic role to define, communicate, and coordinate knowledge management activities	Represents and supports globally all KM activities and initiatives of practices/groups (selected by the KM Community incl. all Knowledge Stewards, headed by a chairperson) Coordinates knowledge management development Defines knowledge management roles, processes, technological requirements and the global KM budget Reports directly to the leadership of the company
Knowledge Advocate (KA)	Top manager and expert	Represents practice/group when communicating the knowledge management requirements Provides support when handling knowledge management activities and global coordination Gives orientation in knowledge management
Knowledge Steward	Operative role carry out knowledge management activities	Carries out the knowledge management process for practice/group Informs practice/group of knowledge management procedures Makes the connection between practice/group and the global knowledge management team (KM Board) Ensures global coordination

Fig. 12.5: Knowledge Management Roles

12.5 The Knowledge Steward

Due to the essential importance of the role of the "Knowledge Steward" the following section describes the main responsibilities:

- report directly to practice/group management
- plan and carry out so-called "case debriefing" meetings in which the project is analyzed critically, and experiences are collected;
- initiate and ensure the creation of knowledge assets;
- initiate and ensure expenses to close knowledge gaps;
- study knowledge assets regarding quality and consistency;

- examine knowledge assets periodically, including updates, corrections, and discards;
- assess the benefits if including items in the knowledge database;
- inform co-workers of new knowledge assets;
- ensure global consistency;
- sanitize sensitive customer data from case studies;
- define and introduce the local roles, processes, and standards of knowledge management;
- hand in requirements to expand the ADL Intranet;
- categorize and store knowledge assets;
- extract erroneously placed items;
- share best practices.

Case Debriefing :
Systematic treatment of project experiences .

Case debriefing" can be divided into three stages: At the beginning of a project data are continuously collected and analyzed. During the project knowledge assets are identified and captured. Debriefing sessions in the end serve the purpose to document all relevant contacts, lessons learned and dos and don'ts. The aim is to reflect project work for more efficient future project management.

Fig. 12.6: Case Debriefing

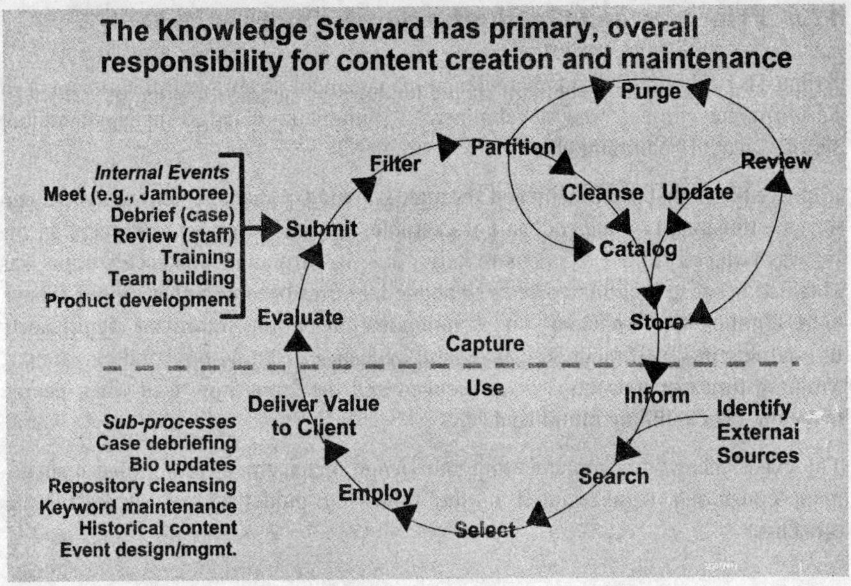

Fig. 12.7: Responsibilities of the Knowledge Steward

The newly created roles at ADL were furnished with evaluation grids. These allow the firm to evaluate the performance in relation to the individual activities. The following table is an example of evaluation criteria for the role of "Knowledge Steward":

Activity in Knowledge Management	Fulfills Requirements	Highly Exceeds Requirements
Acquisition/Capture	Carries out the case debriefing process for the practice or department	Identifies and collects proactively relevant information on industries and practices during the case debriefing process
Dissemination	Informs the practice or department regularly of new content	Contributes continuously to improving productivity and/or quality

Fig. 12.8: Evaluation Criteria for the Knowledge Steward

12.6 Principles and Qualitative Success Stories

Arthur D. Little understands knowledge management as an integral component of its corporate strategy and development: "Without knowledge management our services would be unthinkable."

Improvements that were achieved through knowledge management activities can only be illustrated by anecdotes. For example, the time required to prepare an offer was reduced from 1.5-2 days to half a day. In particular, this acceleration was obtained because lengthy searches or phone inquiries became either more focused, were eliminated or reduced. The communication culture improved significantly through the more efficient and successful exchange of know-how and experience. Training times of new hires were reduced and the know-how loss when people leave the firm could be minimized.

The experiences with implementing knowledge management in a global management consulting firm resulted in the following guidelines of successful approaches:

Global, Global, Global	Priorities and plans have to be defined globally. The implementation occurs on a local level.
Roles, no jobs	Knowledge management roles should be distributed; one person should handle knowledge management and non-knowledge management.
Roles to be taken seriously	Knowledge workers are rated according to their performance regarding knowledge management activities.
No "One Size Fits All"	Each practice/group must decide globally how it implements knowledge management roles and processes.
"In the sweat of one's brow"	Knowledge management is hard work.
Community	We have to be an active community of interests.
Security	The "Network Security Agreement" is our law.
Ownership	We have to respect and protect the rights of ownership.

Consulting firms generally require a systematic management of the corporate resource "knowledge". Arthur D. Little has begun to implement a strategy that is applicable to the entire firm. The strengths of the approach clearly include the definition of the relevant knowledge elements, and the establishment of clear process steps with appropriate roles.

The original idea to have the consultants fill these roles was not carried out. This is where we have identified a potential area for improvement: To secure important experiences, project work should be connected better to knowledge management activities. ADL should not only apply the evaluation criteria of knowledge management to all staff members, but should also design the so-called "soft" aspects of the corporate culture actively. In addition, the application of the desired collaboration tool to support teamwork and the utilization of knowledge suitable for knowledge management activities are supplied at the same time.

Strengths were seen in the definition of relevant knowledge elements, the establishment of process steps of knowledge management, the definition of required roles of knowledge management and the establishment of a catalog of criteria to evaluate performance.

On the other hand side potentials were identified: Better integration into daily business operations and project work, increase in routine processes and better incorporation into the corporate culture.

13 The Aventis Approach to Knowledge Management: Locating Inhouse Expertise for Immediate Anytime, Anywhere Availability

Jürgen Oldigs-Kerber, Alla Shpilsky, Stephen Sorensen[341]

With approximately three billion Euro (2001), Aventis funds one of the largest research and development budgets for prescription drugs in the pharmaceutical industry. The development of a drug, from identifying the active ingredient to approval can take ten to 15 years. A substance with an average to good profitability (for example, € 365 million per year) generates approximately one million Euro each day.

Saving even the smallest amount of time is crucial in drug development. Time is money – money that could be used to conduct additional innovative research.

Within Aventis, it is the associates who possess the most valuable business capital – knowledge. The challenge is tapping into or accessing that knowledge – whenever, wherever or however possible. Since Aventis is a heavily matrixed, global organization, capturing the knowledge of subject experts is especially tricky. Aventis sought to connect people who currently work or have worked on similar problems. By reducing redundancy, productivity can be improved and less re-inventing needs to be invested.

13.1 Facilitating Knowledge Management at Aventis DI&A Using KnowledgeMail

At Aventis Drug Innovation & Approval (a research and development function), the Knowledge Managers have agreed on a practical definition for the term knowledge management (KM). "Knowledge management is the systematic effort to promote connectivity between people and to facilitate the generation, sharing, and use of knowledge to gain competitive advantage."[342]

---

341 The realization of this project in its first phase was only possible through a team approach. Contributing KnowledgeMail team members were Donna Canavan, Claudia Diefenbach, Sébastien Dorey, Mark Feingold, Gabriel Flory, Jean-Marc Girodeau, Robert Irmisch, Anja Klaene, Keith Marin, Jürgen Oldigs-Kerber, Douglas Rush, Alla Shpilsky, Judy Siegel, Kara Smith, Steve Sorensen, Milton Takeguchi, Yung Wang, Hans Wedde and Susann Wolfgram.

342 Aventis Internal KM Global Meeting (Feb. 2, 2001).

One current approach at Aventis – KnowledgeMail – is based on the principle of the employees' willingness for global knowledge exchange. The individual as a knowledge carrier is thereby the focus of KM activities. Locating inhouse expertise (Expertise Location Management) and connecting people (networking) plays an important part in our KM strategy. The individual employee becomes the focus of interest. The goal is to link the various knowledge carriers, in order to use existing knowledge more effectively. The tacit knowledge of the individuals is identified by analyzing the explicit knowledge in their e-mails and other user submitted documents.

Essential to this approach is that it is reinforced by the Aventis culture, which has established core values to provide a framework for the behaviour of the employees. One of these values is networking, which means reaching out beyond boundaries not just to get information and ideas, but also to share ideas and information with colleagues. Positive networking involves breaking down silos. This demands courage, discipline and integrity.[343]

The ease of locating and connecting knowledge carriers adds considerable value to the company with little effort compared to the set-up of "knowledge warehouses" (i.e. documents in databases such as "Lessons Learned", interviews with experts, debriefing papers etc.).

The approach emphasizes quick availability and efficient diffusion of expertise, when needed. Preliminary time-intensive documentation is not required in this case, enabling information to be shared as soon as it is needed. Diffusing the required knowledge as quickly as possible fulfills the urgent demand for knowledge within the company. In all three aspects of this approach – knowledge generation, knowledge retention and knowledge transfer – the knowledge carrier stands in the foreground.

Based on North, Romhardt & Probst (2000) the implementation of KM solutions in general orients itself to the following three strategic areas (cf. Fig. 13.1):

343 Cf. http://www.aventis.com, cited 11-29-2002.

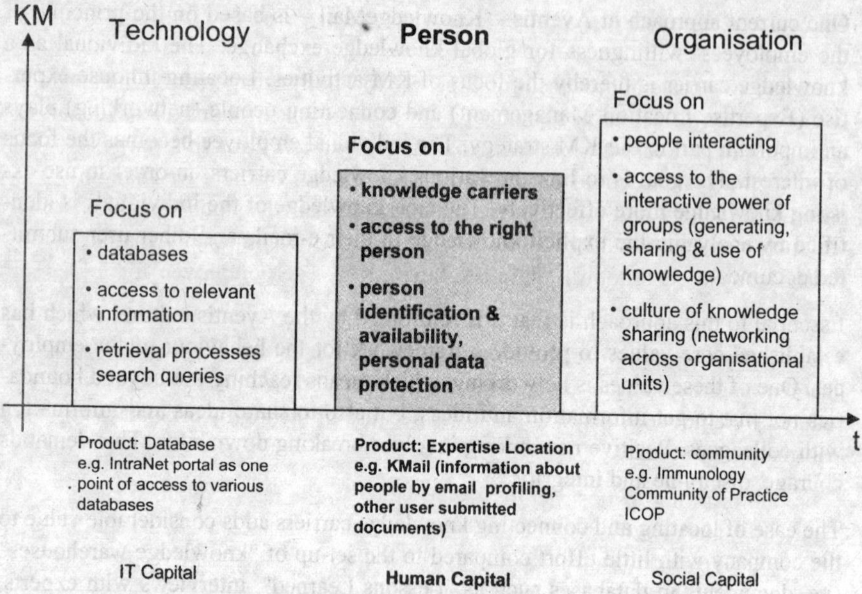

Fig. 13.1: The Involvement of Knowledge Management in Business Processes[344]

In addition to the expansion of IT infrastructure, solutions are consciously differentiated between those that address the individual employee (human capital), and those that aim to support a collaborative network of interacting people (social capital).

The demands for the implementation of new solutions and the buy-in by employees are considered more difficult for solutions that address the human capital and the social capital branch, because they change the way employees work together, address their willingness to share knowledge and to recognize other people's knowledge. A critical mass of users is also essential to make the approaches beneficial to the user.

Taking all this into consideration, Aventis decided in favour of a newly developed tool called KnowledgeMail (KnowledgeMail), marketed by Tacit. KnowledgeMail is an expertise location management software that profiles e-mails and documents submitted by the user to create a user profile detailing proposed areas of expertise. Additionally, the user may enrich this automatically created profile by manually entered description of further expertise fields. Data privacy issues are addressed by specific software security measures.

---

344 Adopted from North, Romhardt and Probst (2000), p. 55.

13.2 Low Workload Expertise Ascertainment (Profiling)

There are two basic approaches for expertise location management. One approach is the creation of a search system of high precision using controlled vocabulary consisting of a taxonomy of terms maintained explicitly for this purpose. However, such a system would not meet the requirements of the ever-changing, fast-paced terminology and topics in pharmaceutical R&D. It is questionable how often such a system would be used because it must be continually maintained by central content (taxonomy) administrators and by each employee for his or her own profile.

The idea was to minimize the workload for individual users for both the establishment and the continuous update of a profile. Yellow Pages or debriefing procedures were not considered alternatives because of the high workload to keep the information up-to-date.

Thus, the profiling of unstructured information in the background, such as documents or e-mails, proves to be the better solution in the context of the Aventis situation. KnowledgeMail supports the creation of a large number of terms per user through the profiling of emails and documents submitted by the user (cf. Fig. 13.2).

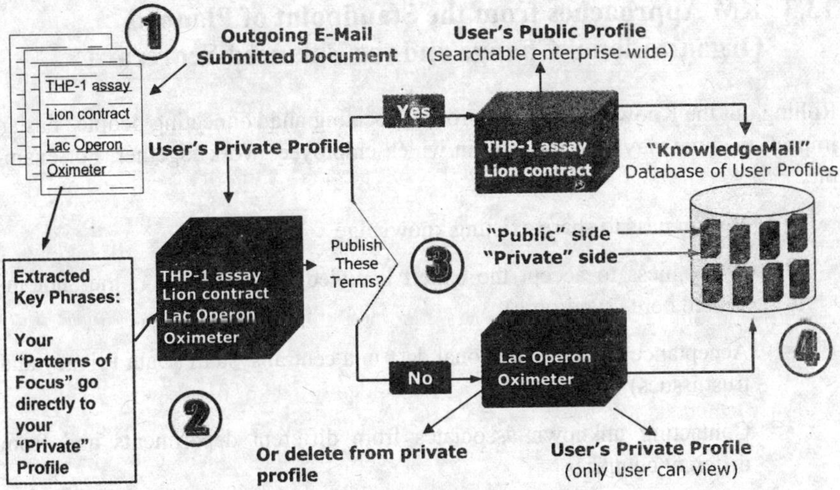

Fig. 13.2: Key Term Extraction from Email or File System by KnowledgeMail

The expert profile is generated in an automated manner by analyzing such texts; along with a description of the job profile in the employee's own words. The profile is based on terms extracted from the documents and then linked to the user's name. The user can freely decide whether and to what degree the profile can be viewed by others. The user profile also consists of a private part (private profile), which can be researched with guaranteed data protection by KnowledgeMail only.

In this case the software contacts the found persons – unknown to the seeker – and the found person is free to respond or not. The identity will never be disclosed unless the found person does it him/herself by responding to the seeker.

The advantages of the KnowledgeMail approach can be summarized as follows:

- Profiles are generated automatically by analyzing emails and other documents submitted by users
- Frequent update of profiles (daily basis)
- Key words are specific to the company business, including internal company codes like AVExxxx
- The search process employed does not only find the right persons but also scores the relevance of the term within the found profile
- Network building amongst employees is promoted
- No documents are stored, only up-to-date user terms based on their emails and other submitted documents
- Data privacy is ensured and participation is voluntary

13.3 KM Approaches from the Standpoint of Planned Organizational Change and the Balanced Scorecard

Rolling out the KnowledgeMail approach "locating and connecting people" has an impact, in some ways, on the way in which employees work together. For example:

- Willingness to share existing knowledge
- Willingness to accept the expert knowledge of others (... the "not invented here" syndrome)
- Acceptance of saving personal data in a central system (Data Privacy and trust issues)
- Contacting unknown associates from different departments and from different countries

The implementation of the KnowledgeMail approach was managed by keeping in mind the organization model of Porras and Robertson[345] for planned organizational changes. Nothing is as practical as a good guiding model.

345 Porras, Robertson (1992), p. 738.

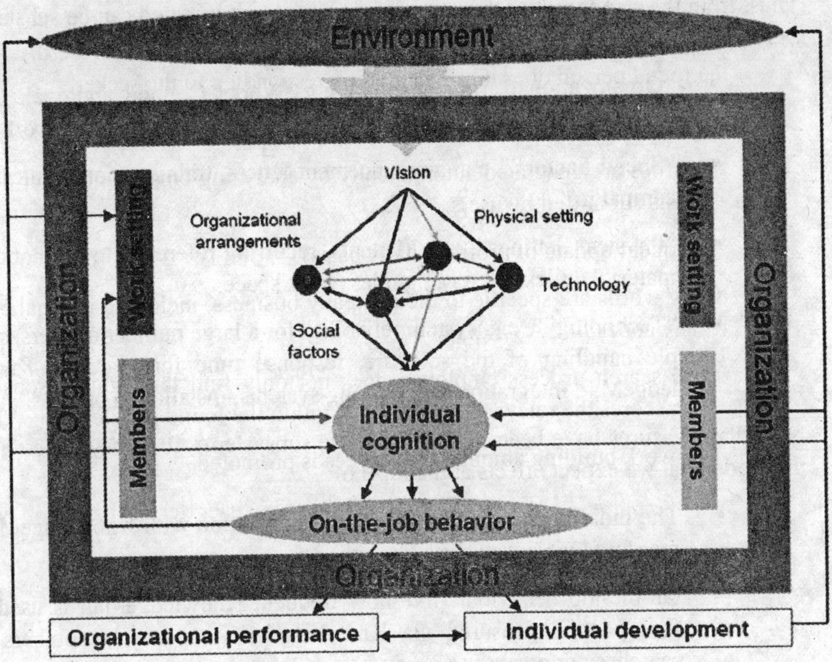

Fig. 13.3: Overview of Organization Model by Porras and Robertson (1992) for Planned Organizational Changes

The figure reflects the following assumptions: "...organizations are contexts within which individuals behave. An organizational work setting comprises four major interrelated subsystems: organizing arrangements, social factors, technology, and physical settings ... Each subsystem consists of specific elements that strongly influence the work behaviour of individual organization members. In turn, individual behaviour is a primary determinant of two categories of organizational outcomes: the level of organizational performance and the level of organization members' individual development. From a research perspective, then, interventions constitute independent variables, and work setting changes, individual behavior and organizational outcomes are dependent variables."[346]

[346] Robertson, Roberts, Porras (1993), p. 620.

Thus, from the standpoint of this model, measures to be initiated can be subsumed as follows:

- "Organizational arrangements", e.g. Works Council approval, Data Protection Officer involvement, obvious top management support.
- "Social Factors", e.g. announcement letters, training, communication campaign.
- "Physical Settings", e.g. Manuals, recurring reference to promotional material, quick reference guides, work space.
- "Technology", e.g. system reliability for a large number of users, simple handling of the software, response time for searches, KnowledgeMail integration into existing systems, installation process.

Now that measures have been initiated in these subsystems, it is easy to see from the model that we expect effects at the level of:

- The individual cognition (like user satisfaction with KnowledgeMail, measured via a web survey rating).
- On-the-job behaviour (like how frequent KnowledgeMail is used to find people measured via KnowledgeMail reports as numbers of searches per one user a year).
- Organizational performance (like increase of productivity measured via cycle time saved through the use of KnowledgeMail).

If we crystallize this logic in a thoroughly designed Balanced Scorecard[347], then Fig. 13.4 represents such a Balanced Scorecard – the KnowledgeMail navigator – to control the rollout of the KnowledgeMail approach at Aventis DI&A.

347 Cf. Kaplan, Norton (1992, 1993, 1996).

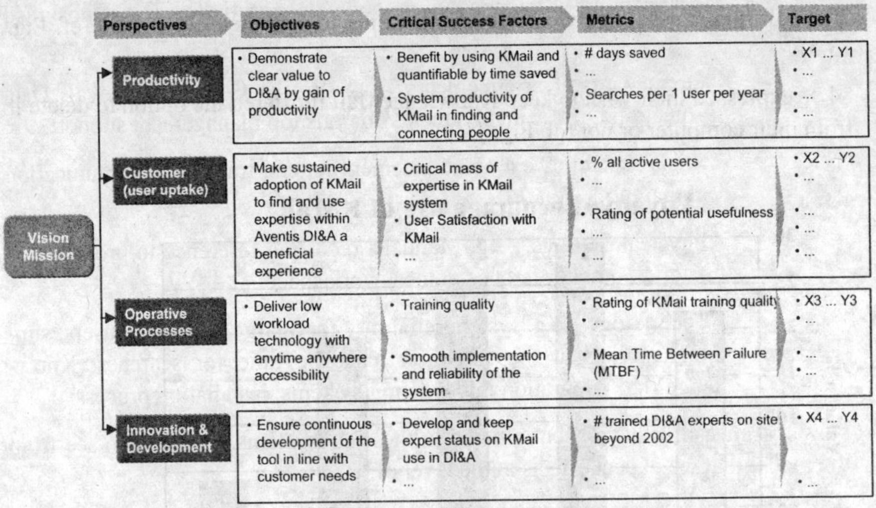

Fig. 13.4: The KnowledgeMail Navigator to Evaluate and Control the Rollout Performance of KnowledgeMail

13.4 User Experiences

The first evaluation in 2001, with over 400 users located in Germany, France, Japan and the US, revealed a high acceptance of KnowledgeMail. With a survey response rate of 32 % the majority of those questioned (62 %) had already been actively using KnowledgeMail (had made a search request) at this point of time. 68 % of these active users assessed the search results as partially relevant or very relevant.

Typical needs with which the employees approached KnowledgeMail were:

- *"I could really use some help setting up this assay."*
- *"I wonder who worked on this compound before?"*
- *"I haven't met this visitor before.*
  What kind of work does he focus on?"
- *"Has anybody worked on this hypothesis before?"*

Those questioned named the following advantages most often:

- *Employee network is expanded*
- *Knowledge about the subject grows*
- *Less time is required to find the appropriate expert*
- *Helps to speed up projects*

75 % of those questioned would recommend the tool to their colleagues (cf. Fig. 13.5).

84 % expressed their wish to keep KnowledgeMail if offered the option to delete it from their computer or not (cf. Fig. 13.6).

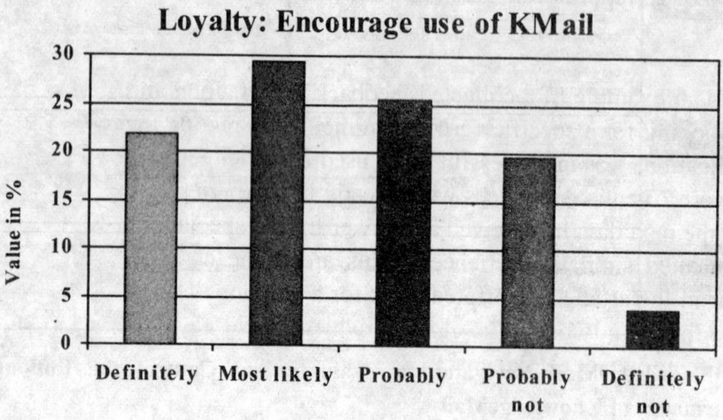

Fig. 13.5: Loyalty Ratings on KnowledgeMail

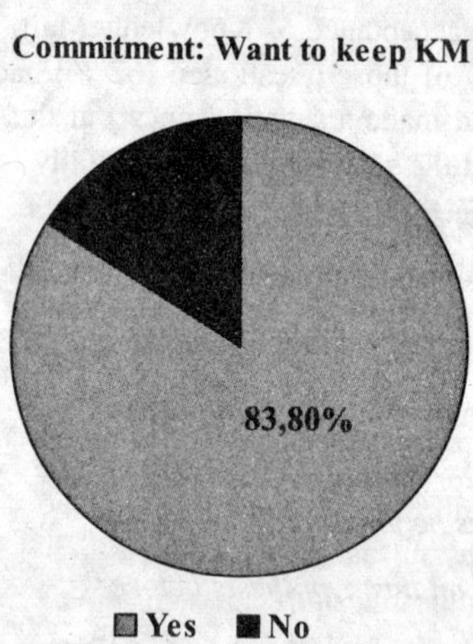

Fig. 13.6: Commitment Ratings on KnowledgeMail

13.5 Business Benefits

Users reported on their experiences with KnowledgeMail in personal interviews. The validated data was used to determine how useful the solution was. A conservative estimate was included when calculating the direct business benefits, and therefore only 10 of 22 reports were taken into consideration.

An example

The following is an example of a collected feedback report: John and his team in the USA were looking for a specific verification method. Creating the method using similar publications was unsuccessful. John used KnowledgeMail to determine whether there were Aventis employees who already had experience with this type of analysis. On the next day, he received an answer from a specialist in Japan who had already collected similar experience in this area. The co-worker informed John of an error in the article and offered hints for a solution. This invaluable information saved the team two months of work, which would have been otherwise required to start their own development.

13.5.1 Calculating the Business Benefits

The direct business benefits were calculated by multiplying the amount of time saved (man days) with the average internal daily cost per associate per day, resulting in an Internal Rate of Return IRR > 200 % over a 5 year period. No other more indirect benefits (sequelae such as faster market entry, reducing the project duration and thereby increasing sales revenue and competitive advantage) were calculated.

The cycle time saved (time span saved ≠ man days) amounted to 6.4 months over a three-month period of usage time for 435 users. The optimistic value was 12.9 months. The indices should make it clear that there is justified hope that this approach can increase productivity and can save a considerable amount of time.

13.6 Summary of Results

Our KM approach KnowledgeMail "Locate and connect people" (building networks) proved considerably successful as early as in the first evaluation phase (December 2001, that is, three months after implementing the software). It showed that:

- it is possible to measure the impact of KM initiatives on the organization, to save a considerable amount of time, to increase productivity, and attain very good results when calculating financial equivalents (ROI > 200 %),
- the users accept the new tool and assess its potential as very high,

- the approach fits the corporate culture and enables employees to network more easily,
- the protection of personal data can be adequately guaranteed; additionally, participation is voluntary,
- there is no considerable additional work for employees (less than 5 minutes maintenance time per month – after some initial effort),
- usage remains high due to the automatic and continual update of information.

14 Cultural Change Triggers Best Practice Sharing – British Aerospace plc.

Peter Heisig, Jens Vorbeck

In 1997, British Aerospace plc. had global sales of 25.6 billion DM (£ 8.6 billion). With its 47,900 employees, it is the fourth largest manufacturer of aerospace and defense systems in the world. In November 1999, British Aerospace merged with Marconi Electronic Systems to form BAE SYSTEMS with sales £ 12 billion and an order book of £ 36 billion. With 100,000 people in nine home markets (UK, USA, Canada, Germany, France, Italy, Australia, Saudi-Arabia, Sweden) across the globe, BAE SYSTEMS is the world's second largest defense company.

In 1998 at the time of data collection for this case study, British Aerospace was divided into two sections: the commercial sector that includes the business unit "British Aerospace Airbus" and the defense sector. Only the defense sector achieved a balanced operating revenue. The commercial sector was able to reduce losses from almost £ 400 mill. in 1992 to £ 20 mill. in 1997. As it is typical of the aerospace and defense systems industry's development of products in long-term international and national projects, British Aerospace plc. participates in 29 larger international partnerships or joint ventures and, additionally, in 50 larger global programs.[348]

BRITISH AEROSPACE

Since 1999: BAE SYSTEMS

http://www.baesystems.com

Industry:	**Aerospace and Defence Systems**
Business Process:	**Management of Improvement and Change**
Employees:	**47.900 (1998)**
Turnover:	**£ 8.6 billion (1998)**
KM Best Practice:	**Cultural Change Program Virtual University Best Practice Centre**

Fig. 14.1: Company Overview – British Aerospace plc., since 1999 BAE SYSTEMS

Changing market conditions – characterized on the demand side by vigorous cuts in national defense budgets since the beginning of the nineties and on the supply side by the merger of the large US aircraft corporations Boeing and McDonnell Douglas – have forced British Aerospace to make internal changes. In addition, the company was operating at a very low margin of profit in the early nineties.

348 British Aerospace Annual Report (1997).

British Aerospace's stock price at that time thus stood at about £ 1 per share. In the meantime, the price per share has risen to £17 (1998). In 1999 a share splitting took place on a one to four ratio. In the early 90's, British Aerospace was still characterized at the management level and in its organizational structures and processes by the holding structure created in the late eighties. "It is the problem of runaway divisional and staff 'fiefdoms'. Lacking sufficient unity and internal coherence, these firms pay too heavy a price for decentralization. (...) BAE's fiercely independent fiefdoms spring from its historical origins in the late eighties as a holding company formed by the merger of disparate pieces of the British aircraft industry."[349]

14.1 The Cultural Change Programme: Benchmark BAE

The former CEO and chairman of British Aerospace plc., Sir Richard Evans, was convinced that the individual divisions of British Aerospace would only be competitive as a synergistic unit, not as excellent but solitary production units. After intense and controversial discussion, the board of British Aerospace turned the initial idea of sharing capabilities into a vision of the "Benchmark BAE Program." The goal of "Benchmark BAE" was to develop the company into a model for the entire industry: "British Aerospace was 'dedicated to working together and with our partners, and to becoming the benchmark for our industry, setting the standard for customer satisfaction, technology, financial performance, and quality in all that we do.'"[350]

Today in 2000, within BAE SYSTEMS the cultural change programme is under review taking into account the British Aerospace cultural change programme and the equivalent initiatives at Marconi Electronic Systems. The following paragraph will focus on the status in 1998 of the cultural change program in British Aerospace only.

14.1.1 From Values to Operational Goals

The next step of the cultural change program in British Aerospace involved discussions with the two or three top managers in each business unit. The managers were to explain which characteristics could be promoted and shared to define the conditions that would contribute to achieving the goals of "Benchmark BAE." These discussions with a total of 130 managers resulted in the following five basic values:

1. people are our greatest strength,
2. customers are our highest priority,

---

[349] Evans (1998).

[350] Evans (1998).

3. partnerships are our future,
4. innovation and technology is our competitive edge, and
5. performance is the key to winning.

In the third step, the 130 top-level managers discussed these values with each other. Then, middle-level management was included in this process.

But, the simplicity of the statements conceals the far more important process of the top managers of the various divisions of British Aerospace, who, with their "cherished histories and strong identities"[351], had just jointly come to a homogenous understanding of their tasks. The goals of the discussion that was carried out on all levels of the company were to break up traditional ways of thinking and to promote questions that eventually led to a new sense of identity and purpose.

New values also require new behaviour characterized by openness and personal candour. These are traits that had previously not widely been seen in the upper management levels because "the fiefdom mentality went so deep that many of them had never met before or even spoken on the phone."[352]

Rather than discussing "values", the lower management levels discussed specific examples of behaviour, specific methods, and specific practices. This led to the establishment of five teams that were each staffed by twelve of the 130 top-level managers. Each team worked on one of the five corporate values. The goal was to develop a company-wide plan of values that would then be disseminated to the different business units. This plan is understood as an ongoing business plan named 'value plan' that contains specific goals for each of the five corporate values. This plan is reviewed continually against the actual performance and adjusted. So the value plan has become an extremely effective tool for managing the business.

For example, an action from the "people" value is to make benefits, which had previously been restricted to the managerial level available to all employees of British Aerospace (personal development plans, car leasing programs, and health insurance). The company produced a video biannually for all the employees that portrays the actual status of their efforts and the respective targets. The personal development plan (PDP) consists of two parts: (1) objectives linked to the business unit value plan and agreed upon with the particular manager, and (2) an individual career plan.

---

351 Evans (1998).

352 Evans (1998).

14.1.2 Monitoring and Coaching

British Aerospace developed and applied an evaluation tool that is based on the predefined corporate values to verify the 130 top managers' pattern of management. The tool is used by the managers themselves to judge whether their behaviour corresponds to and thus promotes the values. The tool was introduced as an Intranet application. A video portrays the self-evaluation of CEO Evans. The result of the evaluation reveals a specific ranking to the respective manager, the CEO, and the project manager. The PDP includes specific coaching measures to improve the individual pattern of management. In a further step, the next lower level of the company's hierarchy with approximately 1,500 managers will be included. For a later stage, British Aerospace aims at 360° feedback.

The BEST program was developed for the Change Management Program "Benchmark BAE" to service the learning and training requirements of the senior management. The program reaches out to all 1,500 managers, and includes the following aspects in five units (length: 1 x 4 days and 4 x 3 days):

- Understand the requirements of the business, the team, and the individual to eventually become a model for the industry;
- Apply efficient managerial abilities in order to manage change;
- Complete a 360° feedback (peer, manager, staff member) before and after the training;
- Show the necessary understanding and the required abilities to coach and support "Team Value Planning";
- Realize the importance of behaviour guided by values and learn the required techniques and tools to support exemplary (benchmark) behaviour;
- Conduct a self-assessment and 360° feedback and receive an EFQM evaluation to improve performance;
- Understand "value planning" and apply the lessons learned in BEST in daily business.

An international program for British Aerospace employees who worked outside the UK was developed and run in 1999. The experience gained in BEST is to be incorporated into management development programs.

The commitment of the top management was ensured with the following measures:

- Active role model of the board members.
- Monitoring of the change by top management itself.

- Integration of the expectations of the 1,500 middle managers by intensive discussions.

14.2 British Aerospace Virtual University

In creating the Virtual University (VU) in May 1997, British Aerospace positioned people, knowledge and know-how at the core of the company's growth, in support of the business strategy to become the Benchmark Company in a highly competitive and consolidating global aerospace sector. Knowledge and innovation embedded in products and customer services are valuable company assets in sustaining business activities, attracting investment, and winning new international business. Today the VU is fully operating as an extended enterprise, twinning academic and business excellence, to build the capability and skills of the workforce through lifelong learning, research and technology and best practice.[353]

Strategic partnerships are crucial as no one has a monopoly on knowledge. Collaboration between the worlds's best in academia and enterprise on clearly articulated business needs, brings leading edge knowledge to the VU programmes, through jointly developed content, best practice and internationally recognized accreditation. The Virtual University includes the Faculty of Learning, the Faculty of Engineering and Manufacturing Technology, the International Business School, Research and Technology Centers, and the Best Practice Center at Farnborough. British Aerospace has appointed experienced managers from the business units as deans and directors of the university.

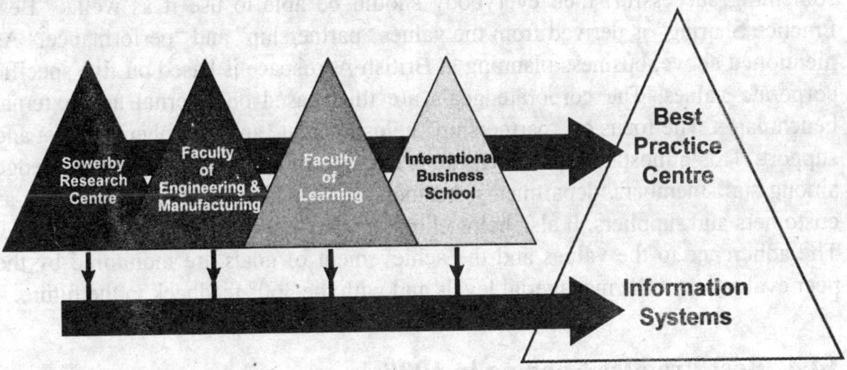

Fig. 14.2: Virtual University in 1998

In May 2000, the BAE SYSTEMS Virtual University has won the U.S.A. Corporate University Xchange Excellence Award, co-sponsored by Financial Times, for its innovative utilization of technology to create a continuous learning environ-

[353] Kenney-Wallace (1999).

ment for all 100,000 employees. The Virtual University has developed intelligent research and retrieval techniques, such as Autonomy, to bring any user a 'one stop shop' of pertinent, timely information needed, saving hours of unnecessary surfing – critical for a large geographically dispersed company. "Creating, sharing and leveraging knowledge is fundamental to competitiveness for any organization in this era of globalization. I am delighted that this Award fully recognizes BAE SYSTEMS' commitment to lifelong learning and the development of talents of all our people and applying technology to accelerate innovation and share business benefit across the Company. These vital components of BAE SYSTEMS drive towards competitiveness and becoming the industry benchmark."[354]

The Virtual University is essentially the organizational umbrella for learning, research and development and best practice and ensures synergy across all areas of the company. In 1998, the Virtual University was staffed with twenty employees; four of whom co-ordinate the best practice center. Today the Virtual University continues within BAE SYSTEMS endorsing the crucial importance attached to developing, sharing and applying knowledge throughout the new global company.

14.3 Best Practice Sharing at British Aerospace

The understanding of knowledge management at British Aerospace can be described as a best practice approach: "To capture and share best or good practices throughout the company". We understand this concept of best practice in the following way: "If you have a practice or process that has worked (i.e., has created something successful) then everybody should be able to use it as well." "Best Practice Sharing" is derived from the values "partnership" and "performance." As mentioned above, business planning at British Aerospace is based on five specific corporate values. The corporate goals are thus based on internal and external benchmarks. The focus on "partnership" helps to create an atmosphere of trust and support. This atmosphere promotes the exchange of information and knowledge among staff members, departments, business units, and external partners, such as customers and suppliers. It also helps eliminate the "not-invented-here" syndrome. The adherence to the values and the achievement of goals are monitored by the peer evaluation at the managerial levels and with the 360° feedback in the future.

14.4 Best Practice Sharing in 1998

Organizationally, knowledge management is incorporated into the BAE SYSTEMS Virtual University. Knowledge management at BAE SYSTEMS is defined through virtual centers of excellence that are the focal points. A virtual center of excellence is characterized by the physical or virtual combination of internal

---

[354] John Weston, CEO BAE SYSTEMS.

and external "elements". This virtual center combines two main dimensions of knowledge. The tacit and implicit dimension as well as internal and external dimension.

Internal elements are the community of experts and best practices. Best practices include the users (staff members of British Aerospace), the processes, knowledge bases (e.g., databases), and other facilities of British Aerospace. External elements or partners are seen in the areas of research and development, in other industrial companies, in external best practices, and in learning from others. The experts with their knowledge represent the tacit dimension. The explicit knowledge is documented in the Best Practice Database, which is accessible via the Intranet (Fig. 14.3).

Virtual centers of excellence are a response to the need to create focal points for learning; the transfer and sharing of best practices are to be channeled internally and externally. In addition, the centers are to function as knowledge centers. The overall goal is to develop and maintain the company's core capabilities at a world-class level.

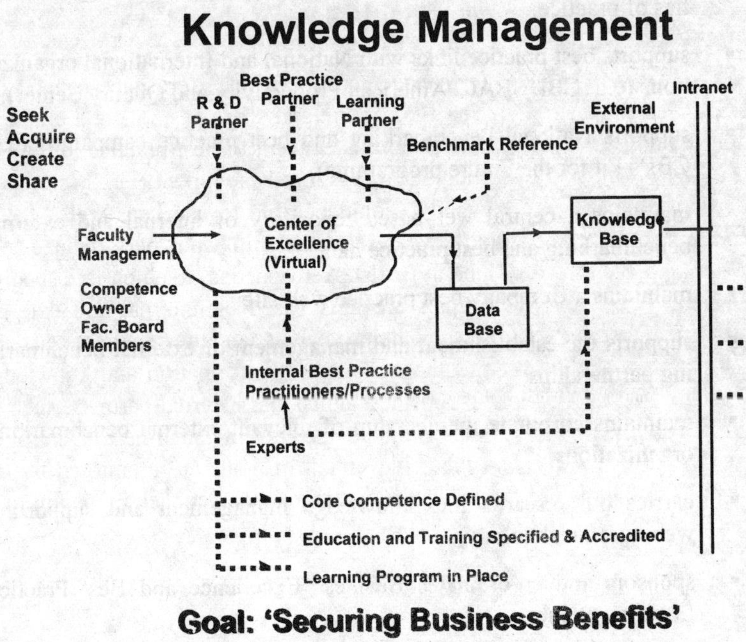

Fig. 14.3: Knowledge Management at British Aerospace in 1998

14.4.1 The Benchmarking & Best Practice Center Acts as Knowledge Broker

In accordance with the understanding of knowledge management, the best practice center acts as a broker of knowledge that identifies and documents best practices both internally and externally. It makes this information accessible through the Intranet and workshops. The application of these best practices is assured by the integration of these benchmarks into the definition of the value plan and the personal development plan (PEP). This process design assures an integrated knowledge management process.

The Benchmarking & Best Practice Center has the following goals:

- maintains an enabling framework to support benchmarking and best practice sharing;
- provides an electronic Benchmarking & Best Practice Library service;
- supports the establishment of special interest groups and communities of practice;
- supports best practice links with National and International organizations (e.g. CBI, SBAC, American Productivity and Quality Center);
- supports National benchmarking and best practice campaigns (e.g. CBI's Fit for the Future programme);
- maintains a central web-based repository of internal and external benchmarking and best practice data;
- maintains a Company best practice web-site;
- supports the establishment and management of external benchmarking partnerships;
- maintains corporate membership of relevant external benchmarking organizations;
- carries out research into knowledge management and supporting web-based technology;
- sponsors and co-ordinates Business Excellence and Best Practice Learning events.

To facilitate the use of the best practice web site, simple electronic forms (best practice templates) and aids (search engines, agents) are offered. In addition, staff members facilitate the exchange of knowledge in workshops where simple questions are asked ("What are you willing to share? What do you need or like to learn?"). The results are published in the intranet to stimulate mutual exchange.

Explicit knowledge is shared via the intranet while tacit knowledge will be shared during the workshops.

14.4.2 How to Structure Knowledge and Best Practices

To guarantee easy retrieval of information, knowledge and best practices a business-oriented structure has been put in place in 1998. To structure the knowledge of internal and external best practices, the center uses acknowledged management-oriented models, e.g., those of the European Foundation for Quality Management (EFQM), Probe of CBI, and "Investors in People" (IiP). The results of EFQM self-assessments are published on the web site to make good practice methods accessible to everyone. British Aerospace supplements this structured information with general contextual data (e.g., what?, how?, why?) in order to promote the transfer of knowledge (Fig. 14.4).

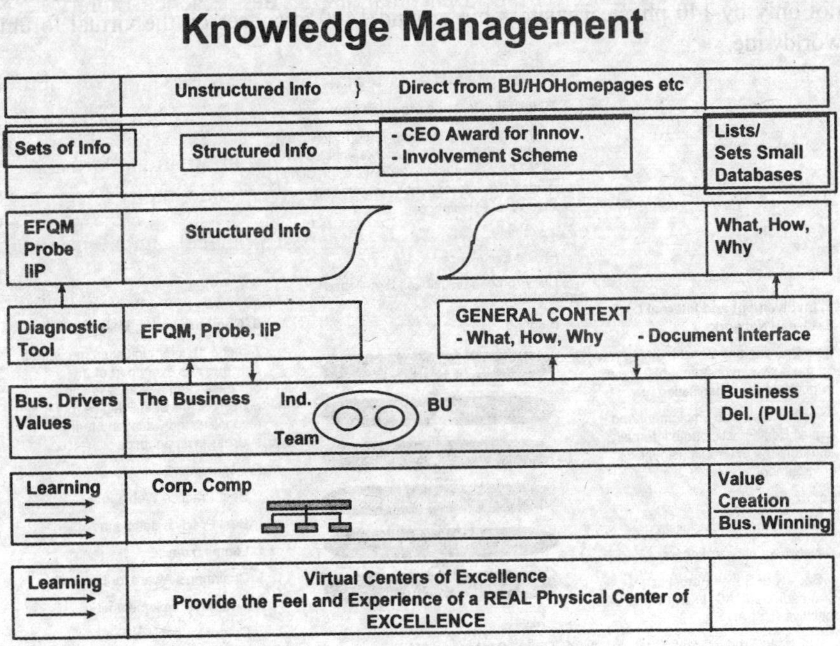

Fig. 14.4: Aspects of Knowledge at British Aerospace in 1998

14.5 Moving forward to 2000

The commitment of BAE SYSTEMS to best practice sharing has been underlined by John Weston, the Chief Executive of BAE SYSTEMS: "Adoption of best practice is no longer optional for business. In fact we believe that a key discriminator for companies is now based on the speed at which they acquire, implement and realize the benefits from best practice in order to support the necessary change demanded by an ever increasingly competitive market place."

BAE SYSTEMS current approach in 2000 to best practice exploitation is a balanced one recognizing that sharing ideas and creating new knowledge is as much about behaviours and culture as it is about process and databases. The approach is based on four key elements (Fig. 14.5): Involvement (Internal/External Networks), External Reference Models, Communication & Recognition, and Tools, Processes & Measures. A six-step process is used to help guide the selection, sharing and implementation of best practices, which have the greatest potential benefit for the company.

By the end of 1999 the best practice database contained over 300 best practice cases and a wide range of activities to support best practice sharing are undertaken. The latest business excellence learning day on April 13, 2000 was attended not only by 140 physical visitors but around 1400 attendees on the virtual forum worldwide.

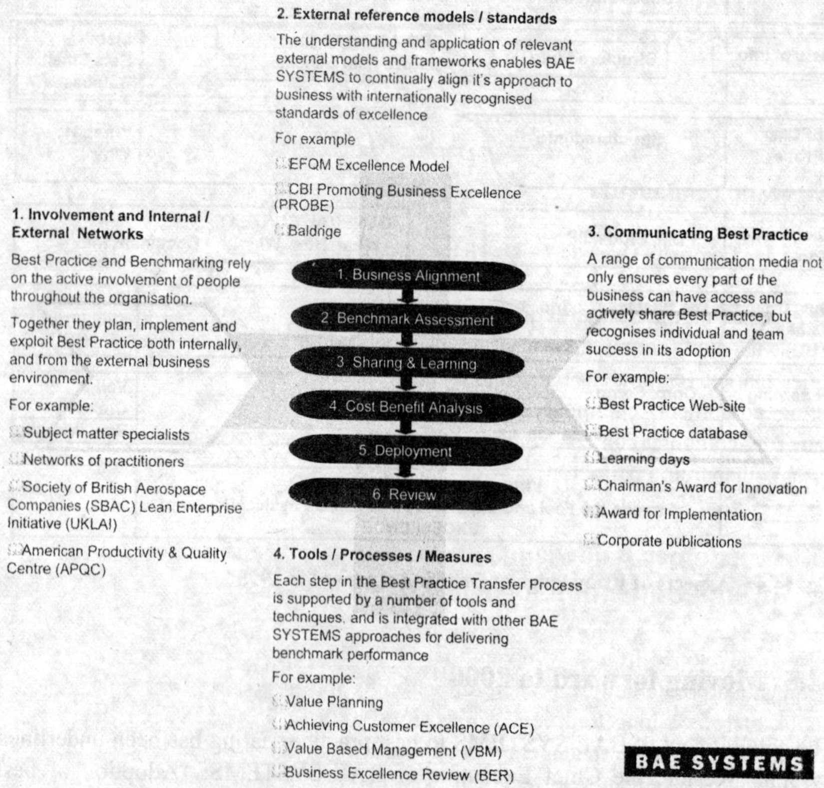

Fig. 14.5: BAE SYSTEMS Benchmarking and Best Practice Framework in 2000

15 Knowledge Management and Customer Orientation Hewlett Packard Austria

Jens Vorbeck, Peter Heisig

Hewlett-Packard Ltd. Austria in Vienna (HP Austria) was founded in 1970. The company is responsible for markets in Austria, Switzerland, Central and Eastern Europe, the Near East, and Africa. In 1997, 444 employees handled a contract volume of € 1.07 billion, mainly in the telecommunications, banking, and gas and oil industries.

HEWLETT® PACKARD

www.hp.com/austria

Industry:	Electronics
Business Process:	Sales & Customer Service
Employees:	420 (HP Austria)
Turnover:	223 Mio ECU (1998)
KM Best Practice:	Culture of Couriousness Customer Voice Internal Knowledge Assessments

Fig. 15.1: Company Overview – Hewlett Packard Austria

HP Austria is a purely marketing company that is divided into eight sections: enterprise account (computers for large customers), computers for HP partners, consumer market (computers for chain stores), medical engineering, measuring and testing equipment, chemical analysis, consulting, and operating business, i.e., managing computer centers – including servicing, repairing, financing, and leasing – as an outsourcing partner.

15.1 Customer Knowledge and Customer Orientation as the Main Motivation for Knowledge Management

HP was developed from a company that originally produced industrial measuring systems. Corporate culture is therefore strongly influenced by measuring. For example, there is a long tradition of measuring many aspects of customer satisfaction. The responsible managers were indeed proud of the traditionally positive comments the company had received from its customers. Then, HP began to notice that competing companies were catching up, and some had even been left HP behind. Research revealed that customer expectations had grown faster than HP itself had developed.

A so-called "10X Program" was initiated to improve product quality 10-fold. Employees noticed that one division had not achieved all the technological improvements planned, but that it still received the best grades by customers. In a worldwide "Reengineering Quality Conference" in 1995, the company reached the logical conclusion from this analysis: The focus on production was replaced by a focus on customers. As one result, the company introduced the program "Quality One On One".

The core of this program is represented by the knowledge of customers. This is made up of three central aspects: a) a deeper understanding of customer requirements, b) the organizational adjustment towards customers, and c) a "passion for customers".

Hewlett-Packard Austria tried to apply this program as profitably as possible. In 1995, the company reached its goal: It was awarded with the HP President's Award for Quality.

The initial situation was characterized by the conviction that a high degree of internal customer orientation was the basis for successfully applying external customer orientation. In internal workshops, individual departments studied neighboring departments as if they were external companies. They analyzed, discussed, and defined customer segments to be observed, products to be offered, and the respective usage of customers. In these programs for quality maturity, departments were allowed to 'hire' an expert from another department who provided his or her knowledge. In regular review meetings the company evaluated the "maturity" of quality. The long-term effect of this approach can be verified by the following anecdote. There was one project that discussed outsourcing hardware repairs. An internal department – against external suppliers, won the public bid for these services. HP Austria attributed the outcome to knowledge management activities and the quality maturity program.

At Hewlett-Packard, external customer feedback has existed for many years. In the beginning of the year 1996 the company put together a so-called "break team" that consisted of six members from six departments. Their job was to analyze existing customer feedback. This "Customer Voice System" will be described in a later chapter as an example of how HP Austria applies knowledge management.

A further example is the SPOC concept (Single Point of Contact), which has existed since 1995. The concept is based on the idea that there is only one entrance to HP. Everything else is then coordinated internally. This is an approach similar to concepts of "one-face-to-the-customer" that are found increasingly in companies worldwide.

Our contact at HP Austria indicated that this approach was very important because it was the basis for project teams that stepped over departmental boundaries to focus consciously on customer-oriented approaches. The responsibilities of these teams included cross-functional tasks, e.g., how to improve the telephone accessibility of staff members. In this case, the team found two solutions: first the system must indicate the presence, not the absence of staff, and secondly the company developed software that allows the central switchboard to connect the call to the correct staff member who signals his or her presence.

15.2 Challenges in the Process of "Understanding Customers and Markets"

The approach to knowledge management at HP Austria is different from the traditional approach. The business process that was analyzed in depth, "Understanding Customers and Markets", focuses on existing activities. Its purpose is to reveal and eliminate deficits in order to promote long-term, autonomous, and accelerated learning in an environment that constantly changes.

The goal is not only to encourage the internal learning of good or better solutions to problems, but to supplement the external product range with services that allow users to manage changes when introducing information technology systems. Customers have criticized the fact that HP supplies excellent soft- and hardware solutions that do not, however, reach their full potential. The reason for this was seen as a lack in attention paid to the so-called "soft factors", especially staff members at customer companies, i.e., the users of the new IT systems.

One barrier in the course of knowledge management was identified in the transfer of knowledge from experts to non-experts. It turned out that experts are not very interested in exchanging or transferring knowledge.
The customer voice system would soon be useless if the organization was not able to react to warning signals: "I've told you twenty times already, and nothing has happened." The organization has to maintain its curiosity.

15.3 Corporate Culture: A Traditional Strength of HP

A lot has been written about the specific features of corporate culture at HP. It is, therefore, not astonishing to read the following statement: "There is no question that HP's corporate culture promotes the development of knowledge. It is quite difficult to implement knowledge management if employees are suddenly confronted with a more or less standardized way of dealing with knowledge. In any case, an open team spirit promotes the development of organizational knowledge, and supports motivation of the staff."[355]

---

[355] Schuller (1998).

The artifacts of HP Austria's culture, i.e., the aspects of corporate culture that are accessible to external analysts can be characterized as open, clear, participative, and communicative. The manager whose desk is located in an open-plan office and who joins all other employees for lunch at the cafeteria is as normal as the communicative atmosphere created in lounges for just this purpose.

These artifacts are obviously compatible to the publicized values of the company. However, HP does point to new ways for the future. For example, with regard to knowledge management, the company confidently assured us that "the pleasure of displaying something new is much greater than the fear of being copied". Such statements are partly due to the promotion of a culture that rewards the reusability of new ideas. This is particularly important if one realizes that the criterion "Innovation" represents a convincing incentive in the specific way the company thinks, and that reusability as such would not be regarded as particularly important if it were not be promoted.

In this context it is noteworthy to mention the so-called "School of Vienna." As the name indicates, this is a specific feature of the Austrian branch. This approach, established in 1997, is based on the question: "How can we develop new business opportunities with all employees"? To allow for "humane change management", HP Austria starts with the premise that people learn slowly, and that corporate activities must always be based on people within the company.

Two examples to examine how HP Austria tried to establish the homogenous culture it desired: 1) When hiring new personnel the recruiters pay particular attention to the integrative capabilities of the applicants. 2) HP developed a particularly exemplary way of treating new employees. To integrate them into the culture described above, they receive detailed information of the company's ideas, purposes, and goals. Among other things, HP shows videos that are prepared for this purpose to familiarize new staff with the processes of the new environment.

A cornerstone of the HP way is the spirit of cooperation that pervades the entire company. This spirit is characterized by a participative management style. This paraphrase of the words of founder David Packard also applies to HP Austria. Even in the stage of defining or strategically aligning the location Austria, management had realized that their ideas could only gain acceptance if they were supported by the employees. The company's hierarchy is correspondingly lean, and there are no barriers if staff persons wish to communicate across departmental or hierarchical levels.

Hewlett-Packard's evaluation system also takes contributions to the internal and external transfer of knowledge into consideration. According to our contact at HP Austria, the most important question is: "What has a staff person done to make his or her knowledge accessible", i.e., to what extent does this person contribute to the transfer of knowledge and information about a customer into the company. For this purpose, HP Austria includes three indicators: the self-assessment of the staff person, the comments of the customer about the staff person, and the evaluation of the superior.

HP does not ascribe specific value to these indicators, but "one of our most important improvements has been to include through this measure the staff's spread of knowledge when calculating incentive payments."

What is true for external customers is obviously also relevant to internal customer orientation. Another aspect of evaluation concerns internal transfer: What has staff recently done to make their knowledge accessible to their teams?"

Cooperation in HP's cross-functional teams is very popular because participation is highly beneficial to one's career. In an internal job search, any evidence of having cooperated in teams increases an individual's chances greatly. In the case of management positions, we might even say that there is a direct relationship between work in cross-functional teams and further career development.

15.4 Customer Voice: If You Want to Know the Customer's Needs – Ask Him and Don't Presume You Already Know the Answer

The term "Customer Voice" refers to an approach at HP Austria to create knowledge about customers. During management of customer responses, all information on the customer is collected systematically, analyzed according to patterns of customer expectations, and transferred into the company while appropriate measures are developed (Fig. 15.2).

This information can reach HP on different paths: as a short note written by a staff person, as a transcript of a telephone message from a customer or a telephone interview with a customer, or directly from the customer through the Internet.

Earlier procedures attempted to reduce customer feedback to central statements and categorize this information according to certain keywords. Today, the collection of information has been separated from its evaluation. The largest sources of information are handwritten and signed notes by staff persons of HP, telephone inquiries, and customer feedback through the Internet and e-mail. 5-7% of all information is collected through letters, 20% through telephone calls, and more than 60% through the handwritten notes mentioned above. Customer surveys through the Internet have the highest growth rates.

About 15-20 staff persons have been trained to analyze incoming comments as to their urgency and the patterns or frequency of certain (sub) processes and information requirements ("frequently asked questions"). Feedback analyzed as urgent is immediately forwarded to the responsible manager. "Customer voice analysts" act not only solely as carriers of knowledge, but also function as a hub of information, as navigators, multipliers, and transmitters of impulses. In contrast to earlier procedures that involved centralized input into customer databases, every department may nowadays add to the database, and may thus contribute to the storage of knowledge about the customer.

Customer Voice : Identify customer knowledge systemically and holistically .

Customer requests and complaints are recorded in full text using plain paper sheets. All information is passed on to customer voice analysts. They categorize the given information and forward long term issues to product development and strategic departments. Short term issues are forwarded to customer support. The separation of information gathering and analysis guarantees unprejudiced judgment and excels most sophisticated pre -categorized schemes.

Fig. 15.2: Best Practice – Customer Voice

JEDER KUNDENKONTAKT – EINE CHANCE

Hast Du Feedback von einem Kunden zu Service & Support erhalten – positiv und/oder negativ –, dann nimm Dir kurz Zeit und teile es uns möglichst wortgetreu mit. Du hilfst uns damit, uns weiterzuentwickeln und zu verbessern!

From: Christina Taborsky

To: Support Feedback

Datum: ______

Bitte kreuze einen der folgenden Bereiche an, auf die sich das Kundenfeedback Deiner Meinung nach bezieht:

Support Bereiche:

- ☐ Anrufannahme
- ☐ Admin. Kundendienstverträge
- ☐ Hardware-Support
- ☐ Software-Support
- ☐ Reparaturzentrum (Bench)
- ☐ Other Services
- ☐ Support Sales & Marketing
- ☐ Other: ______
- ☐ ______

Kundeninformation (optional):

- ○ Name: ______
- ○ Kontakt: ______
- ○ Tel.: ______

Möglichst wortgetreuer Kundenfeedback:

______

HEWLETT® PACKARD

Date ______

To: ______

From: ______

☐ Please approve	☐ Let us discuss
☐ Please handle	☐ For your information
☐ Please comment	☐ For your file
☐ Please call me	☐ As requested
☐ and return	☐ Are you interested?
☐ Please complete/attach:	☐ Please forward to:

Remarks:

HEWLETT-PACKARD Ges. m. b. H.
A-1222 Wien, Liebigasse 1, Postfach 72, Telefon (0222) 25 000-0, Telex 134425

Fig. 15.3: Customer Voice Sheet (Facsimile)

It is very important to strictly separate the collection of data from its analysis. This guarantees that typical distortions and preconceptions are reduced or even eliminated, and that the exact context of the information remains intact. It was often clear that to leave out these two steps leads to distortion. This results in information that includes only data that fits the existing pattern, e.g., regarding customer wishes. This approach shows the importance of the context of data. The "customer voice analyst" retains the context and evaluates the complete information and transforms this info into valuable knowledge.

To spread and apply knowledge, the management teams of the departments meet monthly or quarterly to discuss and act upon internal and external feedback. The connection to medium-term business plans is verified during internal management audits that analyze whether the results of the customer voice program have been linked to the company's business plans.

"Those who listen to the voice of customers on a daily basis, for example, salespeople, but who are too tired at night to write a report, need incentives to document the information and to make it accessible." For this reason, HP Austria handed out simple personalized notepads and developed an Internet/Intranet solution that supports data entry and data retrieval.

Further Approaches to Knowledge Management

In the key account business, HP also analyzed offers in special workshops. The analysis focuses on the twenty most and the twenty least successful offers. Workshops employ the so-called "scenario technique" ("what if ..."). Central aspects include a) the essential requirements of the customer that are hardly ever expressed anymore, b) the price-performance ratio, and c) the attractiveness of Hewlett-Packard. As one result, the company discovered that HP is successful if management techniques are addressed. A further result was the following: "The more integrated HP's approach to the customer is, the faster contract volume grows."

With regard to internal customer feedback, the latest worldwide staff survey revealed that there are deficits in the way internal and external resources are dealt with. A project team of eight pursued these deficits in approximately 40 interviews. The results revealed a lack in up-stream management. Instructions from the headquarters were adopted without critical reflection.

A very interesting approach is the combination of the processes "Understanding Markets and Customers" and human resource management. For example, customer engineering looked for typical clusters of virtues that make a staff person successful. The know-how on the "successful past" of a staff person is then used for personnel recruiting and development purposes.

In close connection to this approach is the attempt to refrain from training in personnel development, and in the return to a focus on "tapping into internal knowledge." For this purpose, HP Austria conducts assessments of internal knowledge to decide which staff person earmarked for development is to be brought into contact with which other staff.

15.5 New Project Management in the Business Field "Professional Services"

Because of its unsatisfactory profit, Hewlett-Packard's business field Professional Services was reorganized. The worldwide hierarchical organization with country-specific orientation was replaced by a committee-oriented structure with logical working teams. In this framework, the company established eight to nine "Fields of Interest" and developed a process for knowledge management called "Structured Intellectual Capital" (SIC).

While developing the SIC approach, an approach that was based to a large degree on the experiences of big consulting firms, HP identified three models suitable for the application of knowledge management: a) learning communities, b) expert communities and c) project snap shots.

- Learning Communities:

Learning communities have the goal to promote the transfer of know-how between "novices" and "experts."

- Expert Communities:

Regular meetings of experts have the goal to identify the "blind spots" of one's expertise, and to initiate the appropriate acquisition of knowledge.

- Project Snap Shots:

In cooperation with the Project Management Institute in the USA, Hewlett-Packard has developed a new project management method. All project managers were trained in this method. The new project management approach is based on the life cycle model. It is divided into six stages: (1) Initiation, (2) Planning & Proposal, (3) Selection, (4) Implementation, (5) Warranty, and (6) Support. It includes 70 to 80 tools. To promote the utilization of existing solutions it is, for example, necessary to check the database about existing offers or offer modules when developing new offers.

The Project Snap Shot is a structured meeting conducted by an external moderator to collect and evaluate all positive and negative experiences. An important responsibility of the moderator is to make sure that the meetings do not focus on typical searches for guilty parties or on mutually apportioning blame. The "lessons learned" are formally entered into the database in the Intranet. Currently, the staff person who enters the data may freely select individual keywords. The company is currently discussing whether to develop a catalog of keywords. The database is accessible to all HP employees.

Project snap shots are developed on the basis of two milestones. The first occurs after the third stage in which the customer concludes the selection of an offer. Here, criteria for success and failure are identified. The customer may be incorporated directly. A second meeting, which is connected to a project meeting, if possible, occurs after the stage of implementation.

In summary, the SIC approach focuses on the development of networks of experts and the availability of knowledge in databases that are accessible through the Intranet.

15.6 An "Evolutionary" Knowledge Management Approach

The case study Hewlett-Packard Austria impressively reveals how comprehensive knowledge management can be connected to customer orientation – how one concept is based on another, and how synergies are created by skillfully networking modern management tools.

Three aspects of this case study are particularly impressive:

The example of HP Austria emphasizes the supporting effects of the design fields, especially of the 'soft' areas of personnel management, management, and corporate culture. The company was able to find an evolutionary way to efficiently produce knowledge management that is appropriate to the characteristics of these design fields. This is evolutionary because the company based its approach on long experience and previous knowledge was systematically reused for further development. HP Austria learned from mistakes and, particularly concerning Customer Voice Analysis, has even taken one step back from complex procedures embedded in categories to simpler but more effective methods.

The example of HP Austria revealed clearly that knowledge management is neither simple nor fast. The costs, e.g., for evaluating the customer voice notes, are hardly insignificant. However, the results seem to justify the expenses.

The example of HP Austria reveals the importance of a unified core process for efficient knowledge management. Not a single activity – creating, storing, spreading, and applying knowledge – is neglected. If we remember how human resource management is incorporated into the business process mentioned above, it seems that HP has even moved one step ahead by connecting several business processes. Perhaps this efficiency can be traced back to the "culture of curiosity" that was quite often mentioned by our contacts.

16 Knowledge Management in a Global Company - IBM Global Services

Jens Vorbeck, Peter Heisig, Andrea Martin, Peter Schütt

IBM is one of the world's leading information technology companies. In 2001, IBM employed 320,000 persons and had revenue of $88,3 billion. IBM Global Services, IBM's business segment responsible for IT services, employs 149,000 persons in 160 countries and had sales of $33,2 billion. In 2001, it was the world's leading provider of information technology services.

IBM®

Industry:	**EDP-Systems, Software and Consulting**
Business Process:	**Produce and deliver services**
Employees:	**307.000**
Sales:	**87.5 Billion USD**
KM Best Practice:	**Intellectual Capital Management Tool** **ShareNet -Conferences** **Category Owner** **Intellectual Capital Seeker**

http://www. ibm.com

Fig. 16.1: Company Overview – IBM Global Services

The business segment Global Services is divided into three units: Business Innovation Services (business strategies and e-business solutions), Integrated Technology Services (design and implementation of infrastructure solutions), and Strategic Outsourcing Services (systems, network and data center outsourcing). The business unit Consulting is assigned to each of these three units. In Germany, the Consulting unit "IBM Unternehmensberatung GmbH" is an autonomous subsidiary of IBM.

This case study describes the experiences and achievements of the unit "Infrastructure & Systems Management" which employs 400 professionals in Germany and 4,000 professionals worldwide. Its responsibilities include carrying out international projects in the systems management and infrastructure area. Another example covers the unit "Automotive Services" which employs 300 professionals in Germany and 3,500 worldwide. They are two examples out of more than 60 communities of practices at IBM.

16.1 Introduction of Knowledge Management

"Competitiveness, profitability, and intellectual leadership" are the keywords to explain the introduction of knowledge management at IBM.

The focus IBM has placed on knowledge management stems from the growing requirements of the company's services. Based on the dynamics of the corporate environment and on the goal to secure and expand competitive advantages, IBM's Consulting Group began to develop a model for knowledge management in October 1994. Under the rubrics "personnel turnover", "mobility", "global customers", and "global competition", IBM raised the questions: "How can we acquire new knowledge? How can we turn local knowledge into globally available knowledge? How can we – in view of staff turnover – preserve knowledge, and utilize knowledge growth?". Because the ability to learn faster than the competition is today's only sustainable competitive advantage (Peter Drucker).

The very first approach only had a lifetime of a few weeks. IBM immediately realized that it is not sufficient to develop databases to store information, experiences, and know-how. "From a technological point of view, the results were successful; however, they were economically disappointing." The assumption that users would come knocking at the door if only technology were made available proved to be wrong.

The second approach based on a holistic framework was much more advanced and became such a success that it is still in use today. The development team consisted of up to fifteen professionals who successfully applied their idea of knowledge management to a Lotus Domino application – this time only to support communication. The idea was called "Intellectual Capital Management (ICM)." "Since 1994, the IBM Consulting Group has employed the ICM effort as part of the company's re-engineering project. The idea of ICM has been to institutionalize and make knowledge management more formal throughout IBM Global Services and Global Industries."[356] Knowledge management at IBM is not an isolated solution. Instead, it is an integral component in a program initiated to optimize customer relations under the title "Customer Relationship Management (CRM)."

From that point on, the development of further activities was based on an international strategy that was adapted to the specific requirements of the individual business segments.

16.2 The Important Dimensions of Knowledge

IBM's knowledge management activities are based on Michael Polanyi's definition of two dimensions of knowledge. Knowledge may be either explicit or tacit,

---

356 Kuan Tsae Huan (1998).

and it may be only available to an individual or to a certain group or even a whole community.

IBM defines explicit knowledge as documented and (ideally) structured knowledge that is fairly easily accessible, and that is available in different media. On the other hand, tacit knowledge exists in the heads of the company's professionals. It includes experiences, ideas, rules of thumb, and tips and tricks. Intellectual capital contains information, explicit and tacit knowledge, and experiences and ideas that are structured in such a way that they can be exchanged and reused to provide the best possible usage for clients (innovation, best practices) and the company (operational effectiveness).

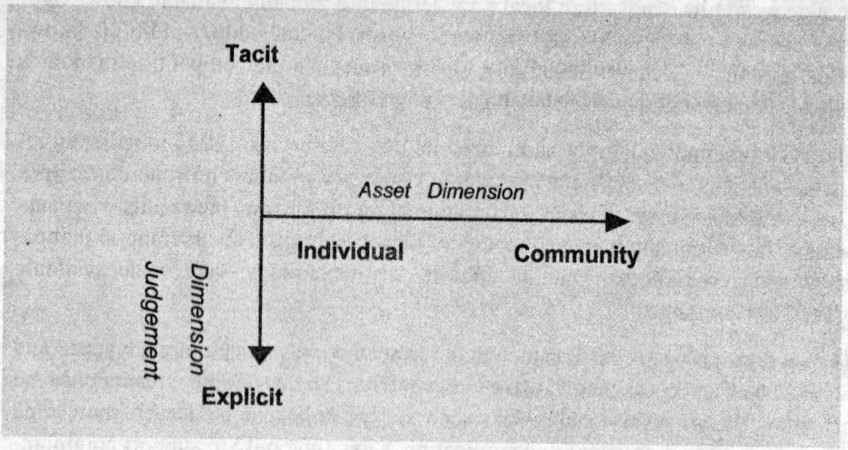

Fig. 16.2: Dimensions of Knowledge

This idea contains two decisive and remarkable aspects. The pragmatic understanding of tacit knowledge that contrasts with an otherwise understanding that refers to tacit knowledge in the sense of mental models and routines, which store knowledge in the human brain. IBM focuses on the management of this pragmatic tacit knowledge. Additionally the two-dimensionality includes possibilities for omni-directional exchanges. IBM's clearly expressed goal is not only the identification of explicit knowledge, but also the retransfer of tacit knowledge into the heads of the professionals through new methods of learning and topics that better relate to daily practice.

The second dimension focuses on the propagation of individual knowledge – tacit and explicit – internally wherever it is relevant to operational performance making it available to a wider audience.

IBM communicates this idea in the following manner: "It is not necessary, realistic or sensible to devote time to defining knowledge. What is important is to achieve an understanding of what it means to use knowledge in contrast to Information. (...) The story I most frequently use to distinguish between knowledge and information is to use the metaphor of a map and a human guide. A map is a set of data organized into a coherent and reusable form – it is information. The guide, on the other hand, is knowledgeable. She does not need to consult a map, takes into account recent experience and has the ability to relate my ability to her knowledge of the terrain. The guide is the fastest way to achieve my objective, provided that I trust her. If I do not have the trust, and am not prepared to take the risk of experimentation, then I will fall back on information – the map. It should also be noted that someone with knowledge of the territory has created the map. If I share the same culture and background as the mapmaker then I am able to use the information. A different background may mean that the map is just data – useless stuff without context."[357]

16.3 IBM's Approach: Intellectual Capital Management

Intellectual capital management (ICM) is a framework of values, processes, people, and technology set up to collect, make available, reuse and further develop knowledge, experiences, and information (Fig. 16.3).

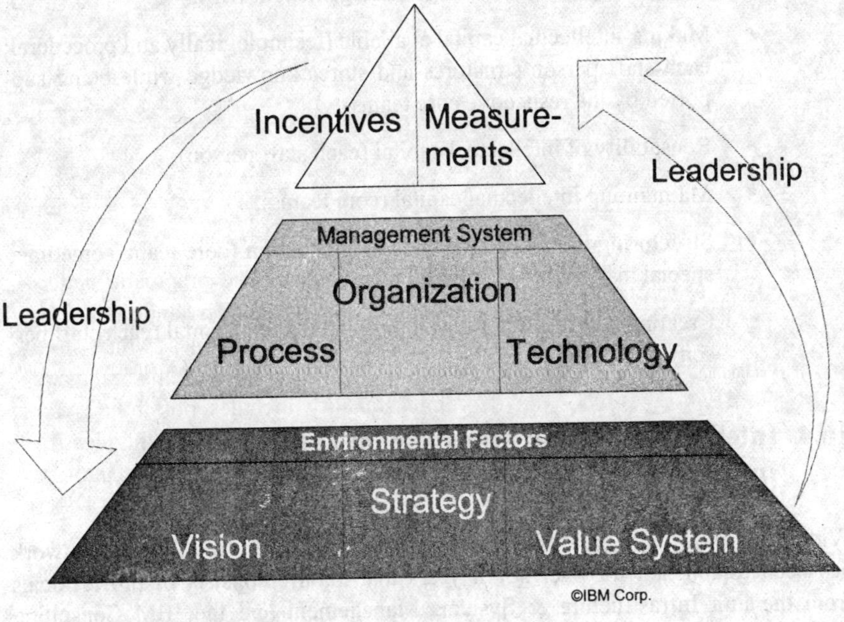

Fig. 16.3: Framework of Intellectual Capital Management

357 Snowden (1998), p. 7.

By using the basis "Vision – Strategy – Values," IBM focuses on homogenizing and adjusting its interests to help customers and do business. Processes fulfill their purpose through consistent standards and methods that support the reusability of knowledge worldwide, and that keep the core process of ICM running. This takes place in an organization that maintains an appropriate balance and relationship between informal and formal networks by using all-inclusive and uniform technology.

„Motivator" may also replace the term „incentives". This aspect is to receive more emphasis. The goal at the top of the pyramid is to motivate staff to actively participate in knowledge management. In this context, the term "incentive", that connotes the allocation of financial assets, would lead in the wrong direction. Finally, it is the managers who have to visibly commit to and promote ICM by implementing active and professional leadership.

IBM's strategy of knowledge management focuses on the following metaphor by Peter Drucker: "In knowledge economy everyone is a volunteer, but we have trained our managers to manage conscripts. ... the real issue is that most of the volunteers still think like conscripts."

The following tasks are part of knowledge management at IBM:

- Making intellectual capital available (technologically and procedural; each staff person structures and stores knowledge while being supported by the respective core teams),
- Reusability of intellectual capital (each staff person),
- Maintaining intellectual capital (core teams),
- Structuring the respective knowledge domain (core team, sometimes special task forces – "solution boards"),
- Creating and/or further developing intellectual capital (each staff person).

16.4 Intellectual Capital Management at the Unit Infrastructure & Systems Management

Within the knowledge area of systems management an informal people network has been established for over seven years that mainly consists of professionals from the unit Infrastructure & Systems Management and the IBM Consulting Group.

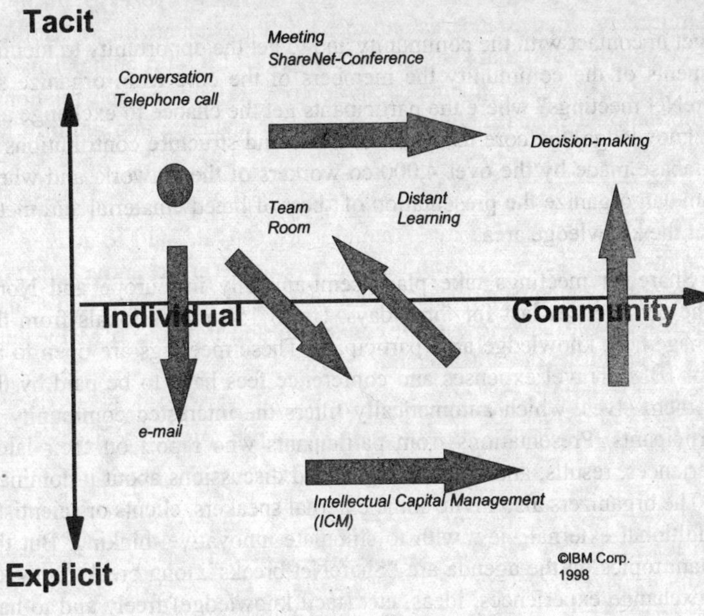

Fig. 16.4: Methods and Tools to Support the Exchange of Knowledge

Actually, it is a cross-organizational network – a so-called Community of Practice – with over 4,000 members worldwide. A Community of Practice Leader together with a core team of about 25 people have form the Community of Practice for systems management. The core team members are working part time for the core team and most of the time as professionals in client engagements.

The manager of the core team is generally allowed to spend 60% of his or her working time on this project. The other core team members dedicate 10% of their time to this job. IBM has not officially regulated how these expenses are accounted for. The core team's job is to:

- Sense and respond to the community's requirements and needs,
- Structure, manage and maintain the knowledge of the community and make it available
- Develop methodologies and harvest knowledge and experiences where these are missing, and
- Communicate to the community about all topics of interest for the Community of Practice.

16.4.1 ShareNet

In order to get in contact with the community and to get the opportunity to identify the requirements of the community the members of the core team organize so-called "ShareNet meetings" where the participants get the chance to exchange and spread tacit knowledge, the core team can evaluate and structure contributions to the ICM database made by the over 4,000 co-workers of the network, and where the core team can organize the presentation of "best of breed" material and methodologies for the knowledge area.

Worldwide ShareNet meetings take place semi-annually in Europe and North America. The conferences last for three days. Up to 150 professionals from the systems management knowledge area participate. These meetings are open to all employees of IBM. Travel expenses and conference fees have to be paid by the participants themselves, which automatically filters the interested community to relevant participants. Presentations from participants who report on their latest project experiences, results, and methodologies and discussions about it dominate the agenda. The organizers also invite some external speakers, clients or scientists, to get an additional external view with to stimulate innovative thinking. But the most important topics on the agenda are "ShareNet breaks", long breaks between sessions to exchange experiences, ideas, etc. (tacit knowledge) freely and to harvest innovative ideas and solutions. The organization of additional national or regional ShareNet meetings takes the cultural peculiarities of a global corporation into account.

A relatively new initiative is the organization of joint ShareNets with related Communities of Practice so that the participants can share and exchange experiences beyond the "borders" of their own Community of Practice with members from a different, but closely related knowledge area.

16.4.2 ICM Tool to Support Knowledge Exchange

The ICM tool, which has received several awards for outstanding knowledge management and commitment to enhance client service supports the movement of the individual tacit knowledge of each of the numerous members of one network to explicit knowledge that is available to all members of the network.

Services in the area of this community of practice follow a self developed "solution life cycle" model that defines the phases "assess, design, deploy and operate" that each product runs through. The ICM database is structured accordingly. It supplies the content of the methods of the "solution life cycle", i.e., specific applications. In particular, the process can be described in the following way:

All members of the network are encouraged to document knowledge (e.g., the description of a project) that they find to be relevant. A "router" is responsible for acknowledging each new entry. He or she must also initially evaluate the data and pass it on to an appropriate reviewer. A reviewer supervises a certain subject area and evaluates the contributions. A subject area may be a phase in the solution life cycle or part of it, which is also represented by a category in the ICM knowledge database.

Evaluation criteria include the following aspects:

- Completeness (Does it contain a summary in English?, Have contacts been identified? Have all relevant documents been attached?),
- Reusability and
- Determination whether the contribution is genuinely new or at least considerably better than any previous information on this topic.

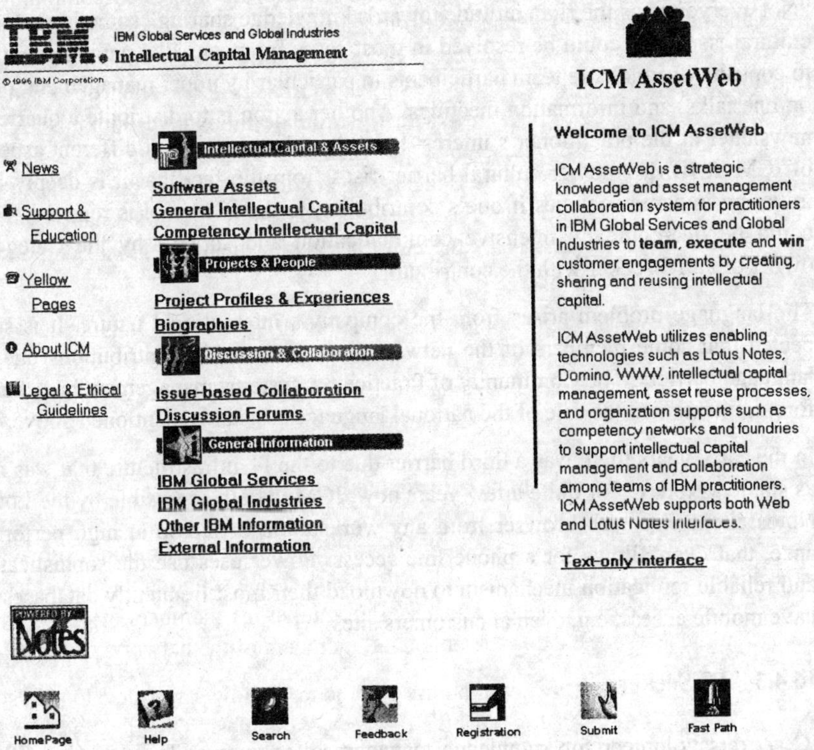

Fig. 16.5: ICM AssetWeb

This means that the reviewer has to remain in touch with the contributor to perfect the document and to achieve the best possible quality. The reviewer can also initiate translations of contributions to this one particular network that are in demand. For each language, the Community of Practice for systems management has identified a national language assistant who gets active when contributions are not in English.

Typically 80% of the contributions are accepted through this method for the knowledge area of systems management. In the meantime, IBM has clearly exceeded the critical minimum amount for such ICM databases. Therefore, it has become very important to remove contributions that are now irrelevant or have now been surpassed by newer and better documents.

The ICM tool is a good example of how IBM – through experience – became conscious of some of the barriers to knowledge management and the reasons for them. Three main classes of barriers have been identified: culture, technology, and language.

"Not everyone has the right attitude towards knowledge sharing" sometimes was a cultural issue that could be resolved in most cases by means like paying attention to contributors and core team participants in particular by upper management, one-on-one talks, and information meetings. Another action is to distribute a quarterly newsletter at the practitioner's interest level that also focuses on different aspects of ICM at IBM. Another cultural barrier rises from the fact that it is deeply humiliating in some societies if one's contribution to the ICM tool is rejected. This could be overcome by intensive communication and support by the Category owner or an IC seeker with the contributor.

The language problem arises from the company's international nature. It is suspected that some members of the network have not sent in contributions due to language barriers. The Community of Practice for systems management has, therefore, implemented the role of the national language assistants mentioned above.

In the early years there was a third barrier due to the IT infrastructure that was not as stable as today. For quite a few years now, ICM is both accessible by the Lotus Notes client or a web browser from any workstation or laptop at high performance, that even allows for a phone line access. Power uses use the sophisticated and reliable replication mechanism to download their most frequently databases to have mobile access, e.g. even at customers sites.

16.4.3 IC Seekers

IC seekers volunteer for knowledge management programs. Their job is to find sources of knowledge and to tap into this potential. They are, so to speak, "scouts for knowledge."

Currently, forty to fifty members of the particular systems management services community regarded do this job on a worldwide basis. These IC seekers attempt to promote the idea of "give and take" when individually talking to participants.

They act as ambassadors for intellectual capital management at IBM. Their achievements are also part of their annual performance evaluation.

To better understand how they work an example: An IC seeker approaches an expert and motivates her or him to share her or his knowledge with colleagues by documenting a project to the ICM tool. Another way would be to tell participants that they could use a certain asset to handle projects more easily. The IC seekers also utilize the ShareNet meetings mentioned above to identify relevant knowledge and the persons related to this knowledge.

16.4.4 Knowledge Cafés

TeamRoom and Qiuickplaces are Lotus applications that promote the cooperation of real and virtual (i.e., sometimes international) teams. In the framework of intellectual capital management, these applications facilitate the transfer of individual tacit knowledge into explicit public knowledge (Fig. 16.4).

16.4.5 Methodology Repository

Within the phases of the solution life cycle for systems management several methodologies are in use. These methodologies are a fundamental part of most of the client engagements that IBM performs in this area of services.

IBM has undertaken a major effort some years ago to harmonize exiting methodologies from different areas of the services business. Today the modules of the unified method labeled the "IBM Global Services Method" are accessible in a Lotus Domino database both with the Notes client and via the IBM intranet.

The IBM Global Services Method is the single repository for best practices approaches that are the nucleus for any business or IT solution provided to IBM's customers. It defines a common language and common mechanisms and procedures to the professionals in order to enable them to reuse and create knowledge assets in a consistent, integrated and best practices way. To protect the IBM Global Service Method from sub-optimal use it is password-protected and the passwords are granted only after training.

The particular systems management related part in the Global Services Methodology is managed by the corresponding knowledge network. They know about the requirements and needs of the community. This is one aspect of the "sense and respond" task that the core teams have. For example, if the community requires a standardized approach for a certain phase of the solution life cycle this may lead to a development project for a new methodology. The development project is ideally performed with client involvement so that the new piece of methodology may be developed and verified in a real world environment before it is included in the overall IBM Global Service Method.

16.4.6 Communications

Another very important aspect of knowledge management within IBM, which has been neglected for a long time, is the bi-directional communication between the core teams and their communities.

Within the knowledge area of systems management several communication means are in place:

- An intranet site for fast-aging news in the systems management area plus a download site with important whitepapers, presentations, etc.
- An internet site with information also accessible for clients and with a download area that holds documents that may be presented at clients.
- A quarterly newsletter (PDF softcopy available for download from the intranet and html for immediate reading) with changing focus areas in each edition that not only spans the systems management knowledge area but also the related knowledge area of networking and connectivity.
- A subscriber service via E-Mail (every 4-6 weeks or when required) offering the latest news in the systems management area, recently published intellectual capital and other news in the ICM area, training and event announcements, hints & tips from successfully completed client engagements, customer reference stories, etc.
- Conference calls with associated online presentations using Lotus Sametime e-meetings about the most recent developments in the systems management area where the listeners also have the chance for questions and answers.
- Newsgroups within the ICM framework where a question and answer section provides the opportunity for the community to fill information gaps.

The benefits of these communication media are verified through the feedback of the community – either online or personally at events such as a ShareNet conference.

In this way all areas of knowledge management at IBM – the ICM tool, ShareNet conferences, methodologies, communication, etc. – are interconnected.

16.5 Intellectual Capital Management at the Unit Automotive Services

The unit Automotive Services has experienced that "the wheel was reinvented several times" while developing software. Because software development requires a lot of preparatory work – due to the necessary compatibility with existing systems –, and because the team of developers is distributed all over the world, the unit has decided to give the introduction of knowledge management priority over the introduction of the re-use of software solutions.

The unit's approach to knowledge management is very similar to the approach taken by the unit Systems Management Services. For example, the TeamRoom is only one tool among many others. The ShareNet is called Solution Board, and is responsible for developing innovative topics and leadership in this field. When topics have been thought through sufficiently, they are published through the ICM process.

A specialty is that the ICM tool is tracked with a tracking tool. This enables the administrators to track the number of accesses, the duration, and number of documents that were read. The tool may even call the attention of users who have not accessed the system for a long time to the possibilities of its use. Incorporating the council into the core team solved objections by members of the works council.

The motivation to participate in knowledge management is based on financial, cultural, and social grounds. IBM has initiated a program that awards active members with up to $5,000. The company has also tried to create a certain composure in its active staff persons that stimulates cooperation. Particularly the IC seekers are able to motivate their co-workers during one-on-one talks. The goal is to escape the necessity of "survival of the fittest", and to experience a colleague who manages knowledge as a friend.

16.6 Achievements

The unit Automotive Services estimated that careful evaluations revealed that each document stored in the ICM tool leads to a time reduction of 30 minutes in the preparation of a new project. An Austrian marketing manager spoke of the invaluable benefits of the ICM tool concerning quality: "I've felt much more competent after preparing a presentation with the Knowledge Network [ICM tool]".

In the knowledge area of systems management the development of the methodologies – especially when they provide templates that are ideal for reuse in any client engagement – has saved a lot of time in client engagements and therefore can create a lot of additional profit when appropriately priced. One methodology can claim that 80% of its contents and templates are ready for reuse in every client engagement and that only 20% have to be adapted to the specific client situation.

Additionally the ICM tool now allows the indication of the time saved through the reuse of a certain piece of intellectual capital. Thus you can document and report how much time has been saved – and the results are amazing: Several sample deliverables saved much more than a week when working on a client engagement. In fact, even the average time that is saved when re-using a piece of intellectual capital is 5 days.

However, it is probably even more important to bear in mind the costs that would occur if knowledge management would not be carried out than to calculate precisely the benefits of the program!

17 Open Minded Corporate Culture and Management Supports the Sharing of External and Internal Knowledge – Phonak

Peter Heisig, Christian Berg, Peter Drtina

Phonak, the world's third largest developer and manufacturer of hearing technology, was founded in 1964 by the engineer Beda Diethelm and the executive Andy Rihs. In its first years, the company copied hearing aids from other manufacturers, while improving the product's quality. However, Phonak soon realized that hearing problems must be treated interdisciplinary. The cooperation of professionals in microelectronics, micromechanics, medicine, psychoacoustics, psychology, physics, and audiology enabled the company to set landmarks on the road to "optimal hearing solutions." Due to the lack of suitable manufacturing equipment, Phonak also developed the appropriate production technology while it produced hearing aids. Nowadays, this development is still reflected in the company's interdisciplinary mix of research, development, and production.

PHONAK

http:// www.phonak.ch

Industry:	**Medical devices**
Business Process:	**Research & Development**
Employees:	**2122 (31.3.2001)**
Sales:	**460 Sfr (1.4.2000 – 31.3.2001)**
KM Best Practice:	**External knowledge sharing Corporate Culture & Management**

Fig. 17.1: Company Overview – Phonak

Today, Phonak is a company with more than 2,100 employees and a global marketing and R&D network. Its main markets are the USA and Europe. The company's sales and profits have risen significantly during the past few years. In 2000, the number of employees has been increased dramatically mainly due to the acquisition of the Canadian hearing instruments manufacturer Unitron which brought 650 employees to the Phonak Group. By internal growth Phonak created an additional 121 jobs. The additional manufacturing and R&D resources in North America now allows the firm to further increase its development effort as well as the output capacity.

Apart from the traditional area of hearing technology, Phonak's communication technology has also gained increasing market shares in security services (surveillance and protection of persons), sports (communication between athletes and coaches), and media (studio technology).

17.1 Knowledge as Part of the Corporate Philosophy

It is hardly possible to speak of an explicit introduction of knowledge management at Phonak. The consciousness of the importance of knowledge and its management is part of its traditional corporate philosophy, and not due to an explicit decision to introduce and apply measures to maintain the intellectual capital of the company. A member of Phonak's management described knowledge as "the tool to deliberately make decisions". The acquisition of knowledge and the best possible communication and cooperation among staff members has always been a key activity. This activity is also reflected in the three following factors of corporate success:

- attracting knowledge creates know-how,
- open minds create superior products, and
- people value creates shareholder value.

The philosophy of the founders also characterizes the current organization. A maximum amount of enthusiasm for research and the motivation to create something new is supported by the best possible environment for all staff members.

17.2 Internal and External Cooperation to Create, Acquire and Share the Best Know-How

Knowledge management at Phonak is carried out consciously, but not explicitly. This conclusion, drawn by one of the top managers, reflects the approach quite well. The following guidelines apply to this idea:

- Interdisciplinary knowledge of hearing is the basis for developing the best possible hearing technology.
- If part of the necessary knowledge is not available internally, it has to be acquired externally. This guideline resulted in close cooperation with several colleges and research institutes. "If we don't know it, perhaps someone else does."
- Knowledge can and should be shared with experts (even if they develop competing products). Sharing knowledge is not regarded as dangerous; it is beneficial to all if an entire field advances. Only in the stage of implementation can we identify differences between suppliers.

When developing a new product, each project initially addresses the following questions:

- In which area do we identify the highest potential for innovation?
- Which resources do we need?
- Which resources are available internally?
- Which resources have to be acquired externally?

An idea that originated at Phonak may therefore be tested and further developed in cooperation with college physicists and engineers. If the result is promising, Phonak produces the solution. Tests of the new product then again occur in cooperation with external institutions, e.g., hospitals, audiologists, psychologists, and customers. Phonak now designs evaluation and diagnostics tools in close cooperation with external partners and clients. This close relation with customers is reflected in the high amount of time spend by R&D staff with clients inside or outside the company.

In this long process, Phonak plays the role of an interface between research and production, and between production and application. The most important goal of these partnerships is always to develop a common language that everybody understands and that results in the best possible solutions. This form of cooperation is profitable for all participants because they can jointly make advances in a certain field of study while they simultaneously pursue individual interests.

17.3 Corporate Culture to Foster Cooperation and Communication

Phonak's corporate culture begins with architecture. The headquarters in Stäfa reflects a philosophy of transparency, openness, and motion. Bright and open offices, isolated stairs, and few doors create an inviting and communicative atmosphere. Communication barriers are avoided wherever possible. The architecture clearly reflects the product of the company: hearing aids, or the intention to enable or improve communication.

Phonak's development from a pioneering company to a medium-sized corporation plays an important specific role. The close cooperation between staff members and appreciation for each individual even now provides Phonak with an internal and external image characterized as "more friendly, more human".

The knowledge that every staff person may ask anyone else questions is another remnant of the philosophy of the founders. This is why hierarchies are as lean as possible. Even managers are easily accessible.

In this sense, the bottom line is primarily understood as a result of successful internal and external cooperation – and only secondarily as a goal.

According to management people at Phonak, the corporate philosophy is best described with the so-called “Silicon Valley Spirit”. This describes a certain subculture of many high-tech companies that is characterized by a highly innovative atmosphere. The company deliberately creates spaces and stimulation for thought, promotes creative and open internal and external discussions, supplies various opportunities to make suggestions, and tries and applies good ideas non-bureaucratically.

“A lot is talked about, and little is written down.” Communication occurs directly and easily at that location where the exchange of knowledge is required. Written documents are reduced to the absolute minimum (“as many as necessary, and as few as possible”). However, rapid growth in the number of employees has put the company in a position in which it can no longer uphold this guideline. The functional “get principle” now has to be supported by the purposeful distribution of information. Therefore, Phonak plans to prepare more documentation to guarantee the successful exchange of information in the future.

The internal openness desired requires that staff members know how to deal with freedom. To get in touch with others actively, to be able to work in teams, and to be able to resolve disputes are the basis of successful projects. Staff members that cannot adjust to this style are – if necessary – asked to leave the company.

Phonak actively maintains a “culture of errors”. The company has realized that errors are an essential factor in the process of learning in research and development. Errors are evaluated in the sense of “lessons learned”, e.g., during debriefings (see below), and are then incorporated into future developments. “Errors can always be made, but please, not twice.”

17.4 Management Actions to Support the Exchange and Generation of Knowledge

The managerial structures also reflect that knowledge management at Phonak is conducted consciously, but not explicitly.

This begins with cooperation in project teams. Project teams correspond to the traditional form of cooperation. They are the result of new ideas, and are reorganized according to current requirements. Constantly changing contacts promote the exchange of knowledge and supply impulses for new ideas.

Phonak pays particular attention to the employment of staff members in individual projects. The guiding question is: “Who should be placed where and why?” Initially, staff members are obviously selected according to their qualifications. However, the managers also pay attention to whether the teams really consist of a new group of people. This is done to avoid “in-groups.“ Management also knows that Phonak’s success during the past years may easily lead to laziness, and to “resting on what has proven to be reliable”. Changing teams are to avoid this. For example, sometimes young staff members with completely different experiences are introduced into a team just to "create some movement".

Furthermore, the management pays attention to whether individual staff members are sole carriers of knowledge for a certain field. The company attempts to set up "know-how cells" of four to five staff members that are jointly responsible for a certain area of know-how. This reduces the danger of losing an entire field of knowledge if one individual leaves the firm.

The composition of committees is also planned carefully. New ideas are discussed and applied without unnecessary amounts of opposition.

Schmitz and Zucker[358] positioned the different measures at Phonak to support the generation and use of internal or external knowledge in the following Knowledge Quadrant. We have added the "Debriefing" and the "Off-Shore-Meetings into their scheme, because the Knowledge Quadrant reflects the essential aspects of knowledge management at Phonak well.

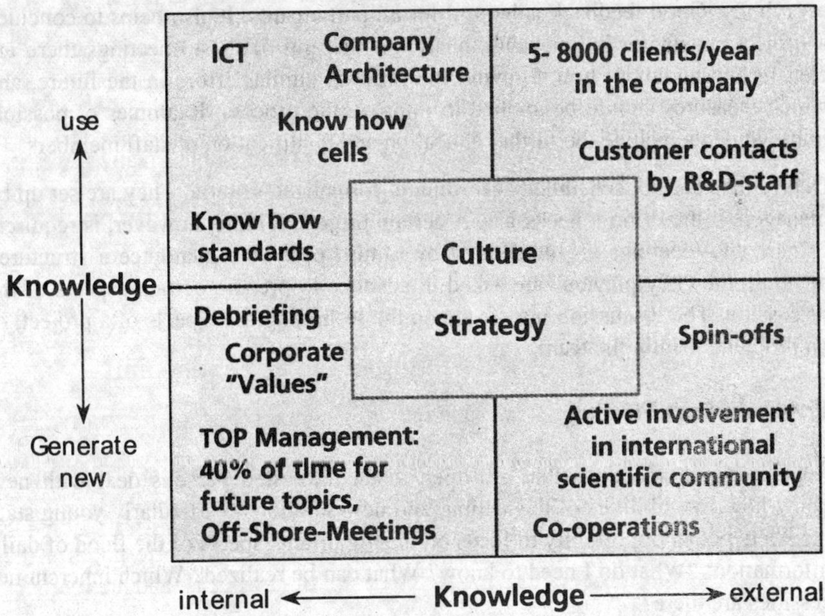

Fig. 17.2: Knowledge Quadrant of Phonak[359]

358 Schmitz, Zucker (1996).

359 Ibid.

17.4.1 Off-Shore Meetings

Once or twice a year, as required, Phonak conducts „Off-Shore Meetings." A certain number of staff members (three to six) are selected to spend several days at another location to discuss new topics and develop ideas. This is easier if the participants are away from their daily job environment. It is easier to come up with new ideas if inherent necessities do not obstruct an individual's viewpoint and thoughts. This is the best possible space to come up with innovative ideas.

17.4.2 Debriefings

Debriefings are an important instrument to integrate identified errors into the continuous process of learning. Phonak has conducted these meetings regularly for ten years. The participants of the workshops evaluate in detail which aspects of a product are good and which should be improved, and why. A clear rule is: "Nobody is guilty!" An individual experience always results in a lesson learned for everybody. Good feedback is helpful for all participants. It also helps to conclude a project satisfactorily and consciously. At the end of these meetings there are clear proposals about how to avoid the same or similar errors in the future, and which measures should be applied to improve the process. Examples of possible improvements include the further education and qualification of staff members.

Debriefings are not scheduled according to formalized criteria. They are set up by managers if they find it necessary. A certain project volume, however, is required. In part, the meetings are moderated by a third party. To guarantee a structured meeting, the "key players" are asked in advance to prepare certain aspects of the discussion. The discussion may focus on the technological aspects of a project, or on problems within the team.

17.4.3 Expert Meetings

Regular expert meetings that are open to all interested persons deal with new know-how, possibilities of the setting, and new solutions. Particularly young staff members get an opportunity to focus on the important aspects of the flood of daily information: "What do I need to know? What can be realized? Which inherent necessities are there?"

17.5 Summary and Outlook

Phonak is an interesting example of a company that carriers out knowledge management activities at different points of the organization without stating it explicitly.

The key element of this approach is the company's traditional consciousness of the importance of exchanging and handling knowledge. This consciousness has implicitly led to an organizational design that promotes an intelligent management of knowledge. The company clearly focuses on promoting opportunities for direct communication, and on intensive partnerships with external experts and institutions. Furthermore this example underlines the importance of management actions to improve the use of knowledge. It shows the advantages small and medium-sized companies could gain; if their founders and managers act as promoters and actively model the supportive roles for knowledge management.

However, it is also obvious that the company is currently – due to its rapid increase in the number of staff members – undergoing radical changes. Measures that previously guaranteed sufficient flow of information and know how have now reached their limits. In the future, it will be necessary to develop a more explicit and more structured approach to knowledge management in order to channel information and know how sensibly.

KM in January 2003 (Annex to the 2nd Edition)

In January 2003 the German economical magazine "Brandeins" contained an article about the corporate culture at Phonak. In terms of knowledge management the author summarized Phonak's approach as follows:[360]

"Knowledge management à la Phonak:

1. The habitat: open architecture

   The form of the principal office uses a lot of glass and light to arrange the asset of the enterprise: its employees. Open-plan offices and optical artifice make sure that no department works isolated. Each employee at any position stands in the center of the enterprise.

2. The organization: flat hierarchies

   Management by objectives: There are functions and objectives, but hardly any instructions. There is an American, polite culture of addressing each other informally. Senior management also works in the open-plan offices. They enjoy no privileges, neither in the canteen nor in meetings. Authority ought to emerge from knowledge.

3. The communication: regular team impulses

   Interdisciplinary teams are formed according to tasks. There are lounges and conference corners at the centerlines of the building. All employees have the same break times, so that the entire staff mingles.

4. The knowledge transfer: narrations

   At regular introductory events, social events and institutionalized professional trainings in the departments the orators change constantly. Before leaving the enterprise pivotal employees pass their knowledge on to all relevant successors in informal chats.

5. Cohesion: cultured culture of dispute

   Probably the most important element: Everything is disputed in every stead at equal eye level. That way no team member dissociates himself from the projects and the professional knowledge becomes more profound. Ideas and innovations pass crucial tests within the enterprise before hitting the market. In case of undissolved conflicts it's the boss who makes the decision."

---

[360] C.f. Grauel (2003), p. 101.

18 Sharing Process Knowledge in Production Environments – Roche Diagnostics – Laboratory Systems

Jens Vorbeck, Ina Finke

The division Roche Diagnostics was created in 1998 after F. Hoffmann-La Roche Ltd. took over the Boehringer Mannheim group. Roche Diagnostics GmbH is divided into four business units: Molecular Biochemicals, Molecular Systems, Laboratory Systems and Patient Care.

http:// www.roche.de

Industry:	**Chemicals and Pharmaceuticals**
Business Process:	**Produce and Deliver Products**
Employees:	**2.700 (Diagnostics -Laboratory Systems)**
Sales:	
KM Best Practice:	**Secure and distribute tacit knowledge** **Team specific culture**

Fig. 18.1: Company Overview – Roche Diagnostics – Laboratory Systems

The main manufacturing centers of the business area are located in the German cities of Mannheim and Penzberg. At Penzberg, the company develops – among other products – product lines for the business area Centralized Diagnostics. Especially immunological reagent components pass through a "scale-up" procedure that works closely with resident researchers. Components are later transferred to routine production sites. In Mannheim, Roche Diagnostics predominantly manufactures products for clinical-chemical analysis. Roche Diagnostics GmbH employs 6,500 people in Germany, from which 2,700 work in the business area Centralized Diagnostics.

18.1 Knowledge Management and Organizational Development: Two Complementing Efforts

The following approach to a new manufacturing concept – connected to team structures developed in the framework of the fractal factory – was initiated in 1995 by the business area Centralized Diagnostics of Roche Diagnostics GmbH at the Mannheim and Penzberg locations.

To meet the increasing demands of markets and customers, the business area Centralized Diagnostics implemented a new and integrated manufacturing concept. The three factors technology, organization, and human resources were to be synchronized optimally to increase productivity on a long-term basis.

Comprehensive organizational development was based on the concept of the fractal factory. The goal of development was to create autonomous working units with clear goals and services.

The specific goals of the new manufacturing concept for the business area Centralized Diagnostics were:

- produce quality products,
- produce cost-effectively and dynamically,
- establish simple and manageable structures and work flow, and
- use flexible technology and new organizational forms.

Units of fractal factories are characterized by a high degree of self-organization and self-optimization. Team structures are developed and established in as many areas of production as possible. The teams enjoy independence, pronounced autonomy, and flexibility. This places great demands on the employees' qualifications and learning capabilities. To encourage and support employees, the business area Centralized Diagnostics initiated training sessions and team-specific CIP groups (Continuous Improvement Process).

Managers perform supervisory and strategic functions while they increasingly delegate responsibilities to teams. The encouragement of team autonomy results in teams that adopt typical management responsibilities. Managers therefore predominantly create and promote space to maneuver so that the staff members are best able to develop their creative abilities. Managers have to visibly support and live by the idea of teams.

Fig. 18.2 illustrates the integrated manufacturing concept and its guiding principle. The optimization of structures and processes also results in adequate spatial developments that integrate new technologies and team structures in the best possible way.

The attempt to create a 'vital' organizational structure that meets the requirements of permanent change connotes the realization of a learning organization.

At the beginning of the redesign of the manufacturing organization, the managers got a general idea of desirable standards in a goal setting workshop. While the product quality was to remain high, manufacturing costs and throughput times were to be reduced. Increasing integration of staff members was to result simultaneously in an increase in flexibility.

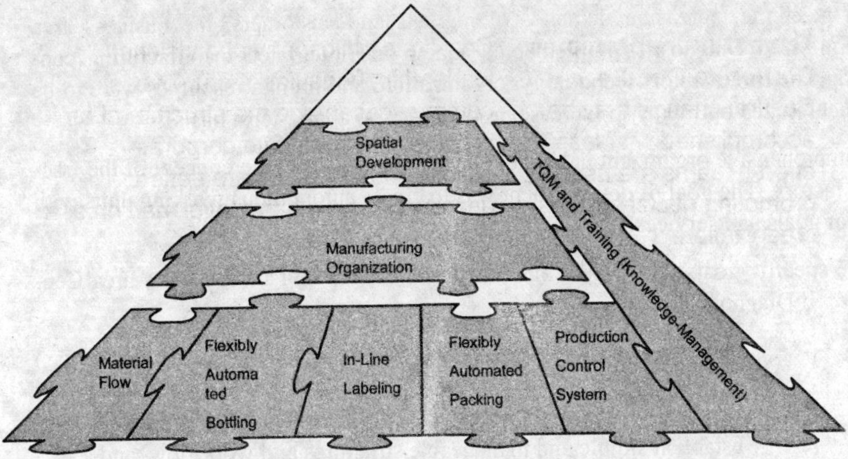

Fig. 18.2: New Manufacturing Concept (Source: Roche Centralized Diagnostics)

Implementation initially occurred in pilot fractals. In the meantime, the concept has been integrated into the daily business operations of almost all manufacturing areas.

18.2 Team Building and Team Empowerment as a Basis for Cultural Change

In summary, Centralized Diagnostics has implemented knowledge management on the basis of an integrated manufacturing concept that is characterized to a great extent by new team structures.

The employees cooperate in manageable units with a high degree of self-management. On the level of teams, the traditional design of knowledge management fields are thus expressed by the fact that the teams manage themselves, and that they autonomously plan and control process sequences. The teams are responsible for sequence planning, operations scheduling, materials purchasing and disposal, documentation, etc.

The business area Centralized Diagnostics placed great emphasis on open communication and the mutual assistance of staff members. The business area regularly conducts team meetings to exchange information. The members are explicitly asked to become involved and participate in the meetings.

A further aspect is the transparency of processes. The motto, "We want a transparent company", is promoted and lived. This becomes particularly apparent in the spatial layout of the firm: Manufacturing areas and administrative offices are accessible and observable by everybody. Diagrams of processes and the continuous process of improvement visualize knowledge.

Team Culture promotes Corporate Culture:
Cultural change may be accomplished with less effort if teams are used as promoting nuclei for the organization.

Corporate philosophy and visions, leadership, behavior of employees, communication and the structure of an organization constitute corporate culture. These issues are better approached on a team level than on a company-wide level.

Fig. 18.3: Best Practice – Team Culture

The business area Centralized Diagnostics revived the original functions of management. The company primarily focuses on transferring responsibilities to employees, i.e., on cooperative management, the promotion of independence, and the creation of space to maneuver and make decisions. This requires a high degree of social competence.

The manager has to be prepared to pass on expertise, and thus to withdraw from the strictly operative level in order to increasingly act in consulting and coaching functions. The decisive aspect of this category is that managers have to live by team orientation consistently and convincingly.

To get rid of old fears, such as the loss of power through the surrender of responsibility, the division conducts training sessions on how to lead teams successfully.

18.3 It is not the Technology that Allows Access to Data but a Trusting Culture

Knowledge has to guide activities. For this to happen staff members need information on context and general conditions. At Centralized Diagnostics, all employees have access to all information on products, customers, goals, and index numbers (e.g., budgeting and performance data) that is relevant to their team. The data is accessible through a PC network. The knowledge and experience profiles of staff members are predominantly paper-based.

Corresponding to the consistent team structures, Centralized Diagnostics also organizes knowledge management in a self-organizing manner.

A single person performs the necessary conceptual coordinating function. This person holds a staff position that is also responsible for additional project management activities.

However, the decisive "promoters" and "multipliers" are the managers in their function as a coach and team manager. The uncoupling of traditional management tasks mentioned above creates room for managing knowledge in teams.

18.4 A Company Specific Conceptual Framework for the Realization of Knowledge Management

Knowledge management at Centralized Diagnostics focuses on the development of the knowledge and experiences of staff members in production areas. The process illustrated in the following figure is an example of how core activities are put into action.

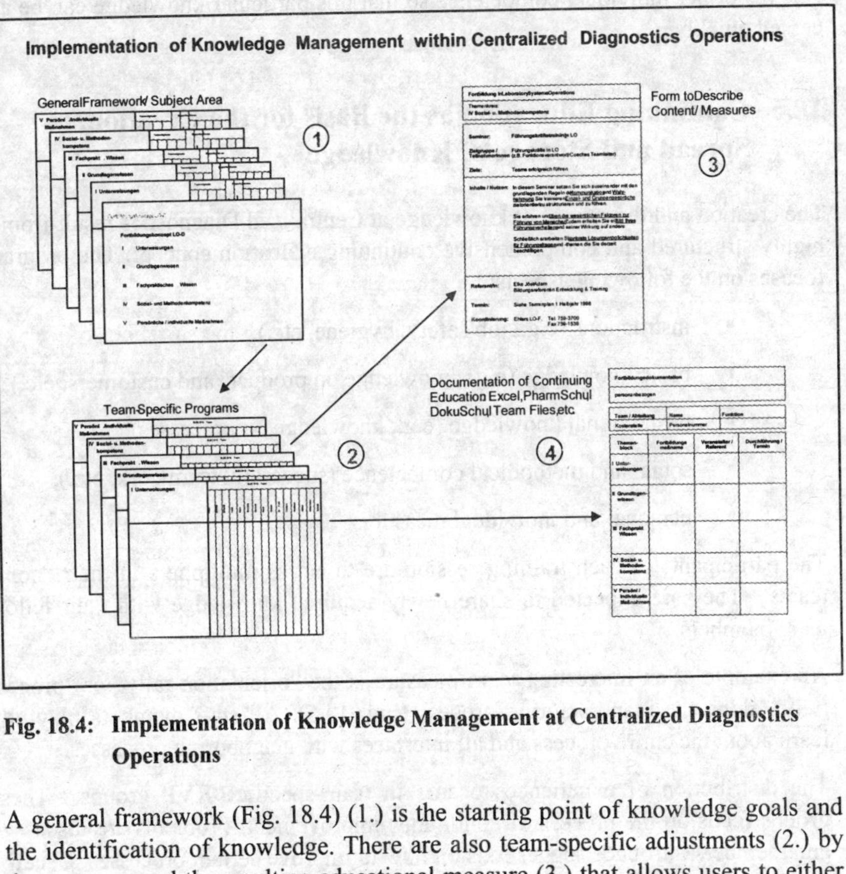

Fig. 18.4: Implementation of Knowledge Management at Centralized Diagnostics Operations

A general framework (Fig. 18.4) (1.) is the starting point of knowledge goals and the identification of knowledge. There are also team-specific adjustments (2.) by the manager and the resulting educational measure (3.) that allows users to either store or spread knowledge (4.). This process – and others as well – are described in the following chapters in detail.

18.5 Informal Interviews for the Identification of Knowledge and the Setting of Knowledge Goals

To improve the knowledge of existing staff potential, the business area Centralized Diagnostics approved a dissertation – submitted for a diploma – on "Knowledge Stock and Development Perspectives of Staff Members." During informal interviews, many employees on various hierarchical levels related their professional knowledge and their ideas of additional development perspectives, e.g., concerning multiple qualifications. Managers were asked about their opinions and wishes regarding opportunities for the development of individual staff members.

In teams, specific knowledge is visualized in tables. Each team member can indicate his or her individual competence so that this particular knowledge can be accessed quickly.

18.6 Continuing Education as the Basis for the Creation, Spread and Storage of Knowledge

The creation and the spread of knowledge at Centralized Diagnostics result from a highly structured and comprehensive continuing education concept. The program focuses on the following aspects:

- instructions (e.g., job safety, hygiene, etc.),
- basic knowledge (e.g., knowledge on products and customers, etc.),
- professional knowledge (e.g., knowledge on processes),
- social and methodical competence (successful teamwork, etc.),
- personal and individual measures.

The participants of each training session are to act as multipliers in their "home teams". They are expected to share newly acquired knowledge with their fellow team members.

An example of an interesting team measure is the "orientation rally" or "process rally" along the manufacturing process (Fig. 18.5). All staff members playfully learn about the entire process and all interfaces with neighboring processes.

The distribution of experiences occurs in team-specific "KVP groups." These groups focus on the process of continuous improvement. Problems are discussed and members introduce suggestions on how to improve certain processes. Usually, the groups nominate one member who is responsible for implementing ideas. The groups meet every two weeks.

Process Rally: A playful attempt to communicate process orientation on all levels of an organization.

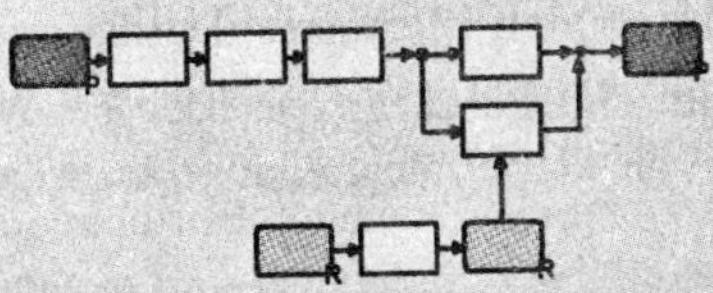

The process is visualized. This visualization is used to analyze the process for critical stations (e.g. interfaces). The process is mapped in a simulation game where all employees have to solve tasks along the process chain. The process rally takes approximately half a day and is fit for 10-30 "players".

Fig. 18.5: Best Practice – Process Rally

The various teams are connected by a "round table." Here, members from different teams meet to discuss professional problems. At the moment, teams meet once a month.

In addition, the company sets up bulletin boards to inform staff members about plans, on-going projects, and results.

The teams document their knowledge of staff competence and staff training in databases. Each team member can access the database.

Due to the rigid requirements of product and process documentation in the pharmaceutical industry, process parameters are recorded regularly. Interview partners in manufacturing areas use the "remark" field of the software to document their process-specific know-how. These "in-process remarks" enable staff members to turn their tacit knowledge into explicit knowledge that is then applicable by others. Professional know-how is thus accessible to all team members. This results in optimal manufacturing processes.

The central idea, however, is to store knowledge in the heads of the staff members and to illustrate to other staff members (team members) to which extent which knowledge is hidden in which brain.

At the beginning of 1999, the business area Centralized Diagnostics introduce a team file that contains guidelines, goals, team-specific information, working time models, products, customers, specific measures, etc. Especially newcomers can use this file as an initial source of information.

The rigid documentation requirements force staff members to follow process records carefully. This guarantees that stored know-how is spread and applied.

The various methods to create and spread knowledge allow all team members to access knowledge easily. Knowledge can, therefore, be applied quickly.

18.7 Roche's Experience with Knowledge Management at the Production Level

Long-term restructuring towards teamwork is now being carried out based on firm convictions and comparatively good experiences with the pilot project. Measures to make knowledge clear and manageable on all levels are being intensified.

The success of the first teams in Mannheim and Penzberg has been so overwhelming that the company may now move on to the integration of knowledge management into daily business operations.

Potential has been found in the following areas: content-based team networks, incorporation of new members into the team, and the re-qualification of managers for the new requirements.

The most important aspect is – according to our contacts – the re-orientation expected from managers.

Apart from many details and minor initiatives, there are four aspects that are remarkable:

The unity of the core process of knowledge management: Focusing knowledge management activities at the level of teams was one of the reasons why Centralized Diagnostics was able to conclude the core process successfully. In this context, it was noteworthy that the company took particularly important aspects of knowledge goals and the identification of knowledge into account.

Many companies are challenged by the fields "Culture", "HRM" (Human Resource Management), and "Management." The example of Centralized Diagnostics shows that these fields can be designed at the level of teams by providing appropriate general conditions. However, top managers must be committed to providing these conditions. At Centralized Diagnostics, this was the case.

As part of our second remark, we would like to point out how people at Centralized Diagnostics were willing to hand over responsibilities and to take staff problems seriously. For example, each team member is able to moderate and initiate the CIP process of his or her scope of work. The trust and confidence implied here results in highly motivated employees who are willing to take over responsibility.

The centralized coordination of all knowledge management activities in the form of the production management of a department attracts attention with creative and unusual ideas on how to create new knowledge or how to influence design fields in the teams. These ideas include the "process rally" mentioned above, outdoor events to promote team spirit, and other activities.

19 KnowledgeSharing@MED – Enabling Knowledge Sharing by Turning Knowledge into Business

Manuela Müller

19.1 Corporate Context

Siemens Medical Solutions is one of the world's largest and most diversified suppliers of innovative products, services and fully integrated solutions to the health-care industry. The division offers products, solutions and services for integrated health care. Its offerings extend from innovative technologies for fast diagnosis to services that optimize processes and increase efficiency in hospitals and doctor's offices. The Group has some 29.000 employees worldwide and in the fiscal year 2001/2002 has posted sales of approx. 7,6 billion EUR.

This case study demonstrates MED's knowledge management approach, the overall concept, the implementation of dedicated solutions along the value chain and the integration of these solutions within the processes and within daily routine.

Innovation and process improvement are the driving factor for Medical Solutions' continued growth in sales and earnings. Siemens Medical increases the efficiency of health care by combining innovative medical engineering with information and communications technology. Today, two thirds of the products on the market are less than three years old, and the division applies for a new patent almost every day. Growth in information technology for health care applications is in the double-digit range.

In mass production, material flows sequentially from one process to the next. Operations aiming at supporting knowledge processes are more complex. The ubiquity of the KM solutions – the fact that anyone can link with anyone else – makes it potentially possible for a contributor to choose knowledge in order to introduce new products and services. As a platform for innovation, knowledge is the way to produce and distribute new combinations of products, services or solution modules. Companies could become magnets by introducing new business models that take advantage of the interactive capabilities of their employees. And clearly, innovation will enforce competition if companies choose to create new value through global collaboration by providing something that was previously furnished by someone else.

Siemens Medical Solutions continues to secure short-term success and long-term viability. A particular knowledge management objective, in support of whichever strategy the division pursues, is to leverage the best knowledge available to make Siemens Medical acts as effectively as possible. More and more, Siemens Medical Solutions manages its knowledge systematically in order to increase its competitive capabilities – although many knowledge management activities and functions may be implicit in each employee's and department's daily work and practice.

19.2 Strategy and Approach of Knowledge Management at Siemens Medical Solutions

KnowledgeSharing@MED was founded in October 2000 as a global initiative being an integral part of an effective strategy that will help to generate, capture, share, disseminate and apply knowledge and information that is relevant to the organization's mission. In the first months, the team worked on the basic framework, which was aligned, to the company's strategy. Based on this, several fundamental solutions have been developed and integrated after a pilot phase in the organization.

The organizational structure of the KnowledgeSharing@MED team was characterized by a very flat hierarchy, a project-oriented management and a milestone-based roadmap with the superior goal of an early integration of future knowledge-intensive solutions in the operating processes. Apart from the solutions, the idea was to develop a corporate culture of knowledge sharing within networks – in real time from person to person, in order to find faster and better solutions for the customers. The basic assumption was that knowledge created by experts all over the world should be made available for global reuse. The solutions have to be supported by an user-friendly, web-based infrastructure that facilitates capturing and finding the required knowledge that fosters global joint learning and collaboration – not only in theory, but on a daily basis, every single day at work – across all MED divisions and boarders, at all times and MED-wide.

Building best-in-class knowledge sharing capabilities required a holistic and MED-wide approach. Siemens Medical has taken a number of concrete steps to encourage cooperation and knowledge sharing. A powerful concept has been developed in an extensive benchmarking effort of leading knowledge management companies before areas of interventions have been identified. This concept has been applied and tailored to MED's specific needs. It includes the integration of existing initiatives in the overall approach.

Within MED, a successful knowledge management approach consists of 6 pillars:

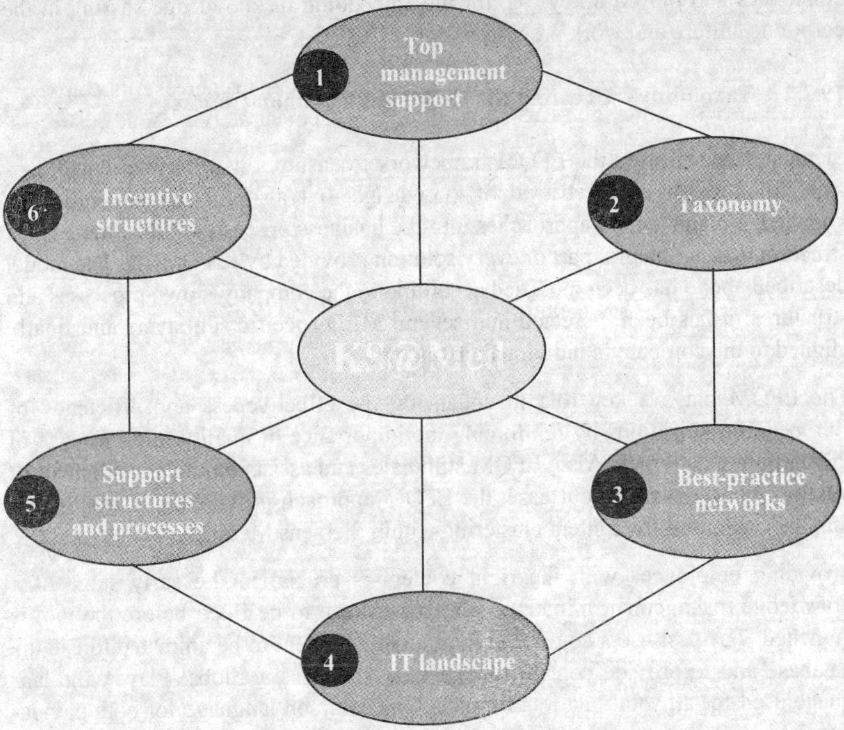

Fig. 19.1: Six-Pillar Approach of KnowledgeSharing@MED

19.2.1 Strong Top Management Integration and Support

A strong and visible top management support, demonstrating the high priority of the topic and the leverage to overcome roadblocks, was needed to emphasize transparency and strong support for this topic. Management helped to communicate the idea of the initiative across organizational levels and functional departments to ensure its value added was understood and applied. Practice Area Leaders have been nominated for all main topic areas and for each process of the value chain. Practice Area Leaders for Marketing and Sales have been appointed in the first place. They have been responsible of controlling the KS topics in their respective area.

Management at all levels had to be won and involved in various meetings or by individual convincement. Later on, a structured reporting process was introduced for top executives in order to motivate management to maintain the support for the initiative. The core team discussed the initiative's progress, success stories, problems, open issues and priority changes in detail with the management. Once convinced, the management acted as an additional communication channel for the employees and played a key role for the embedding of knowledge sharing in the corporate culture and work processes.

19.2.2 Taxonomy – Creating the Right and Common Context

Siemens Medical used the EFQM framework to ensure a workflow-oriented content structure for easy retrieval of knowledge, but also for the integration of knowledge management approaches into the business processes. Thus, the EFQM structure was an integral part of every solution provided by KS@MED. The model described above has been modified according to the company's own processes and structures; it has been dissected into several MED-specific sub-layers and finally aligned to the company's individual taxonomy.

The EFQM plays a key role in enhancing the effectiveness and efficiency of European organizations by reinforcing the importance of quality in all aspects of the corporate activities. Also, EFQM stimulates and assists on the development of quality improvement. In our case, the EFQM approach provided a clear structure that has been used for content clustering within Siemens Medical.

Providing employees with the right content is crucial in the early stage of a knowledge management initiative – databases have to be filled before the tool is launched. The first users could find high quality content to be animated to use the database and to provide content themselves. Acting as a Global Player, the language used for all solutions was English. One common language for both navigation and content allowed to reduce the complexity of knowledge management. Many barriers occurred during the first implementation phase, but the level of frustration dropped once users discovered that sharing crucial knowledge about a customer in English was easy and valuable. Where the level of frustration reached a high level, the Global Editing Team helped to translate their papers and demands. Further, a specially designed incentive system was introduced to eliminate obstacles and barriers. Also, it provided valuable incentives, for example special language courses (e.g. English) that made knowledge sharing and creation attractive.

19.2.3 Best Practice Networks – Working in Multiple Collaborative Environments

The complexity of the twenty-first century world has speeded up the pace of evolution in healthcare. Processes of learning, adapting, and changing will dictate the company's success. The integration of Best Practice networks in decentralized structures make those companies smarter, faster, more innovative, and more agile. This is why Best Practice networks are often called the living engine of each knowledge management solution. They focus on users' needs and high impact initiatives. Before Best Practice networks were integrated, the team attended to human needs and dynamics. Investigations were made on how people exchange information and communicate with one another in everyday company life. Mutual trust, care, identity and the link to core processes of the value chain play an important role when it comes to sharing knowledge and skills within a CoP (Community of Practice). CoPs aim at linking employees not only regarding their work issues but also socially and emotionally. Members developed a feeling of responsibility for the entirety of the Knowledge Community. Within CoP's, knowledge was transferred from one member to the whole community and ideas were openly communicated. Trust is a basic prerequisite for any exchange of knowledge. Selected employees have served as pilots, and online communication suites for these pilots have been implemented as a first step.

19.2.4 Creating a Flexible and User-Friendly IT Landscape

Knowledge management is not about technology. Excellent technology alone is not sufficient, since knowledge management is a people business. It is probably important to stress this issue because we are dazzled by technical solutions. IT-solutions have to be integrated in a structure that recognizes individual contribution. Thus, within knowledge management solutions, IT technology is considered as a medium of transport rather than a solution in itself. Experts share knowledge if their relationship is part of a knowledge enriching culture.

At Med, a customized and flexible, web-based IT architecture facilitated access and provided intuitive navigation. The users' feedbacks were integrated immediately into the solutions where it seemed reasonable. Engaging people in the debate about the compromises necessary in order to find practical solutions was a key success factor at an early stage of the solutions' implementation. Additionally, easy access to the intranet was ensured, single user profiles and the use of meta directories have been established, existing applications have been made available to other business units and local companies. MED′s staff had the opportunity of using all of the provided tools on a voluntary basis.

19.2.5 Designing Support Structures and Processes

Training/familiarization programs have been a very important element of the overall process. These programs have been intensified during the local implementation of the solutions. The process of knowledge generation, storage and utilization, as well as the way in which this process is nowadays embedded in the current workflow, have been defined. Criteria for evaluation of content have been set up and updated on a regular basis. This includes also a complex and far-reaching process for quality assurance. In this context, a global editing team ensured high-quality content before the documents were published in the database. At the same time, the solution allows the user of a certain document to rate the quality and reusability of this document from his own perspective.

Part of this pillar was the decision to manage this initiative by a central core team that incorporated clear roles and responsibilities to push the initiative forward. The core team centered the solutions according to Med's strategy and integrated them in core processes by evaluating what types of knowledge fostered the solution strategy. They tried to integrate several solutions in the core processes, so that knowledge management became an integral part of the daily workflow. One example is the foundation of the Global Market Research network, part of the Community of Practice solution. This network enhances the core process marketing by exchanging information and leveraging market and customer knowledge (e.g. market trends for specific regions) to all decision makers so that they could respond in a more flexible way to the changes of the market place. The core team clearly focused on the aspect in quickly bringing up a solution before they tried to introduce the whole range of possible applications.

19.2.6 Integrating Incentive Structures

A reward and recognition concept for the initiative has been developed in order to accelerate adoption and overcome initial barriers. The concept was aligned with the existing HR performance processes and systems where it makes sense. The incentives mainly fostered individual knowledge exchange and transnational collaboration. In order to handle the time issue and to overcome reservations against KS@MED, the system incentivizes givers and takers of high-quality knowledge. The contributors of knowledge are rewarded by collecting points depending on the quality of the respective input. The points can be converted into non-monetary rewards. A key success factor in implementing this concept was that all facets have been well orchestrated and balanced at the same time. Only then can the benefits of knowledge sharing be captured. Converting information into knowledge is a dynamic process, and the needs of receivers as well as the sources available to them are constantly changing. Consequently, optimization is a persistent process and this task was solved faster and the decision-making turned out to be better with each interaction.

19.3 Turning Ideas into Practice – The KnowledgeSharing@MED Solutions

It has been elaborated above that Siemens Medical's activities for knowledge management are driven by its goals and its strategy. The strategy is the rationale for a set of operating activities, and as such, the goal-setting and strategic planning processes are the central element of the operation. The company selected some solutions to be introduced globally. The solutions were positioned as a true value-adding factor that helps to solve relevant problems in employee's daily work. With the creation of knowledge management solutions, a business driven portfolio of the initiative has been implanted globally, which has the potential to show short-term benefits to the organization. The solutions considered for a global roll-out within Siemens Medical were:

19.3.1 Knowledge Square

The Knowledge Square distributes the expertise of an expert to all MED-employees. Concerning the evaluation of all entries, the Knowledge Sharing Team consulted experts from various groups. As compared to the Intranet, the Knowledge Square provides employees with current know-how, which is organized according to topics, linked to the value chains and across organizational boundaries. This allows for the employees to dispose of the source quickly and efficiently in order to use their working time more efficiently. Once optimized methodologies, processes and structures for corporate business are developed and standardized, they can easily and with little adaptation be reused in future business dependencies across the globe. The same applies to market and evaluation know-how and strategic approaches in similarly developed markets. It gives the opportunity to pass on the know-how to other employees and allows to leverage the local knowledge globally.

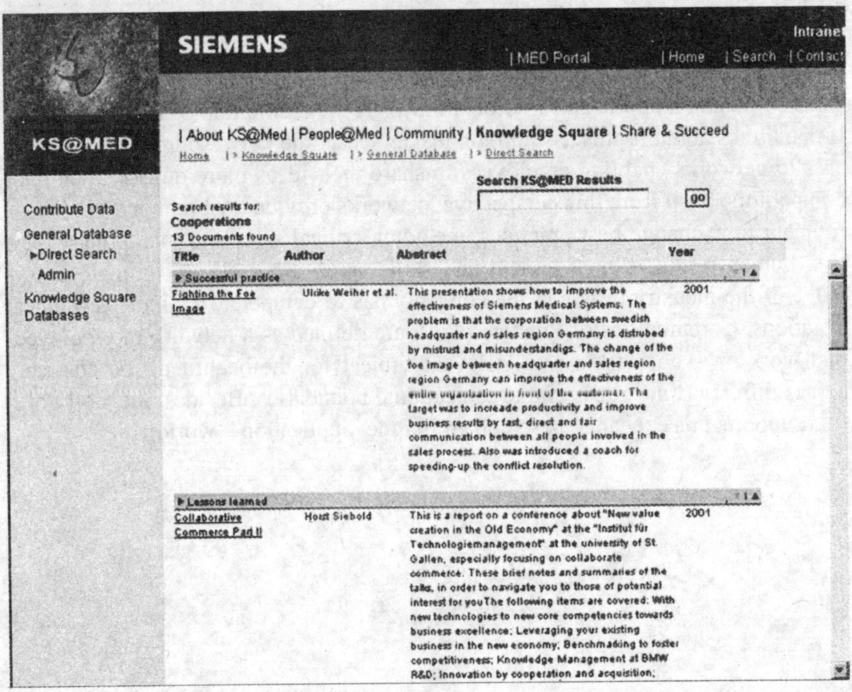

Fig. 19.2: Screenshot Knowledge Square

19.3.2 People@MED

People@MED is a network developed to foster faster networking between employees within Siemens Medical. Employees insert their knowledge profiles, their tasks, their function and personal data that can be called up at any point in time by other employees worldwide. In total, the network provides instant access to experts and their contact data.

Since the purpose behind this solution was to foster knowledge exchange through expertise location and an easy-to-use means of accessing skills and experience across the company, it was important to ensure its keeping up with the company's universe of users. When a MED-employee is looking for a subject matter expert, e.g. on competitors in the US market for clinical information systems, or if he is trying to locate someone who can answer product-related questions, People@MED lists all knowledge profiles and current scopes of responsibilities of experts as entered by them. A transparent process ensures the availability of up-to-date information – People@MED is linked to several yearly performance talks that were held with every employee of Siemens Medical.

19.3.3 Communities of Practice

It was discovered, however, that the principal value-adding activity underlying the solution-business-approach for Global Player is the facilitation of network access. If one thinks about it, this assumption makes sense: the key resource needed is access to networks enabling employees to share their ideas more quickly than their competitors. Seen from this perspective, networks provide a vibrant and vital environment to exchange best practices, including critical sales and marketing expertise, human resource management and skills, and, above all, an intimate knowledge of the industry in which the company has to compete. At Siemens Medical Solutions, Communities of Practice (CoP) are defined as a network of employees that work on a business-driven, relevant subject for the benefit of the company across different time zones and organizational boundaries. Besides, the CoP Solution supports this exchange based on an Intranet application – worldwide.

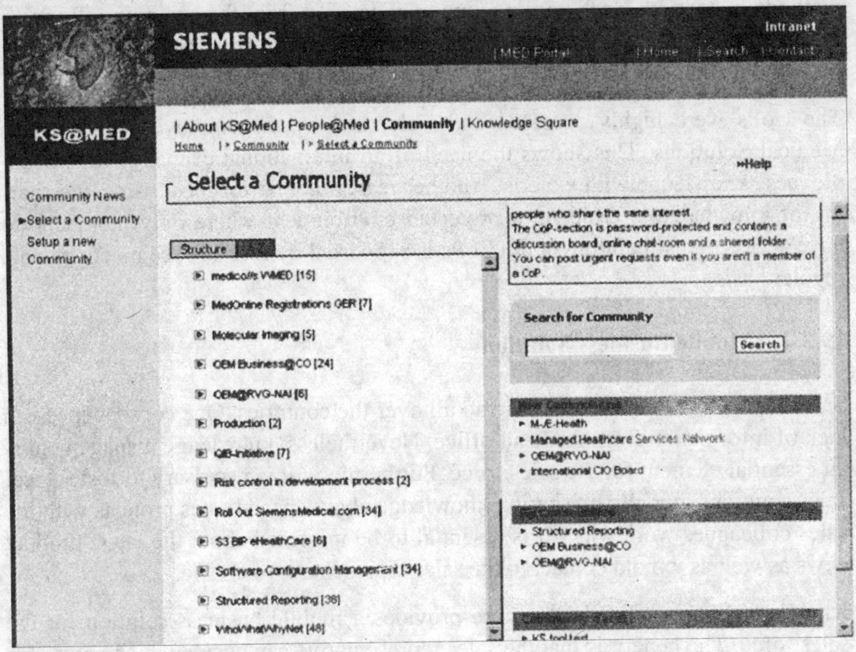

Fig. 19.3: Screenshot of Communities of Practice within MED

But what good are processes of learning without a re-use of the experiences resulting from them? Defining team goals, roles and responsibilities were the most prominent tasks of many successful CoP's. Having a common view of a project's intended outcome made it easier for team members to focus on the goals, establish a common vocabulary and maximize the amount of effort spent on project activities. Without a sense of clarity, it is difficult for team members to know exactly what they were trying to accomplish and how they could evaluate their progress along the way. As a member of a CoP each employee is familiarized with subject matter from experts in other groups who are involved in the same or similar projects.

The employees also had access to several partners for sharing questions and know-how that enabled them to handle the daily workflow faster and more effectively with regard to costs. CoP's are designed with a password-protected area on the Knowledge Sharing application that can be accessed by members only. This solution provides the possibility to exchange knowledge in discussion boards and chat rooms as well as to use a common calendar and a shared document folder where knowledge is shared in one place. The members can also store knowledge according to different topics in a common document archive in order to make it available to other CoP members.

Prior to the launching of such a network, each CoP owner has the opportunity to attend a number of one-on-one training sessions and be informed about tools and methods in order to work with maximum efficiency from the beginning. Running through potential problems during that phase helped the employees to prepare for potential difficulties and prevent breakdowns in team performance.

The CoPs were highly accepted and welcomed by the employees of Siemens Medical Solutions. This shows the need of an international exchange of the employees' knowledge with experts. Another reason is that for most users, the sharing of knowledge is easier in a protected environment where only members do have access. Trust building in the CoPs is supported by face-to-face meetings on a regular basis.

19.3.4 Mobile Business Solution

Sales representatives are often spread all over the countries. One consequence is a lack of information from the head office. Nevertheless, knowledge management is an essential element for the sales force. Furthermore, it is necessary to foster local sales solutions globally by sharing knowledge about special sales projects with the sales colleagues worldwide. It is essential to be informed about the latest product news as well as to hold contact to the sales representatives on site.

KnowledgeSharing@MED therefore provides a mobile business solution for the sales force. The benefit is that the sales representative can decide which news s/he is interested in and download it from a local server onto her/his organizer. The marketing specialists usually provide the data of the different divisions of Siemens Medical Solutions. Normally the data are also generated out of a sales system. The sales system contains information about products, competitive news, status of the delivery of sold products and the contact data of the different product specialists in the head quarter. With the mobile solution of KnowledgeSharing@MED the sales representatives have fast and easy access to the necessary information without having to search for it in different systems.

19.3.5 Balanced Scorecards (BSC)

The Balanced Scorecards is a solution that helps to translate vision and strategy into measurable actions. It enables the executive management not just to focus on capital allocation, but also to emphasize non-financial, critical success factors. Therefore a BSC system reverts to internal data sources and visualizes them in a graphical overview. The data displayed in the solution are defined by the so-called Scorecards. Scorecards cover the following areas: financial aspects, customer loyalty, business processes and the learning of the organization. They are determined by the business strategy of the organization. Starting with the goals the organization wants to achieve, it has to be determined what the company has to change to be successful with its strategy. By knowing what should be changed, it is possible to determine how to measure long-term the development of the change by the so-called Balanced Scorecards.

While tangible values (e.g. travel or training costs) could be easily captured short-term, the hidden value of tacit knowledge flows or the future project values were extremely difficult to calculate. None of the models for measuring the impact of knowledge management were regarded as an optimal stimulus to operate with. It was recognized by the core team, that the successful implementation of such an initiative would require a critical monitoring of the global rollout for each individual country or business unit. It was the main intention to concentrate short-term on measuring the acceptance and the use of all solutions in each individual country instead of showing the ROI for knowledge management. An overview indicated the status of the initiative in each country. During the implementation phase these early indicators helped to concentrate on countries with poor use and exchange of knowledge. First success stories that met a business need from countries with excellent progress were handled in an open way and communicated to the whole community. The result: more than 8000 users from over 65 countries are today an active part of this global network, exchanging ideas or using best practices and proven project insights for repeatable businesses.

19.4 Globalize Local Knowledge and Facilitating Change

19.4.1 Knowledge Sharing Managers Network as Multipliers

Striking the right balance between global and local implementation continues to be a challenge for today's organizations. One the one hand, many successes had been reported in order to control or supervise KM-related activities from the headquarters. One the other hand, many decentralized organizations have shared great success stories from grassroots efforts to promote knowledge sharing and re-use in their local companies only. Siemens Medical Solutions combined both tactics and triggered the global Rollout strategy for more than 40 countries worldwide. New roles and responsibilities have been defined. The Knowledge Sharing Managers ensured the rollout and support for the initiative in the respective countries chosen in the first phase of the rollout. They act in their new role as multiplier and facilitators for all activities and solutions in the local cultural environment and share roll-out-related information with the core team members. They also support local contributors in capturing knowledge and experiences and enforce the development of reusable knowledge. Without the network of the Knowledge Sharing Manager, the rollout of KnowledgeSharing@MED would not have been possible. The Knowledge Sharing managers were trained and prepared for their task. The backbone of such an initiative is the user, which is contributor at the same time. As the main driving force behind the initiative, the Knowledge Sharing managers supported the idea of leveraging local innovations on a global scale. Once selected by their top management for this task, they have actively participated in setting the strategic goals and plans for the initiative itself. They had to show results and have been involved from the first beginning in the development of the solutions. The core team integrated their feedback into the solutions as far as possible. The feedback from the Knowledge Sharing managers helped the core team to reduce barriers and to implement change management measures. Concerns and conflicts have been openly discussed in larger teams in order to elaborate alternatives. Decision making regarding project cancellations or new programs were based on a common agreement. Generally, solutions had to meet a business need. This added enormously to the motivation of Knowledge Sharing managers. In fact they had the impression that the solutions agreed upon were developed by themselves.

19.4.2 Brand Building for the Initiative

Putting it clearly, the real benefits of knowledge sharing played an important role in the rollout of KnowledgeSharing@MED. In bigger companies, a lot of databases, content management systems and intranet applications already exist. The core team developed a marketing strategy that pointed out the differences of these applications and the benefits for the user. Establishing a good reputation for the initiative was as important as offering individual support for users. Most of the used tools were not actually self-explaining, even if the solution was as user-friendly as possible. Therefore, individual hotline support was necessary in order to help the users and to animate them to provide knowledge.

For long time, acceptance success stories of the solution have been marketed within the organization. The team found out what benefits the users gained and marketed them in all internal communication media in the organization.

19.4.3 European Award 'Solution of the Year 2002'

Through these solutions, Siemens Medical is developing into a learning and fast changing organization – an organization that shares Best Practices internally and absorbs ideas from outside by regularly benchmarking its activities against those of its strongest competitors.

Siemens Medical's success in the area of knowledge management has been recognized by several knowledge management assessments and benchmark activities around the world. In the end of September 2002, KnowledgeSharing@MED registered more than 7.500 active users from more than 65 countries.

In September 2002, the company received the European Award 'Solution of the Year 2002' from the AIIM (Association for Information and Image Management). The award was designed for organizations that are particularly successful in the area of document and process management services, innovations or knowledge management. The initiative KnowledgeSharing@MED was chosen as Best Practice solution not only because of the highly valuable contribution to the company's success, but also for being able to improve and foster all the core processes in the organization.

Part IV

KM – Made in Europe

20 Building Communities

Organizational Knowledge Management within the European Commission's Information Society Technologies Programme

Paul Hearn[361], *Agnes Bradier*[362], *Anne Jubert*[363]

20.1 Introduction

"Knowledge management" means different things to different people. At its conception, the juxtaposition of the terms "knowledge" and "management" contrasted, among other things, with the more common concept, at the time, of "information management". In this way, it was intended to clarify that there is an indeed distinction between "knowledge" and "information". This distinction is still today the subject of much debate in international conferences: *What is the difference between knowledge and information? Can knowledge be managed?* Within the European Commission's IST Research Programme "knowledge management" has more to do with the radical transformation of organizations than with the desire to "get the right information to the right people at the right time". Such information or library science approaches to the management of knowledge – though much quoted and analyzed in the early days of knowledge management, and still, unfortunately, to some extent today – represent only a partial, very limited – one might argue reductionist – view on the challenge of managing knowledge within enterprises.

The management of information benefited from insights from, for example, artificial intelligence, case and model-based reasoning and other "systems" approaches. The argument was that such information filtering or data analysis tools, when working in conjunction with tools providing some kind or contextual and/or location-based information, would lead to improved results and efficiency. The information overload that many employees suffer is perhaps the clearest evidence that

361 European Commission, DG Information Society, B-1049 Brussels, Paul.Hearn@cec.eu.int

362 European Commission, DG Information Society, B-1049 Brussels, Agnes.Bradier@cec.eu.int

363 European Commission, DG Information Society, B-1049 Brussels, Anne.Jubert@cec.eu.int

such tools are indeed necessary. The information revolution has allowed us to access more and more information, but for many at the cost of increased stress in managing information overload.

Within IST, our research in knowledge management attempts to take the different approach of focusing on knowledge as a "social construction process" or activity rather than as a machine-processable product as such. In this setting, knowledge management is seen as "a movement, which places people, and exchange between people, in the center of the picture". Seen in this way, we believe that knowledge management offers profound lessons to organizations. Those organizations that see the opportunity and need for organizational and cultural change in the knowledge economy can, using the learnings of knowledge management, implement radical transformations of their organizations. Those willing to put people and their abilities and competencies first have the most to gain from the knowledge "revolution" which is upon us.

20.2 The History of Knowledge Management Research at the European Commission

The continued dedicated effort in the area of knowledge management – first under the Esprit Research Programme "Learning and Training in Industry" theme (1994-1998),[364] and later under action line II.1.2 Knowledge Management in Framework Programme V (1998-2002)[365] – has led to the establishment of an European movement which we refer to as "Knowledge Management Made in Europe". The critical mass of researchers, academics, businesses and knowledge workers has been reconfigured, restructured and consolidated into a coherent framework to address current and future research needs. Many research and industrial players have been invited to provide input to the European research agenda in the knowledge management area.

We realized early on that most innovation arises from the ability to shift perspective. This is why in the management of our research portfolio we put equal emphasis on tracking developments within our own research projects as in bringing in the outside world. We see clear opportunities to take IST research in knowledge management on to higher levels of innovation through constant external challenge and critique.

Some 40 projects have been funded under the "Knowledge Management" banner in Framework Programme V. These projects can be classified into 3 main groups:

---

364 Cf. LTI. IT for Learning and Training in Industry, http://www.cordis.lu/esprit/src/ltihome.htm, cited 12-4-2002.

365 Cf. IST. Knowledge Management for eCommerce & eWork, http://www.cordis.lu/ist/ka2/al12.htm, cited 12-4-2002, cited 12-4-2002.

First Generation KM (1998)

- Information portals – tools and methodologies integrating to a large or lesser extent information necessary for back and front office processes in organizations. A reflection of early movements by information systems suppliers into knowledge management, such first generation knowledge management approaches suffered from a lack of a holistic framework or people or community-centered approach. Since 1999, we have been moving away from this arena towards a more holistic treatment of primarily tacit knowledge in organizations.

Second Generation KM (1999-2000)

- Knowledge processes to business processes – tools and methodologies linking knowledge and business processes.
- Assessment or measurement-type projects – which attempt to measure and benchmark knowledge management implementation within and between organizations, and to manage and measure impact of knowledge life cycles within the enterprise.
- Collaboration and innovation spaces – tools, methodologies and good practices that accelerate creative exchange between people working within and across organizations. The end objective of such projects is to support the transition of organizations into knowledge-based communities.

Third Generation KM (2001-2002)

- Knowledge and innovation ecologies – tools, methodologies and good practices which identify contextual barriers and enablers of absorptive and innovative capacities of organizations and attempt to replicate co-creation abilities across the enterprise or network.
- Human-centered knowledge management – focus on people as unique holders of knowledge, and exchanges between people as primary generators of new knowledge for innovation.
- Knowledge networks and working groups – which attempt to build critical mass within and outside the IST programme.

20.3 Major Challenges Ahead in the Field of Knowledge Management

The overriding objective, or (wo-)man-on-the-moon vision, for knowledge management is to support organizational innovation through increased collaboration, flexibility and openness. We see our role, and the role of our research projects, as that of facilitating linking of ideas, people and projects to this end.

It is interesting to see the journey of Europe into the knowledge economy through an "ecology" lens. At a macro level therefore, we see the challenge for organizational leaders and regional planners to be to identify those conditions which, when aligned into the right ecology or environment, make innovation and co-creation possible, and to replicate those conditions to as many industry groups as possible.

At a micro or organizational level, the fundamental challenge ahead is to convince organizations of the need for a major cultural transformation of business, towards more open, inclusive, communicative and collaborative working environments, which encourage rich exchange between people, and provide the organizational and technological conditions necessary to make such exchanges possible.

This is critical. Organizations that will survive in the future will be those in which people are holding rich and meaningful exchanges leading to greater collective sense-making on the issues that concern them. Organizations which are not able to create an environment where new ideas can flourish or in which creative knowledge workers feel they can make a real and meaningful contribution may lose key staff and key markets and may ultimately disappear.

20.4 Building Strong Constituencies and Networks

In the organizational knowledge management area, we began to focus on building constituencies at the very start of Framework Programme V around our "KM Made in Europe" movement. Since 1999 we have launched a number of cluster projects, working groups and discussion groups to leverage our work and to provide for the possibility of wide collaboration and impact.

- Working group on **Multidisciplinary Research** and its application to knowledge management. This working group is expected to spin-off a number of new project ideas in the future. This is one of the several expert groups who advise us on new inputs to our work, which convene periodically.
- The **European Knowledge Management Forum** is a cluster project and web site comprising all KM-related projects. In addition, it is open to free contribution from outside IST, and some 4,000 organizations working in the knowledge management space have become members of this cluster, in 90 countries worldwide. The forum follows a self-organizing approach. On a content level, the project aims at supporting identification of commonality in KM terminology, application and implementation.

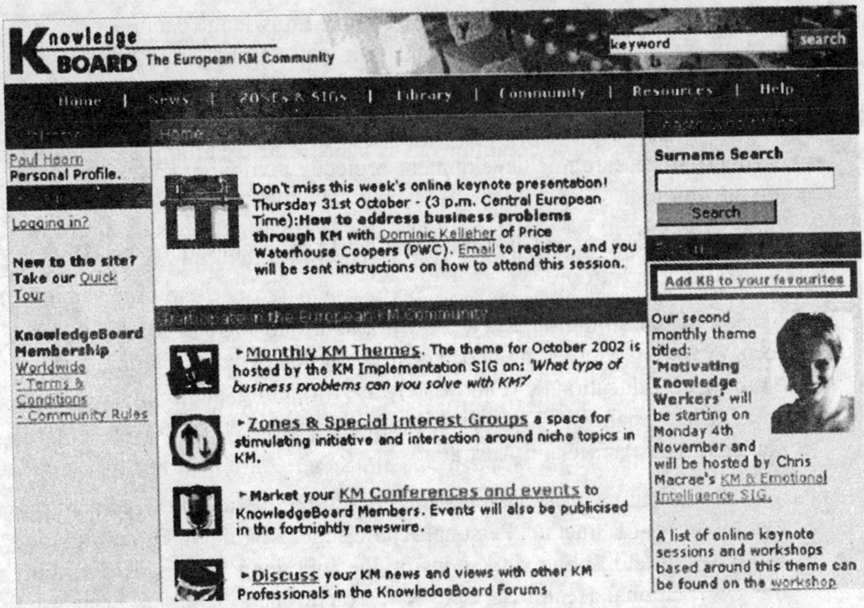

Fig. 20.1: The Portal of the European Knowledge Management Forum[366]

Examples of Special Interest Groups nurtured at the Forum [367]

- KM Visions and Strategy
- Frameworks and Standards
- KM Implementation
- KM in the Public Sector
- Communities of Practice
- KM in SME's
- Innovation
- Space
- Emotional Intelligence
- Research and Action in the Learning Society

Special Language Groups

- Le coin des francophones (French)

---

366 At Knowledge Board, http://www.knowledgeboard.com, cited 12-4-2002.

367 Cf. Knowledge Board. The European KM Community, http://www.knowledgeboard.com/community/zones/index.html, cited 12-4-2002.

- Saber latino (Spanish)
- Campo dei Saperi (Italian)

Seed Projects Launched

- Around 40 research & development projects, involving several hundred organizations.[368]
- CEN/ISSS working group on KM Standards, involving more than 100 organizations in Europe.[369]
- Various discussion groups at a pre-working group stage dealing with subjects including Motivating and Incentivizing Knowledge Sharing, KM in Training & Education, KM and Enterprise Modeling, and regional groups of actors in Spain/Latin America, European accession countries, France-Canada, Asia Pacific and Australia.
- Various internal workshops held in 2001-2002 on Narrative Techniques, Communities of Practice, Personal Knowledge, KM in NGOs and KM in Engineering and online workshops on the following themes: KM within the UK National Health Service, KM Practitioners Toolkits, Scenarios for the Knowledge Economy, Branding and Marketing a KM Initiatives, The human dimensions of KM, The Role of Ontologies in Knowledge Management, Games and action: an interactive approach towards KM, and Fun as a Business Proposition.[370]

In the IST Programme, the KM team has worked hard to create an environment in which new ideas can emerge and flourish. In order to achieve this, we believe it is important to create an environment in which people and their ideas can be reconfigured in myriad ways, and in which social capital and trust can be established. Ideas tend to flow when such environmental conditions are right.

In August 2002, we were lucky enough to have the opportunity to test our constituency-building approach against that of leading global organizations, as part of a Global Benchmarking Study[371]. One of the eloquent conclusions of this study was that we were indeed showing the organizations and people involved that "if we share, we can win". By sharing in multiple different contexts, the participants

---

368 Cf. Knowledge Board. IST Project Corner, http://www.knowledgeboard.com/item/69381, cited 12-4-2002.

369 Cf. CEN/ISSS. Workshop, http://www.cenorm.be/isss/Workshop/km/Default.htm, cited 12-4-2002.

370 Cf. Knowledge Board. Workshop, http://www.knowledgeboard.com/workshop/index.html, cited 12-4-2002.

371 Cf. APQC. Global Benchmarking Study, http://www.apqc.org/proposal/6506, cited 12-4-2002.

of our research programme have found many ways to contribute to and learn from the ideas of their peers.

20.5 Future and Promising Research Areas to be Addressed in Framework Programme VI

We have identified a number of major challenges associated with Knowledge Management in the Future. The uniting theme is the move to more holistic appreciation of the nature of knowledge and its management:

- **Organizational innovation:** Research into the enablers and blockers of effective conversion of knowledge into value. At a macro level, this includes: research into the culture of entrepreneurship and innovation in Europe, enablers and blockers of cross-European flow of ideas and knowledge, carriers/facilitators/blockers of free-flowing exchange, the role of trust and opening up of trust bases, creative work, European-wide incentive guidelines to foster knowledge conversion and sharing processes. At a microlevel, this includes research into -identification and co-creative replication of effective conditions ("ecologies") that enable maximum creativity and innovation within and across organizations.
- **New organizational forms, networks and connections:** Research will explore and prototype emerging organizational forms, networks and connections; re-configuration or new uses of existing organizational forms, networks, and connections that maximize knowledge sharing as enablers of organizational innovation and creativity. Research will focus on internal incubators, informal constellations, groups and communities, formal alliances, partnerships, and consortia of organizations or groups, and mechanisms, tools and methodologies for collecting; enhancing and synthesizing distributed intelligence and tacit knowledge of workers. Consideration will also be given to human resource issues including skills and competency development, new roles and learning requirements, and reward systems and recognition mechanisms for knowledge contributors and knowledge workers.
- **Communities of Practice:** Research will focus on what has been tried and learned from COPs thus far, the promise and limitations of COPs, new innovations, scalability and interaction between communities, new organic organizational forms and supporting tools emerging from our COP experiments that will provide us with more powerful ways to connect and create shared meaning, managing communities and leveraging their potential. Technology research will investigate new platforms and tools, which can support and enhance the work of COPs.

- **Global Connectivity:** Research on how ICTs can enable organizations and communities to connect anything to anything, anyone to anyone, in ways that facilitate high value knowledge creation and sharing through connections that create meaningful context, but avoid meaningless complexity. Research will build on multidisciplinary approaches to psychology, biology, anthropology, ethnography, sociology, linguistics and other disciplines. The focus will be on finding effective means to enhance collective intelligence of organizations, communities, and other groups through connections that form patterns of meaning, and tackle questions such as how to create sense-making patterns of connections, and how to recognize existing patterns that provide for sense-making.

True to the nature of knowledge management as a multidisciplinary research area, the above work will be influenced by state-of-the-art developments in areas including socio-economics, social and management science, social psychology, complexity theory, work on complex adaptive systems, social capital, new growth theory, quantum behaviour in dense networks, organizational living systems, narrative research, development of algorithm-driven knowledge-cognition software for the synthesis of knowledge in operational environments and across networks, research on experience-driven, narrative and other knowledge exchange technologies considering conversation is a core competency.

Finally, as is often the case, the simplest ideas are the hardest to implement. We will be truly successful in knowledge management when we are able to show organizations and the individuals that work in them that "if they share, they can win" – when we can say we have contributed to the establishment of more open and collaborative business and organizational cultures.

20.6 Research Views

To better structure and clarify the research topics in the future, one might describe KM along the following four views:

20.6.1 Strategic View

The focus is on building a world beating innovation and co-creation culture within and between enterprises, and, at a macro level through the aggregation of such innovative organizations, developing benchmarks and indicators for regional, national and European knowledge ecologies, including:

- Understanding of interplay of organizational maturity and innovation or co-creation ability;

- Methodologies which help to identify facilitators and catalyzers of innovation and creativity within and between organizations, and attempt to model and transfer such features to other settings;
- Multi-disciplinary research on the social, ethical, sociological and economic facets of new organizational collaborative forms;
- Multi-disciplinary research focusing on the individual as unique carrier of knowledge, and exchange between individuals as unique mechanism for creation of new knowledge;
- Organizational design principles for the set-up of co-creation sites;
- Analysis of macro-level interdependencies and challenges.

20.6.2 Organizational View

The focus here is on seeing the organization as an organic entity, with emergent properties, interoperating in distinct knowledge ecologies:

- Exploration of concepts such as complexity and quantum theory applied to the organization activity at an operational level;
- Effective leverage of tacit knowledge and management of people as knowledge contributors for the collective benefit of the enterprise, including mechanisms to identify and trace the value-added contribution of each participant, and responsibility and liability within the knowledge ecology in which he or she resides;
- Exploration of the limits of communities of practice as vehicles for organizational innovation;
- Implementing and measuring impact of new organizational forms at operational level, and analysis of the individual as the fundamental element of value creation and sustainability.

20.6.3 Product/Service View

This view sees the delivery of new products and services as a collaborative, cross-organizational and multi-organizational challenge, which needs to take into account richer multidisciplinary lessons and multiple stakeholder views:

- Incorporation of truly multidisciplinary views to product/service design which lead us away from binary user requirements/customer-focus approaches to truly holistic lifestyle and experiential solutions;
- Rich exchange benefiting from global connectivity and inter-organizational exchange.

20.6.4 Infrastructure View

This view promotes technology as an enabler of rich exchanges, collaboration and interaction across time and space:

- Definition and implementation of next generation technology platforms and tools for knowledge sharing, and co-creation;
- Research into next generation community portals and co-creation spaces supporting new organizational forms;
- Application of enabling technologies from many ICT fields where useful, including semantic web and language engineering, open-source of components or infrastructures, taxonomies and shared ontologies, open architectures, grid technologies, agent technologies for automated interactions, self-adapting modules and components, multi-agent architectures, P2P, etc.

20.7 The Road Ahead

When we launched our research activities in KM back in 1994, although, like most other organizations, we did not call it KM! From 1998, in the IST Programme, we saw the opportunity to reach beyond our programme by promoting European research under the brand "KM Made in Europe". In the last four years we have frequently been surprised, encouraged and humbled to see how European businesses and research institutes have been able to rise to the challenge of meeting and exceeding the demands of the European research agenda in KM. Our constituency of research projects and the wider KM community in Europe are, together, setting the context for the creation of Europe as the most dynamic and competitive knowledge economy in the world by 2010, the challenge set by our political leaders at Lisbon.[372]

True to what we have learned in KM over the past years, we believe that to be the most competitive knowledge economy, Europe will need also to be the most collaborative, and the IST Programme has shown examples of research collaboration at its best.

The Information Society Technologies Programme is about supporting world-class research into the management and leveraging of knowledge for organizational performance enhancement. Through our research projects in Framework Programme

---

[372] Cf. European Commission. eEurope initiative, http://www.europa.eu.int/information_society/eeurope/action_plan/index_en.htm, cited 12-4-2002.

V we have striven to promote excellence in methodologies and tool development to make KM a reality for business organizations in Europe and further afield, whether they are large established players or networks of smaller enterprises.

We believe that Europe can show the way forward in knowledge management, and through our research and standardization activities in Framework Programmes V and now VI, we hope to continue to show how Europe can lead in knowledge management in the period 2003-2007.

20.8 Abstract

Under the IST Programme, the European Commission is attempting to build community around the subject of Knowledge Management Made in Europe. In the past four years almost 40 research-, development-, and take-up projects have been launched which are tackling various pieces of the organizational knowledge management puzzle. In addition, the cluster project known as the European Knowledge Management Forum is attempting to build a sustainable network of knowledge management theoreticians and practitioners who are interested in the journey into the knowledge economy, and what knowledge management methods and tools can contribute to this journey. The Forum is also attempting to support commonality in KM terminology, application and implementation in Europe. This paper gives an overview of the current work, which has taken place under the IST Programme during Framework Programme V, and attempts to highlight some interesting research themes on the way to Framework Programme VI.

Finally, in order for Europe to remain competitive in the knowledge economy, the authors argue that European organizations will need to show they are capable of effective and open collaboration.

List of Figures

References

- Abecker, A., Hinkelmann, K., Maus, H., Müller, H.J., (eds.) (2002): Geschäftsprozessorientiertes Wissensmanagement. Effektive Wissensnutzung bei der Planung und Umsetzung von Geschäftsprozessen. Springer, Berlin, Heidelberg, New York
- Allee, V. (2000): Knowledge Networks and Communities of Practice. OD Practitioner, Vol. 32, No. 4, cited 28-07-2002, available at http://www.odnetwork.org/odponline/index.html
- Allweyer, Th. (1998): Modellbasiertes Wissensmanagement. In: Information Management, 1, 37-45
- Amelingmeyer, J. (2000): Wissensmanagement. Analyse und Gestaltung der Wissensbasis von Unternehmen. Gabler Verlag, Wiesbaden
- American Productivity & Quality Center (APQC) (ed.) (1996): Knowledge Management. Consortium Benchmarking Study Final Report. American Productivity and Quality Center, Houston.
- American Productivity & Quality Center (APQC): Global Benchmarking Study. Cited 12-4-2002, http://www.apqc.org/proposal/6506
- American Productivity & Quality Center (APQC), Elliott, S. (eds.)(2000): Knowledge Management: A Guide for your Journey to Best-Practice Processes. American Productivity & Quality Center, Houston
- Andriessen, D., Tissen, R. (2000): Weightless Wealth. Find your real value in a future of intangible assets. Financial Times Prentice Hall, London
- ARC Seibersdorf research GmbH (ed.) (2001): Wissensbilanz 2001. Wissen schafft Zukunft. ARC Seibersdorf research GmbH, Wien
- ARCHmatic-Glossar und –Lexikon, cited 12-12-2002, http://www.glossar.de
- Argyris, C., Schön, D. (1978): Organizational Learning. A theory of Action Perspective. Addison-Wesley Publishing, Reading (Mass.)
- Arthur Andersen, The American Productivity & Quality Center (eds.) (1995): Knowledge Management Assessment Tool. Internal Paper
- Asch, S. (1987): Social Psychology. Prentice Hall, New York
- Atkinson, J. (1988): Einführung in die Motivationsforschung. Motivation und Leistung. Klett-Cotta, Stuttgart
- Aventis (2001): Aventis Internal Global KM Meeting, cited 11-29-2002, http://www.aventis.com
- Bach, V., Vogler, P., Österle, H. (eds.) (1999): Business Knowledge Management. Praxiserfahrungen mit Intranet-basierten Lösungen. Springer, Berlin, Heidelberg, New York

- Bach, V.; Österle, H., Vogler, P. (eds.) (2000): Business Knowledge Management in der Praxis. Prozessorientierte Lösungen zwischen Knowledge Portal und Kompetenzmanagement. Springer, Berlin, Heidelberg, New York
- Bachmann, T., Finke, I., Klaus-Stöhner, V. (forthcoming 2003): Handbuch Barrierentypen.
- Bain, J. (1968): Industrial Organization. Wiley, New York
- Bair, J. (1998): Developing a Scaleable Knowledge Management Architecture and IT Strategy. In: Building the Knowledge Management Framework. The New IT Imperative. 1st & 2nd July 1998, Business Intelligence, London
- Bair, J. H., O'Connor, E. (1998): The state of the product in knowledge management. Journal of Knowledge Management 2, (2), 20-27
- Barros, B., Verdejo, F. (1998): Designing Workspaces to Support Collaborative Learning. Cited 12-12-2002, http://sensei.lsi.uned.es/ ~bbarros/
- Beckman, Th. J. (1999): The Current State of Knowledge Management. In: Liebowitz, J. (ed.): Knowledge Management Handbook. CRC Press, Boca Raton 1, (1), 1-21
- Bergmann, K.: Wissenslandkarten. Das flexible Unternehmen. Cited 12-12-2002, http://www.flexible-unternehmen.de/fb990708.htm
- Berner, S. (1999): Reaktionen der Verbleibenden auf einen Personalabbau. Dissertation. Bamberg
- Bennet, D., Bennet, A. (2003): The Rise of the Knowledge Organization. In: Holsapple, C.W. (ed.): Handbook of Knowledge Management. Springer Verlag, Berlin, Heidelberg, Vol. 1, 5-20
- Bierly, E., Daly, P. (2002): Aligning Human Resource Management Practices and Knowledge Strategies. In: Choo, C.W., Bontis, N. (eds.): The strategic Management of Intellectual Capital and organizational Knowledge. Oxford University Press, Oxford, New York, 277-295
- Boisot, M. (1999): Is your Firm a creative Destroyer? Competitive Learning and Knowledge Flows in the technological Strategy of Firms. In: Zack, M. (ed.): Knowledge and Strategy. Butterworth-Heinemann, Boston, 251-273
- Brauner, Elisabeth (1994): Soziale Interaktion und mentale Modelle: Planungs- und Entscheidungsprozesse in Planspielgruppen. Waxmann, Münster
- British Aerospace (1997): Annual Report 1997
- Bromme, R. (1992): Der Lehrer als Experte. Zur Psychologie des professionellen Wissens. Verlag Hans Huber, Bern, Göttingen, Toronto
- Brooking, A. (1996): Intellectual Capital. Core asset for the third Millennium. Internat. Thomson Business Press, London

- BSI (ed.)(2001): Knowledge Management PAS 2001: A Guide to Good Practice. London (British Standards Institution)
- Bukowitz, W. R., Williams, R. L. (1999): The Knowledge Management fieldbook. Financial Times Prentice Hall, London.
- Bullinger et al. (2002): Business Communities. Professionelles Beziehungsmanagement von Kunden, Mitarbeitern und B2B-Partnern im Internet. Gallileo Press GmbH, Bonn
- Bullinger, H.-J., Wörner, K., Prieto, J. (1997): Wissensmanagement heute. Daten, Fakten, Trends. Fraunhofer IAO, Stuttgart
- Carbon, M., Heisig, P. (1993): Verbesserung der Prozeßtransparenz durch konstruktive Veränderungen. In: Bolte, A., Martin, H. (ed.): Flexibilität durch Erfahrung. Computergestützte erfahrungsgeleitete Arbeit in der Produktion. Verlag Institut für Arbeitswissenschaft, Kassel, 71-77
- Carter G., van der Spek, R. (2002): Successful knowledge management strategies. Report on a best practice screening survey. European Foundation for Quality Management and CIBIT Consultants | Educators, Brussels
- CEN/ISSS (2002): Knowledge Management Workshop Business Plan, Brussels Version 2.1, cited 25-06-2002, http://www.cenorm.be/isss/Workshop/km/Default.htm
- Center for the New Engineering: What are clustering algorithms. Cited 12-12-2002, http://cne.gmu.edu/modules/dau/stat/clustgalgs/clust1_bdy.html
- Chevron (ed.) (1997): Best Practice Transfer in Chevron. San Francisco
- CIBIT Consultants | Educators, cited 12-4-2002, http://www.cibit.nl
- Cohen, D., Prusak, L. (2001): In Good Company. How Social Capital Makes Organizations Work. Harvard Business School Press, Boston
- Collins, D., Montgomery, C. (1999): Competing on Ressources. Strategy in the 1990s. In: Zack, M.: Knowledge and Strategy Butterworth Heinemann, Boston
- Collison, C., Parcell, G., (2001): Learning to fly – Practical lessons from one of the world's leading knowledge companies. Capstone Publishing Ltd., Oxford
- Comelli, G. (1985): Training als Beitrag zur Organisationsentwicklung. Handbuch der Weiterbildung für Praxis in Wirtschaft und Verwaltung. Vol. 4, Hanser Verlag, München
- Commission of the European Community (1994): An Industrial Competitiveness Policy for the European Union. Bulletin of the European Union, Supplement 3

- Content Management Systems Europe. Cited 12-12-2002, http://www.contentmanager.eu.com/history.htm
- Cranfield School of Management, Information Strategy, The Document Company Xerox (eds.) (1998): Europe's State of the Art in Knowledge Management, November Cranfield
- Csikszentmihalyi, M., Aebli, H. (1999): Das Flow- Erlebnis. Jenseits von Angst und Langeweile: Im Tun aufgehen. Klett- Cotta, Stuttgart
- Davenport, T., Prusak L. (1998): Working Knowledge. How Organizations manage what they know. Harvard Business School Press, Boston
- Davenport, T., Probst, G. (2001): Knowledge Management Case Book. Publicis MCD and John Wiley & Sons, Munich, Berlin
- Davenport, T. (1999): Knowledge Management and the Broader Firm: Strategy, Advantage and Performance. In: Liebowitz, J. (ed.): Knowledge Management Handbook. CRC Press, Boca Raton, 2-1 – 2-11
- Davenport, T., Jarvenpaa, S., Beers, M. (1996): Improving Knowledge Work Processes. Sloan Management Review, Summer, 53-65
- De Hoog, R. (1997): CommonKADS: Knowledge Acquisition and Design Support Methodology for Structuring the KBS Integration Process. In: Leibowitz J., Wilcox, L. C. (ed.): Knowledge Management and Its Integrative Elements. CRC Press, Boca Raton, New York, 129-141
- DELPHI GROUP (ed.) (2002): Taxonomy & Content Classification. Market Milestone Report. Delphi Group Whitepaper
- Denning, St. (2001): The Springboard. How Storytelling Ignites Action in Knowledge-Era Organizations. Butterworth-Heinemann, Boston, Oxford
- Despres, Ch., Chauvel, D. (2000): A Thematic Analysis of the Thinking in Knowledge Management. In: Despres, Ch., Chauvel, D. (eds.): Knowledge Horizons. The Present and the Promise of Knowledge Management. Butterworth-Heinemann, Boston, Oxford, 55 – 86
- Diebold Deutschland GmbH (1993): Geschäftsprozesse im Zentrum der modernen Unternehmensführung. Diebold Deutschland GmbH, Eschborn
- Drucker (1999): Management in the 21st Century (2nd Ed.). Econ Verlag, Düsseldorf
- Earl, M., Scott, I. (2000): What do we know about CKOs? In: Despres, C., Chauvel, D.: Knowledge Horizons. Butterworth-Heinemann, Boston, 195-203
- Eck, C.D. (1997): Wissen – ein neues Paradigma des Managements. Die Unternehmung, 3, 155 – 179
- Edvinsson, L., Brünig, G. (2000): Aktiv Posten Wissens Kapital. Unsichtbare Werte bilanzierbar machen. Gabler, Wiesbaden

- EFQM, APQC, KMN (eds.) (1997): Knowledge Management and the Learning Organisation. Results of a joint EFQM/APQC/KMN Benchmarking Study Project, Brussels
- Ehms, K., Langen. M. (2002): Holistic Development of Knowledge Management with KMMM ®. Cited 01-20-2003, http://www.knowledgeboard.com/doclibrary/knowledgeboard/kmmm_article_siemens_2002.
- Ellis, D. (1989): A behavioural approach to information retrieval design. Journal of Documentation 46, 318-338
- El Sawy, O., Josefek, R. (forthcoming 2003): Business Process as Nexus of Knowledge. In: Holsapple, C.W. (ed.): Handbook of Knowledge Management. Springer Verlag, Berlin, Heidelberg, Vol. 1, 425 – 438
- Eppler, M. (2003): Making Knowledge Visible through Knowledge Maps: Concepts, Elements, Cases. In: Holsapple, C. (ed.): Handbook of Knowledge Management. Springer Verlag, Berlin, Heidelberg, Vol. 1, 189-205
- Eppler, M., Röpnack, A., Seifried, P. (1999): Improving Knowledge Intensive Processes through an Enterprise Knowledge Medium. In: Proceedings of THE 1999 ACM SIGCPR Conference, Managing Organizational Knowledge for Strategic Advantage. The Key Role of Information Technology and Personnel
- Erlach, C., Hausmann, I., Mandl, H., Trillitzsch, U. (2000): Knowledge Master – a collaborative learning program for Knowledge Management. In: Davenport, T., Probst, G. (2000): Knowledge Management Case Book. Siemens Best Pratices. Publicis MCD Verlag, Erlangen, Munich, 179-197
- European Commission. eEurope initiative, cited 12-4-2002, http://www.europa.eu.int/information_society/eeurope/action_plan/index_en.htm
- European Foundation for Quality Management, cited 12-4-2002, http://www.efqm.org
- Evans, Sir R. (1998): A Benchmark that brought Managers in from the cold. In: Dauphinias, G., Price, C. (eds.)· Straight from the Clothe world's Business Leaders reveal Ideas that every manager can use. Nicholas Brealey Pub., 121-128
- Festinger, L., Irle, M., Möntmann, V. (1978): Theorie der kognitiven Dissonanz. H. Hubert, Göttingen
- Firestone, J., McElroy, M. (2002): Generations of Knowledge Management. Source Executive Information Systems. Cited 12-12-2002, http://www.dkms.com/white_papers.htm
- Fischer, M. (1998): Visualisierung von Management-Informationen. Roderer Verlag, Regensburg
- Fleig, J., Schneider, R. (1995): Erfahrung und Technik in der Produktion. Springer Verlag, Berlin, Heidelberg, New York

- Forkel, M. (1994): Kognitive Werkzeuge – ein Ansatz zur Unterstützung des Problemlösens. Hanser Verlag, München
- Foster, K. (1999): Justifying Knowledge Management Investments. In: Knowledge and Process Management 6, (3), 154-157
- Frank, U., Döring-Katerkamp, U., Trojan, J. (2002): Knowledge Management in der Praxis. Bd 1: Aktivierung der Mitarbeiter im Rahmen von Knowledge Management. Institut für e-Management e.V. (IfeM), Köln
- Fraunhofer AiS: Textmining. Cited: 12-12-2002, http://ais.gmd.de/KD/textmining.html
- French, W., Bell, H. (1994): Organisationsentwicklung. Sozialwissenschaftliche Strategien zur Organisationsveränderung. 4th Ed., Haupt Verlag, Bern
- Fried, J., Süßmann, J. (2001): Revolutionen des Wissens – Eine Einführung. In: Fried, J., Süßmann, J. (Eds): Revolutionen des Wissens: Von der Steinzeit bis zur Moderne. Verlag C.H. Beck, München, 7 – 20
- Frey, B. (1997): Markt und Motivation: Wie ökonomische Anreize die (Arbeits-) Moral verdrängen. Vahlen Verlag, München
- Frey, B., Osterloh M. (2002): Managing Motivation. Gabler Verlag, Wiesbaden
- Frey, B., Osterloh, M. (1997): Sanktionen oder Seelenmassage? Motivationale Grundlagen der Unternehmensführung. In: Die Betriebswirtschaft (DBW) 57, (3), 307-321
- Gabler Wirstchafts-Lexikon (1994), Rev. 13th Ed., Gabler Verlag, Wiesbaden
- Gattermeyer, W., Ayad, A. (2001): Changemanagement und Unternehmenserfolg. Grundlagen-Methoden-Praxisbeispiele. 2nd Ed., Gabler Verlag, Wiesbaden
- Gebert, D., von Rosenstiel, L. (1996): Organisationspsychologie. Person und Organisation. 4th Ed., Kohlhammer, Stuttgart
- Göbler, T. (1992): Modellbasierte Wissensakquisition zur rechnerunterstützten Wissensbereitstellung für den Anwendungsbereich Entwicklung und Konstruktion. Hanser Verlag, München
- Gödicke, P. (1992): Wissensmanagement – aktuelle Aufgaben und Probleme. In: io Management Zeitschrift, 61, (4), 67-70
- Grant, R. (2000) : Shifts in the World Economy. The Drivers of Knowledge Management. In: Despres, C., Chauvel, D. (eds.): Knowledge Horizons. The Present and the Promise of Knowledge Management, Butterworth-Heinemann, Boston, 27 – 53
- Grauel, R. (2003): Die Sprachfabrik. In: Brand Eins 03, (1), 96-101

- Grewe, A. (2000): Implementierung neuer Anreizsysteme. Rainer Hampp Verlag, München, Mering
- Grey, D. (1999): Knowledge Mapping: A Practical Overview. Cited 12-12-2002, http://www.smithweaversmith.com/knowledg2.htm
- Gruber, T. (1993): A Translation Approach to Portable Ontology Specifications. In: Knowledge Acquisition 5, (2), 199-220
- Grundstein, M. (2000): From Capitalizing on Company's Knowledge to Knowledge Management. In: Marey, D., Maybury, M., Thuraisingham, B. (eds.): Knowledge Management. Classic and Contemporary Works. MIT Press, Cambridge, 261 – 287
- Guretzky, B. von: Schritte zur Einführung des Wissensmanagements: Wissenskarten – Gelbe Seiten – Teil B. Cited 12-12-2002, http://www.community-of-knowledge.de/cp_artikel.htm?artikel_id=39.
- Hammer, M., Champy, J. (1993): Reengineering the Corporation. Campus, Frankfurt, New York
- Hansen, M., Nohria, N., Tierney, T. (1999): What's your Strategy for Knowledge Management. In: Harvard Business Review, March-April, 106-116
- Hedlund, G. (1994): A Model of Knowledge Management and the N-Form Corporation. Strategic Management Journal 15, 73-90
- Heisig, P. (1998a): KVP und Wissensmanagement. In: Howaldt, J., Winther, M.: KVP: Der Motor der lernenden Organisation. Wirtschaftsverlag Bachem, Köln, 214 – 229
- Heisig, P. (1998b): Erfahrung sichern und Wissen transferieren: Wissensmanagement im Projektmanagement. Projektmanagement 9, (4), 3-10
- Heisig, P. (2000): Benchmarking Knowledge Management und wissensorientierte Gestaltung von Geschäftsprozessen. In: Bühner, R. (ed.): Loseblattwerk Organisation, Landsberg/Lech, 38 pages
- Heisig, P. (2001a): Business Process Oriented Knowledge Management. In: Mertins, K., Heisig, P., Vorbeck, J. (eds.) (2001): Knowledge Management. Best Practices in Europe. Springer, Berlin, 13-36
- Heisig, P. (2001b): Wissensmanagement: Die Synchronisation von Menschen, Prozessen und IT. Proceedings of the X. Produktionstechnisches Kolloquium – PTK 2001 "Unternehmenswerte durch Technologie" Berlin, 27.-28. September 2001, 269-275
- Heisig, P., Finke, I., Jaitner, A. (forthcoming 2003): Wissensmanagement-Kompetenz-Check (Knowledge Management Skill Test). In: Rosenstiel, L., Erpenbeck, J. (eds.): Handbuch der Kompetenzmessung. Schäffer-Poeschel Verlag, Stuttgart.

- Heisig, P., Vorbeck, J. (2001): Benchmarking Survey Results. In: Mertins, K., Heisig, P., Vorbeck, J. (eds.) (2001). Knowledge Management. Best Practices in Europe. Berlin: Springer, 97-123
- Helbig, H. (2001) : Die semantische Struktur natürlicher Sprache: Wissensrepräsentation mit Multinet. Springer, Berlin, Heidelberg, New York
- Herzberg, F. (1968): Work an nature of man. Staples Press, London
- Hiebeler, R. (1996): Benchmarking Knowledge Mangement. In: Strategy & Leadership. March / April , 22-29
- Hinkelmann, K., Karagiannis, D., Telesko, R. (2002): PROMOTE – Methodologie und Werkzeug für geschäftsprozessorientiertes Wissensmanagement. In: Abecker et al. (2002): Geschäftsprozessorientiertes Wissensmanagement. Springer Verlag, Berlin, 65-90
- Hoffmann, I. (2001): Knowledge Management Tools. In: Mertins, K., Heisig, P., Vorbeck, J.: Knowledge Management. Best Practices in Europe. Springer, Heidelberg, 74-94
- Holsapple, C. (ed.) (2003): Handbook on Knowledge Management. Vol. 1. Knowledge Matters, Vol. 2. Knowledge Directions, Springer Verlag, Berlin, Heidelberg
- Hopfenbeck, W., Müller, M., Peisl, T. (2001): Wissensbasiertes Management in der New Economy. Moderne Industrie, Landsberg/Lech
- Huang, K. (1998): Capitalizing Collective Knowledge for Winning, Execution and Teamwork. Unpublished IBM Working Paper
- Huang, K., Lee, Y., Wang, R. (1999): Quality information and Knowledge. Prentice Hall, London
- Hylton, A.: A KM Initative is Unlikely to Suceed Without a Knowledge Audit. Cited 12-12-2002, available at: http://www.knowledgeboard.com/library/the_need_for_knowledge_audits.pdf
- ISO-2788 (1986) : Documentation – Guidelines for the establishment and development of monolingual thesauri. 2nd Ed.
- Issing, L. J. (1988): Wissensvermittlung mit Medien. In: Mandl, H., Spada, H. (eds.) : Wissenspsychologie. Psychologie Verlags Union, München
- IST. Knowledge Management for eCommerce & eWork. Cited 12-4-2002, http://www.cordis.lu/ist/ka2/al12.htm
- Jaitner, A. (2001): Abschied vom Taylorismus. Wissensmanagementsysteme in Unternehmensberatungen als Modelle für den öffentlichen Sektor. Verwaltung & Management 5, (7), 304-310
- Jäkel, F., Arroyo Pinedo, J. (2000): Development of a Demonstrator for Modelling and Simulation of Global Distributed Enterprises. In: Mertins, K.;

Rabe, M. (eds.): The New Simulation in Production and Logistics. 9th ASIM Dedictaed Conference on Simulation in Production and Logistics, Fraunhofer IPK, Berlin, 375-384

- JOANNEUM REASEARCH Forschungsgesellschaft m.b.H. (ed.) (2002): JR Explorer – Wissensbilanz. 2001
- Kaiser, T., Vogler, P. (1999): PROMET®I-NET: Methode für Intranet-basiertes Wissensmanagement. In: Bach, V., Vogler, P., Österle, H. (ed.): Business Knowledge Management. Praxiserfahrungen mit Intranet-basierten Lösungen. Springer, Berlin, Heidelberg, New York, 117-129
- Kamphusmann, T. : Methoden und Konzepte. PowerPoint Präsentation. Cited 12-9-2002), http://www.do.isst.fhg.de/wm/events/pdf_dateien/Texmining.pdf
- Kaplan, R., Norton, D. (1992): The Balanced Scorecard – Measures that Drive Performance. Harvard Business Review, (Jan/Feb), 71-79
- Kaplan, R.., Norton, D. (1996 a): Using the Balanced Scorecard as a Strategic Management System. Harvard Business Review, (Jan/Feb), 75-85
- Kaplan, R., Norton, D. (1996): The Balanced Scorecard. Translating Strategy into Action. Harvard Business School Press, Boston (Mass.)
- Kaplan, R., Norton, D. (1993): Putting the Balanced Scorecard to Work. Harvard Business Review, Sep-Oct, 134–142
- Kaps, G., Nohr, H. (2001): Erfolgsmessung im Wissensmanagement mit Balanced Scorecards. nfd Information – Wissenschaft und Praxis, 52, (2), 89-97 (part 1) and 52, (3), 151-158 (part 2)
- Karagiannis, D., Junginger, S., Strobl, R. (1996): Introduction to Business Process Management System Concepts. In: Scholz-Reiter, B., Stickel, E. (eds.): Business Process Modelling. Springer, Berlin, Heidelberg, New York, 81-106
- Katz, W. (1982): Introduction to Reference Work. Vol. 1. McGraw-Hill, New York
- Kemp, J., Pudlatz, M., Perez, Ph., Munoz Ortega, A. (2002): KM Framework. Research paper of the European KM Forum (IST Project No 2000-26393)
- Kenney-Wallace; G. (1999): The Virtual University. In: Defence Management Journal 3, (1)
- Kleinbeck, U. (1996): Arbeitsmotivation. Entstehung, Wirkung, Förderung. Juventa, München
- Knowledge Board: The European KM Community. Cited 12-4-2002, http://www.knowledgeboard.com/community/zones/index.html
- Knowlegde Board: IST Project Corner. Cited 12-4-2002, http://www.knowledgeboard.com/item/69381

- Knowlegde Board: Workshop. Cited 12-4-2002, http://www.knowledgeboard.com/workshop/index.html
- Kosiol, E. (1968): Grundlagen und Methoden der Organisationsforschung. Duncker &Humblot, Berlin
- Krause, O. (2000): Management Knowledge Engineering – A Toolkit to Engineer Adaptive Management Systems. In: Neely, A. (ed.): Performance Measurement Past, Present and Future, Centre for Business Performance, Cranfield School of Management, 307-314
- von Krogh, G. M. Venzin (1995): Anhaltende Wettbewerbsvorteile durch Wissensmanagement, Die Unternehmung , 417 – 436
- Krogh, G. von, Ichijo, K., Nonaka, I. (2000): Enabling Knowledge Creation. How to Unlock the Mystery of Tacit Knowledge and Release the Power of Innovation. Oxford University Press, Oxford, New York
- Kuznets, S. (1966): Modern Economic Growth: Rate, Structure, Spread. Yale University Press, New Haven, Cited in: Teece, D. (1998): Capturing Value from Knowledge Assets: The New Economy, Markets for Know-how, and Intangible Assets, California Management Review 40, (3), 55-79
- Langen, M., Ehms, K. (2000): KMMM. Eine Methodik zur Einschätzung und Entwicklung des Reifegrades im Wissensmanagement. In: KnowTech 2000 – Knowledge Engineering, Management, Consulting & Training, Leipzig 5.-8. September 2000
- Lee, L. (2001): Knowledge Sharing Metrics for Large Organizations. In: Morey, D., Maybury, M., Thuraisingham, B.: Knowledge Management. The MIT Press, Cambridge (Mass.), 403-419
- Leitner, K.-H., Bornemann, M., Schneider, U. (2002): Development and Implementation of an Intellectual Capital Report for a Research Technology Organization. In: Bontis, N. (ed.): World Congress on Intellectual Capital Readings. Butterworth-Heinemann, Woburn
- Lev, B. (2001): Intangibles. Management, Measurement and Reporting. Brookings Institution Press, Washington DC.
- Lev, B., Zarowin, P. (1999): The Boundaries of Financial Reporting and How to Extend Them. Journal of Accounting Research 37, (29), 353-385
- Levinthal, D., March, J. (1993): The Myopia of Learning. Strategic Management Journal 14, (2), 95-112
- Liebowitz, J., Rubenstein-Montano, B., McCaw, D. (2000): The Knowledge Audit. In: Knowledge and Process Management 7, (1), 3-10
- Lindemann, U., Aßmann, G., Freyer, B. (1998): Vernetzte Information zur Handhabung von Entwicklungswissen. Zeitschrift für wirtschaftlichen Fabrikbetrieb (ZWF) 93, (9), 386-389

- Lorsch, J. (1985): Strategic myopia: Culture as an invisible Barrier to Change. In: Kilman,R., Saxton, M., Serpa, R. at al. (eds.): Gaining Control of the Corporate Culture. Jossey-Bass, San Francisco, 84-102
- LTI: IT for Learning and Training in Industry, cited 12-4-2002, http://www.cordis.lu/esprit/src/ltihome.htm
- Lühker, M., Vaanholt, S. (1994): Motivation: Mehr als nur ein Mythos? In: Personal 46, (5), 230-235
- Luhn, G. (1999): Implizites Wissen und technisches Handeln am Beispiel der Elektronikproduktion. Meisenbach Verlag, Bamberg
- Lullies, V., Bollinger, H., Weltz, F. (1993): Wissenslogistik. Über den betrieblichen Umgang mit Wissen bei Entwicklungsvorhaben. Campus Verlag, Frankfurt/Main, New York
- Maier, Ronald (2002): Knowledge Management Systems. Information and Communication Technologies for Knowledge Management. Springer Verlag, Berlin, Heidelberg
- Martin, H. (1995): CeA-Computergestützte erfahrungsgeleitete Arbeit. Springer, Berlin, Heidelberg, New York
- Mason, E. (1939): Price and Production Policies of Large-Scale Enterprises. American Economic Review 29, (1), 61-74
- Mavrinac, S., Siesfeld, A. (1998): An Explanatory Investigation of Investors' Information Needs and Value Priorities. In: OECD/Ernst&Young Center for Business Innovation: Enterprise Value in the Knowledge Economy. Measuring Performance in the Age of Intangibles. New York, 49-72
- McGinn, C. (2001): Wie kommt der Geist in die Materie? Das Rätsel des Bewußtseins. Verlag C.H. Beck, München
- Mehrwald, H. (1999): Das Not-Invented-Here-Syndrom in Forschung und Entwicklung. Dt. Uni-Verlag, Wiesbaden
- Mertins, K., Heisig, P., Vorbeck, J. (eds.) (2001): Knowledge Management. Best Practices in Europe. 1st Ed., Springer Verlag, Berlin
- Mertins, K., Heisig, P., Finke, I. (2001): Wissensmanagement-Audit: Benchmarks für den Umgang mit Wissen. In: Schwuchow, K., Gutmann, J. (eds.): Jahrbuch Personalentwicklung und Weiterbildung. Luchterhand Verlag, Neuwied, 157-162
- Mertins, K., Jochem, R. (1999): Quality-oriented design of business processes. Dordrecht, Norwell
- Mertins, K., Schallock, B., Ganz, W., Moll, K., Metz, A., Hornung, (1991): Handbuch der humanen CIM-Gestaltung. Fraunhofer Institut Produktionsanlagen und Konstruktionstechnik IPK Berlin, Juni 1991 (Projektbericht „Aufbe-

reitung von HdA-Gestaltungswissen für das Beratungsangebot der CIM-TT-Stellen")

- Mertins, K., Schallock, B., Carbon, M., Heisig, P. (1993): Erfahrungswissen bei der kurzfristigen Auftragssteuerung. In: Zeitschrift für wirtschaftlichen Fabrikbetrieb 88, (2), 78 – 80.
- Metternich, J. (2001): Wissen als Grundlage von Wettbewerbsstrategien. Ein Modell zur Analyse und Planung von Wissensstrategien. Shaker Verlag, Aachen
- Mintzberg, H. et al. (1998): Strategy Safari. A Guided Tour Through The Wilds Of Strategic Management. Prentice Hall, London
- Morris, C. (1977): Pragmatische Semiotik und Handlungstheorie. Suhrkamp, Frankfurt am Main
- Murry, J., Hammons, J. (1995). Delphi: A Versatile Methodology for Conducting Qualitative Research. The Review of Higher Education 18, (4), 423-436.
- Netherlands Ministry of Economic Affairs (October 1999): Intangible Assets. Balancing Accounts with Knowledge. The Hague: Ministry of Economic Affairs
- Nonaka, I., Takeuchi, H. (1995): The Knowledge-Creating Company. Oxford University Press, Oxford
- North, K. (2001): Wissensorientierte Unternehmensführung. Gabler Verlag, Wiesbaden
- North, K., Romhardt, K., Probst G. (2000): Wissensgemeinschaften – Keimzellen lebendigen Wissensmanagement. io Management 7/8, 52-62
- O'Dell, C., Grayson, C. (1998): If Only We Knew What We Know. The transfer of Internal Best Practice. The Free Press, New York
- Oracle (ed.) (2002): J2EE and Microsoft .NET. An Oracle White Paper.
- Organization for Economic Cooperation Development (ed) (1996): Employment and Growth in the Knowledge-based Economy. OECD, Paris
- Organization for Economic Cooperation Development, cited 12-4-2002, – http://www.oecd.org
- Pfeifer, J., Sutton, R. (2000): The knowing-doing gap. Harvard, USA
- Platon (1981): Theätet. Reclam Verlag, Ditzingen
- Porras J., Robertson P. (1992): Organizational Development: Theory, Practice and Research. In: Dunnette, M., Hough, L. (eds.): Handbook of Industrial and Organizational Psychology, Consulting Psychologists Press Inc, Palo Alto – California, 719-822.

- Porter, M. (1980): Competitive Strategy: Techniques for Analyzing Industries and Competitors. Free Press, New York
- Pribilla, P., Reichwald, R., Goecke, R. (1996): Telekommunikation im Management. Schäffer-Poeschel, Stuttgart
- Probst, G., Büchel, B., Raub, S. (1998): Knowledge as a Strategic Resource. In: Krogh, G., Ross, J., Kleine, D. (eds.): Knowing in Firms. Understanding, Managing and Measuring Knowledge. SAGE Publications, London, New Delhi, 240- 252
- Probst, G., Raub, St., Romhardt, K. (1998): Wissen managen. Wie Unternehmen ihre wertvollste Ressource optimal nutzen. Vol. 2., Gabler Verlag, Frankfurt/Main, Wiesbaden
- Prusak, L. (2002). The death and transfiguration of knowledge management. Presentation, given at the First International Conference on the Future of Knowledge Management, Berlin, March 8-10, 2002.
- Rabrenovic, O. (2001): Die Rolle der Organisation des Wissensmanagements im Unternehmen. Ein Entscheidungsmodell zur Optimierung des Wissensmanagements am Beispiel des Auftrags- und Auftragsabwicklungsprozesses in der produzierenden Industrie. Shaker Verlag, Aachen
- Rao, R., Sprague, R. jr., (1998): Natural Technologies for Knowledge Work: Information Visualization and Knowledge Extraction. Journal of Knowledge Management 2, (2), 70-80
- Rehäuser, J., Krczmer, H. (1996): Wissensmanagement im Unternehmen. In: Schreyögg, G., Conrad, P. (eds.): Managementforschung. Vol.6: Wissensmanagement., De Gruyter, Berlin, New York,1- 40
- Reinmann-Rothmeier, G., Mandl, H. (1998): Wissensmanagement. Eine Delphi-Studie (Forschungsbericht Nr. 90). Ludwig-Maximilians-Universität, Lehrstuhl für Empirische Pädagogik und Pädagogische Psychologie, München
- Reinmann-Rothmeier, G., Mandl, H. (2000): Individuelles Wissensmanagement. Strategien für den persönlichen Umgang mit Wissen am Arbeitsplatz. Verlag Hans Huber, Bern, Göttingen
- Remus, U. (2002): Prozessorientiertes Wissensmanagement. Konzepte und Modellierung. Dissertation Universität Regensburg
- Rickson, R. (1976): Knowledge Managment in industrial Society and environment Quality. In: Human Organization 35, (76)
- Robertson, P., Roberts, D., Porras, J. (1993): Dynamics of planned organizational change: Assessing empirical support for a theoretical model. Academy of Management Journal 36, (3), 619-624
- Romhardt, K. (1998): Die Organisation aus der Wissensperspektive: Möglichkeiten und Grenzen der Intervention. Gabler, Wiesbaden

- Roos, J., Roos, G., Dragonetti, N., Edvinsson, L. (1998): Intellectual Capital. Navigating the New Business Landscape. New York University Press, New York
- Rosenstiel, L. von (1999): Motivationale Grundlagen von Anreizsystemen. In: Bühler, W., Siegert, T. (1999): Unternehmungssteuerung und Anreizsysteme. Kongressdokumentation 52. Deutscher Betriebswirtschafter-Tag 1998. Schäffer-Poeschl Verlag, Stuttgart
- Ruggles, R. (1997): Knowledge Management Tools. Butterworth-Heinemann, Boston, Oxford
- Rühli, E. (1993): Unternehmensführung und Unternehmenspolitik. 3rd Ed., Paul Haupt, Bern
- Ruth, S., Shaw, N., Frizzell, V. (2003 forthcoming): Knowledge Management Education: An Overview of Programs of Instruction. In: Holsapple, C. (ed.): Handbook of Knowledge Management. Vol. 2, Springer-Verlag, Berlin, Heidelberg, 581-603
- Sarvary, M. (1999): Knowledge Management and Competition in the Consulting Industry. California Management Review 41, (2), 95-107
- Scheer, A. (1997): Wirtschaftsinformatik. Referenzmodelle für industrielle Geschäftsprozesse. 7th Ed., Springer Verlag , Berlin, Heidelberg
- Schmitz, C., Zucker, B. (1996): Wissen gewinnt: Knowledge-Flow Management. Metropolitan Verlag GmbH, Düsseldorf, München
- Schnurr, H.-P., Staab, St., Studer, R., Stumme, G., Sure, Y. (eds.) (2001): Professionelles Wissensmanagement. Erfahrungen und Visionen. Konferenz, Baden-Baden (Germany) 14.-16. März 2001, Shaker, Aachen 2001
- Scholl, W., König, C., Meyer, B. (2002): Delphi-Studie zur Zukunft des Wissensmanagements. In: 43. Kongress der Deutschen Gesellschaft für Psychologie, Berlin, 24. September 2002
- Scholl, W., König, C., Meyer, B., Heisig, P. (forthcomming): The Future of Knowledge Management – An international Delphi Study.
- Schreiber, A., Hoog, R., Akkermans, H., Anjewierden, A., Shadbolt, N., Velde, W. (2000): Knowledge Engineering and Management. The Common-KADS Methodology. The MIT Press, Cambridge, London
- Schüppel, J. (1996): Wissensmanagement: Organisatorisches Lernen im Spannungsfeld von Wissens- und Lernbarrieren. Deutscher Verlag, Wiesbaden
- Schütt, P. (2000): Wissensmanagement. Falken, Niederhausen/Ts.
- Schuller, F. (1998): Wissensaufbau erfordert eine offene Lernkultur. Personalwirtschaft, 5, 27-30

- Schwermer, M. (1998): Modellierungsvorgehen zur Planung von Geschäftsprozessen. IPK, Berlin
- Shapiro, C. (1989): The Theory of Business Strategy. Rand Journal of Economics 20, (1), 125-137
- Shristava, P. et al. (1992): Resource-based View of the Firm: Advances in Strategic Management. JAI Press, Greenwich (Conn.), London
- Siebert, G. (1998): Prozess-Benchmarking – Methode zum branchenunabhängigen Vergleich von Prozessen. IPK, Berlin
- Siemens AG (ed.): Investor Relation Reports. Cited 12-12-2002, http://www.siemens.com/index.jsp?sdc_p=pfs3ndcmut4o1026936l&sdc_sid=33161385733
- Skandia (ed.) (1997): Intelligent Enterprising. Intellectual Capital Supplement to Skandia's 6-month interim report
- Skyrme, D. (2000): Developing a Knowledge Strategy: From Management to Leadership. In: Morey, D., Maybury, M., Thuraisingham, B. (eds.): Knowledge Management. Classic and Contemporary Works. The MIT Press, Cambridge, London, 61-83
- Skyrme, D. (2003 forthcoming): Commercialization: The Next Phase of Knowledge Management. In: Holsappel, C. (ed.): Handbook of Knowledge Management. Vol. 2, Springer, Berlin, 639 – 655
- Skyrme, D., Amidon, D. (1997): Creating the Knowledge-Based Business. Business Intelligence, London, New York
- Smith, A., Rupp, W. (2002): Communication and loyalty among knowledge workers: a ressource of the firm theory view. In: Journal of Knowledge Management 6, (3), 250-261
- Snowden, D. (2002): Complex Acts of Knowing – paradox and descriptive self-awareness. IBM Global Services. Cited 07-09-2002, http://www.1.ibm.com/services/files/complex.pdf
- Snowden, D. (1998): A Framework for Creating a Sustainable Programme. In: Rock, S. (ed.): Knowledge Management. A Real Business Guide. Caspian Publishing Ltd., London
- Spek, R. Van der, Hofer-Alfeis, J., Kingma, J.(2003): The Knowledge Strategy Process S. In: Holsappel, C. (ed.): Handbook of Knowledge Management. Vol. 2, Springer, Berlin, 443-446
- Spek, R. van der, Kingma, J. (1999): Achieving successful knowledge management initiatives. In: CBI/IBM (eds): Liberating knowledge, business guide of Confederation of British Industry. Caspian Publishing, London

- Spek, R. van der, Kingma, J., Hofer-Alfeis, J. (2002): The knowledge strategy process. In: Hollsapple, W. (ed.): Handbook of knowledge management. Vol. 2, Springer, Berlin, 443-467
- Spek, R. van der, Spijkervet, A. (2002): Knowledge Management. Dealing intelligently with knowledge. CIBIT, Utrecht
- Spur, G., Mertins, K., Jochem, R. (1993): Integrierte Unternehmensmodellierung. Beuth Verlag, Berlin, Wien, Zürich
- Standards Australia (HB 275 – 2001): Knowledge Management. A framework for succeeding in the knoweldge era. Standards Australia, Sydney
- Stewart, T. (1997): Intellectual Capital. The New Wealth of Organizations. Bantam Books, London
- Stewart, T. (1997): Der vierte Produktionsfaktor. Wachstum und Wettbewerbsvorteile durch Wissensmanagement. Carl Hanser Verlag, Munich
- Stroebe, J. (1990): Einstellung II- Strategien der Einstellungsänderung. In: Stroebe, W., Hewstone, M., Codol, J., Stephenson, G. (eds.): Sozialpsychologie. Eine Einführung. Springer, Berlin, Heidelberg, New York
- Studer, R. et al. (2001): Arbeitsgerechte Bereitstellung von Wissen – Ontologien für das Wissensmanagement. Forschungsgruppe Wissensmanagement Institut AIFB Karlsruhe, cited 12-09-2002, http://www.aifb.uni-karlsruhe.de/WBS
- Sullivan, P. (1998): Profiting from Intellectual Capital. Extracting value from innovation. John Wiley & Sons, New York
- Sullivan, P. (1998): Introduction to Intellectual Capital Management. In: Sullivan, P. (ed.): Profiting from Intellectual Capital. Extracting Value from Innovation. John Wiley & Sons, New York, 3-18
- Süssenguth, W. (1991): Methoden zur Planung und Einführung rechnergeführter Produktionsprozesse. Hanser-Verlag, München
- Sveiby K. (1997): The Intangible Assets Monitor. Journal of Human Resource Costing & Accounting 2, (1), 73-97
- Sveiby, K. (1987): Managing KnowHow. Bloomsbury, London
- Sveiby, K. (2001): Methods for Measuring Intangible Assets. Cited 12-12-2002, http://www.sveiby.com/articles/IntangibleMethods.htm
- Sveiby, K. (1997): Wissenskapital das unentdeckte Vermögen, Immaterielle Unternehmenswerte aufspüren, messen und steigern.Verlag Moderne Industrie, Landsberg
- TFPL Ltd. (1999): Skills für knowledge management – building a knowledge economy. TFPL Ltd., London

- The Business Intelligence and Data Warehousing Glossary. Cited 12-12-2002, http://www.sdgcomputing.com/glossary.htm
- The Danish Trade and Industry Development Council (May 1997): Intellectual Capital Accounts: Reporting and Managing Intellectual Capital. Copenhagen
- Tiemeyer, E. (1997): OrgTools. Vol. 3: Projektmanagement. Schäfer-Poeschel, Stuttgart
- Tiwana, A. (2000): The Knowledge Management Toolkit. Prentice-Hall, Upper Saddle River
- Transfer Center for Technology Management, cited 22-01-2003, http://www.tectem.ch
- Tünschel, L., Hille, T., Jochem, R. (1998): Geschäftsprozessmodellierung – Werkzeug für das Management des Wandels. io Management, 5, 66-74
- Ulich, E. (1998): Arbeitspsychologie. 4th Ed., Schäffer-Poeschel Verlag, Stuttgart
- Vallabhajosyula, S. (2001): Accounting Rules and Regulations for Intangibles. In: Lev, B.: Intangibles. Management, Measurement, and Reporting. Brookings Institution Press, Washington D.C., 135-154
- Victor, B., Boynton, A. (1998): Invented Here. Maximizing Your Organization's Internal Growth and Profitability. Harvard Business School Press, Boston
- Vogel, E. (1999): Wissensmanagement bei den Helvetia Patria Versicherungen – Ein Vorgehen zur Bewertung des Ist-Stands und zur Entwicklung eines Grobkonzepts. In: Ralph Schmidt (ed.): 21. Online-Tagung der DGI: Aufbruch ins Wissensmanagement. Frankfurt am Main, 18. bis 20. Mia 1999, Proceedings /– Frankfurt am Main: Deutsche Gesellschaft für Informationswissenschaft und Informationspraxis (DGI), 117 – 128
- Volberda H. (1998): Building the flexible Firm. How to remain competitive. Oxford University Press, Oxford, New York
- Vopel, O. (2001): Lehrjahre einer Wissensorganisation – das Beispiel Ernst & Young. Wissensmanagement 3, (3), 4-7
- Vorbeck, J., Finke, I. (2001): Motivation and Competence for Knowledge Management. In: Mertins, K., Heisig, P., Vorbeck, J. (eds.): Knowledge Management. Best Practices in Europe. Springer, Berlin, 37-56
- Warnecke, G., Gissler, A., Stammwitz, G. (1998): Referenzmodell Wissensmanagement – Ein Ansatz zur modellbasierten Gestaltung wissensorientierter Prozesse. Information Management 1, 24-29
- Weber, F., Pawar, K. (2000): ICE 2000. Proceedings of the 6th International Conference on Concurrent Enterprising. Toulouse, France, 28-30 June 2000, 161-164

- Weber, F., Wunram, M., Kemp, J., Pudlatz, M., Bredehorst, B. (forthcoming): Towards Common Approaches and Standards for Knowledge Management in Europe.
- Weggemann, M (1999): Wissensmanagement. Der richtige Umgang mit der wichtigsten Unternehmens-Ressource. MITP-Verlag, Bonn
- Wernerfelt, B. (1984): A resource-based View of the Firm. Strategic Management Journal 2, (5), 171-180
- Westphal, R. (1998): Globale Sammelleidenschaft. BasicPro 1998, (1)
- Widhalm, R., Mück, T. (2002): Topic Maps: Semantische Suche im Internet. Springer, Berlin, Heidelberg, New York
- Wiig, K. (1995): Knowledge Management Methods. Practical Approaches to Managing Knowledge. Vol. 3., Schema Press, Arlington
- Willke, H. (1998): Systemisches Wissensmanagement. Lucius und Lucius, Stuttgart
- Wolford, D., Kwiecien, S. (2003): Driving Knowledge Management at Ford Motor Company. In: Holsapple, C. (ed.): Handbook of Knowledge Management. , Vol. 2, Springer, Berlin, Heidelberg, 501-51
- Woods, E., Sheina, M. (1998): Knowledge Management: Applications, Markets and Technologies. Ovum Ltd., London

Recommended Further Readings

- Argyris, C. (1990): Overcoming organizational defense. Allyn & Bacon, Boston
- Boisot, M. (1998): Knowledge Assets. Securing competitive advantage in the information economy. Oxford University Press, Oxford
- Bontis, N. (2001): World Congress on Intellectual Capital Readings. Butterworth Heinemann, Woburn
- Bontis, N., Choo, C. (2002): The Strategic Management of Intellectual Capital and Organizational Knowledge. Oxford University Press, Oxford
- Brooking, A. (1999): Corporate Memory. Strategies for Knowledge Management. International Thomson Business Press, New York
- Bullinger, H., Rüger, M., Koch, A. et al. (2001): Anreizsysteme im Wissensmanagement. Knowledge meets motivation. CD-Rom Fraunhofer IAO. Fraunhofer IRB Verlag, Stuttgart
- Cepis, Informatik/Informatique (eds.) (February 2002): Knowledge Management and Information Technology 3, (1)

- Cole R. E. (1998): Special Issue on Knowledge and the Firm. California Management Review 40, (3)
- Cross, R., Israelit, S. (2000): Strategic Learning in a Knowledge Economy. Butterworth- Heinemann, Boston, Oxford
- Davenport, T. (1998): Wenn ihr Unternehmen wüsste, was es alles weiss...: Das Praxisbuch zum Wissensmanagement. Verlag Moderne Industrie, Landsberg/ Lech
- Eccles, R., Herz, R., Keegan, E., Philipps, D. (2001): The Value Reporting Revolution. John Wiley & Sons, New York
- Edvinsson, L., Malone, M. (1997): Intellectual Capital: Realizing Your Company's True Value by Finding Its Hidden Brainpower. HarperBusiness, New York
- Edvinsson, L., Sullivan, P. (1996): Developing A Model For Managing Intellectual Capital. In: European Management Journal 14, (4), 356 – 364
- Emery, F. (1959): Characteristics of socio-technical Systems. Tavistock Institute of Human Relations, Document No. 527
- Erpenbeck, J., Sauer, J. (2000): Das Forschungs- und Entwicklungsprogramm "Lernkultur und Kompetenzentwicklung". In: Arbeitsgemeinschaft Qualifikations- und Entwicklungs-Management, Geschäftsstelle der Arbeitsgemeinschaft Betriebliche Weiterbildungsforschung Kompetenzentwicklung (ed.) (2000): Lernen im Wandel – Wandel durch Lernen. Vol. 5.: Waxman, Berlin, 289-332
- Fitz-ens, J. (2001): The ROI of Human Capital. Measuring the economic value of Human Capital. Amacom, New York
- Fopp, L., Schiessl, J. (1999): Business Change als neue Management Disziplin. Wie der Chief Change Officer den Unternehmenswandel mitgestaltet. Campus Verlag, Frankfurt
- Heckhausen, H. (1989): Motivation und Handeln. Lehrbuch der Motivationspsychologie. Springer, Berlin
- Hudson, W. (1993): Intellectual Capital. How to Build It, Enhance It, Use It. John Wiley & Sons, New York
- Klein, D. (1998): The Strategic Management of Intellectual Capital. Butterworth-Heinemann, London
- Krickl, O., Milchrahm, E. (2000): Integrativer Ansatz zur Wissensbewertung. In: Knorz, G., Kuhlen, R. (eds.): Informationskompetenz – Basiskompetenz in der Informationsgesellschaft, Proceedings des 7. Internationalen Symposions für Informationswissenschaft, Univ.-Verl. Konstanz, Konstanz, 113-125

- Maier, Ronald (2002): Knowledge Management Systems. Information and Communication Technologies for Knowledge Management. Springer Verlag, Berlin, Heidelberg
- Marchand, D., Kettinger, W., Rollins, J. (2001): Making the Invisible Visible. How companies win with the right information, people and IT. John Wiley & Sons, Chichester, New York
- Morey, D., Maybury, M., Thuraisingham, B. (2001): Knowledge Management. The MIT Press, Cambridge (Mass.)
- Morey, D., Maybury, M., Thuraisingham, B (eds.) (2000): Knowledge Management. Classic and Contemporary Works. MIT Press, Cambridge, London
- Myers, P. (1996): Knowledge Management and Organizational Design. Butterworth- Heinemann, Boston, Oxford
- Park, J. (2002): XML topics maps: creating and using topic maps for the Web. Addison-Wesley, Boston
- Pepels, W. (ed.) (2000): Motivationsmethoden für Wirtschaftstudierende. Sich selbst und andere motivieren. Cornelsen Verlag, Berlin
- Petrash, G. (1996): Dow's Journey To A Knowledge Value Management Culture. European Management Journal 14, (4), 365-373
- Prusak, L. (1998): Knowledge in Organizations. Butterworth- Heinemann, Boston, Oxford
- Rümler, R. (2001): Wissensbarrieren behindern effektives Wissensmanagement. In: Wissensmanagement 5, (01)
- Schneider, U. (2001): Die 7 Todssünden im Wissensmanagement. Kardinaltugenden für die Wissensökonomie. Frankfurter Allgemeine Zeitung Verlagsbereich, Frankfurt am Main
- Stewart, T. (1995): Trying to Grasp the Intangible. Fortune Magazine, october 2nd, 1995
- Stewart, T. (2001): The Wealth of Knowledge. Intellectual Capital and the Twenty-first Century Organization. Currency, New York
- Sullivan, P. (2000): Value driven Intellectual Capital. How to convert intangible corporate assets into market value. John Wiley & Sons, New York
- Sullivan, P., O'Shaughnessy, J. (1999): Valuing Knowledge Companies. Les Nouvelles 34, (2), 83-89
- Sveiby, K. (1997): The new oranizational wealth. Managing and measuring knowledge-based assets. Berrett-Koehler, San Fransico
- Sveiby, K., Lloyd, T. (1990): Das Management des Know-how. Führung von Beratungs-, Kreativ- und Wissensunternehmen. Campus, Frankfurt/Main

- Trojan, J., Döring-Katerkamp, U. (2002): Motivation und Wissensmanagement- Eine praktische Perspektive. In: Franken, R., Gadatsch, A.: Integriertes Knowledge Management. Konzepte, Methoden, Instrumente und Fallbeispiele. Vieweg Verlag, Braunschweig
- Weinert, A. (1998): Organisationspsychologie. Psychologie Verlags Union, Weinheim
- Wiig, K., de Hoog, R. et al. (1997): Knowledge Management (special issue) Expert Systems with Applications 13, (1)
- Wiig, K. (1995): Knowledge Management. The Central Management Focus for Intelligent-acting Organizations. Schema Press, Arlington

Journals and Newsletters

- **Journal of Intellectual Capital;** published by MCB University Press
- **Journal of Knowledge Management;** published by MCB University Press
- **Knowledge Management Review;** published by Melcrum Publishing
- **Knowledge Organisation** – Official Quarterly Journal of the International Society for Knowledge Organisation. International Journal devoted to Concept Theory, Classification, Indexing and Knowledge Representation; published by Ergon Verlag, Würzburg
- **Knowledge and Process Management** – The Journal of Corporate Transformation. The Official Journal of The Institute of Business Process Re-Engineering; published by John Wiley & Sons
- **Knowledge Management** – The Magazine for Knowledge enabled Enterprise; published by Bizmedia Ltd.
- **Knowledge Management;** published by arkgroup
- **Wissensmanagement** – Das Magazin für Führungskräfte; published by docu-line Verlags GmbH
- **The International Knowledge Management Newsletter;** published by Management Trends International

Studies

- Ashton, C. (1998): Managing Best Practices. Transforming business performance by identifying, disseminating and managing best practice. London, Business Intelligence
- Chase, R. (1998): Creating a Knowledge Management Business Strategy. Delivering Bottom-Line Results. Lavendon. Management Trends International

- Chauvel, D., Despres, C. (2002): A review of survey research in knowledge management: 1997-2001. In: Journal of Knowledge Management 6, (3), 207 – 223
- ILOI (1997): Knowledge Management. Ein empirisch gestützter Leitfaden zum Management des Faktors Wissen. Internationales Institut für Lernende Organisation und Innovation, München
- Kluge, J., Stein, W., Licht, T. (2001): Knowledge unplugged. The McKinsey & Company global survey on knowledge management. Palgrave, New York
- KPMG (ed.) (August 2001): Bedeutung und Entwicklung des multimedia-basierten Wissensmanagements in der mittelständischen Wirtschaft. Studie im Auftrag des Bundesministeriums für Wirtschaft und Technologie, Schluss-bericht 20
- Maier, R., Klosa, O. (1999): Knowledge Management Systems 99: State-of-the Art of Use of Knowledge Management Systems, research paper no. 35 of the Department of Business Informatics III, University of Regensburg
- META Group (ed.) (2002): Der Markt für Knowledge Management in Deutschland. http://www.metagroup.de/studien/_km2001/
- Skyrme, D. (1998): Measuring the Value of Knowledge. Metrics for know-ledge based business. Business Intelligence, London
- The Economist Intelligence Unit (1996): The Learning Organisation. Manag-ing knowledge for business success. Written in Co-operation with the IBM Consulting Group. The Economist Intelligence Unit, Research Report, New York
- Bottomley, A. (1998): Enterprise Knowledge Management Technologies. An Investment Perspective. Durlacher Research Ltd., London
- Mühlbauer, S., Versteegen, G. (2000): Wissensmanagement. Empirische Untersuchung, beste Praktiken und Evaluierung von Werkzeugen. It Research GmbH, Höhenkirchen
- Seifried, P., Eppler, M. (2000): Evaluation führender Knowledge Management Suites. Wissensplattformen im Vergleich. Benchmarking Studie. NetAcademy Press, St. Gallen

Editors

Prof. Dr.-Ing. Kai Mertins

Prof. Dr.-Ing. Kai Mertins has been the Director for Corporate Management at the Fraunhofer Institute for Production Systems and Design Technology (IPK), Berlin/Germany since 1988. After completing Control Theory studies in Hamburg as well as Economy and Production Technology at the Technical University of Berlin, he became a member of the scientific staff at the University Institute for Machine Tool and Manufacturing Technology. He has held a Ph.D. in production technology since 1984 and has more than 25 years experience in the design, planning, simulation and control of flexible manufacturing systems (FMS), manufacturing control systems (MCS), computer integrated manufacturing (CIM), business reengineering and enterprise modeling. He was General Project manager in several international industrial projects and gives lectures and seminars at the Technical University Berlin and several other universities. His special field of interest is strategic development of organizations, modeling and planning for production systems, simulation, BPR and production management. Kai Mertins is a member of the editorial board of the journal „Production Planning and Control" and of the „Business Process Management Journal". He is Vice Chairman of the IFIP-working group "Production Management".

Professor Mertins is responsible for the Representative Office of the Fraunhofer Society (FhG) in Jakarta. He has led several projects in Asia since 1985.

Peter Heisig

Peter Heisig is Head of the Competence Center Knowledge Management at Fraunhofer IPK. He studied Social Sciences at the Universities of Göttingen, Vienna and Bilbao, and has carried out research projects in Spain and Argentina. He started applied research at Fraunhofer IPK in 1990 with a project on the tacit knowledge of experienced workers and designers in metal companies. He was responsible for several projects in cooperation with industrial companies, in the fields of factory planning and business process engineering, teamwork, continuous improvement, benchmarking and knowledge management. In 1996 he assumed responsibility for the Information Center Benchmarking (IZB). Since summer 1997 he has been in charge of the area of knowledge management at the Fraunhofer IPK. In autumn 1997 he initiated the first German consortium-benchmarking study on knowledge management and in autumn 2001 the First Global Delphi Study on "The Future of Knowledge Management". Since 2000 he is the Chairman of the Global Benchmarking Network (GBN). He is researching the integration of knowledge management into daily business tasks and processes.

He has been invited to over 60 conferences and workshops about Knowledge Management in China, Germany, Italy, Sweden, Switzerland, Spain, United Kingdom and Mexico since autumn 1998.

In autumn 2002 he was appointed by the European Commission DG Information Society and the CEN as a project team member with other seven European KM experts to work on the "European Guide to Good Practice in Knowledge Management".

Jens Vorbeck

Jens Vorbeck was senior researcher at the Fraunhofer IPK, Berlin/Germany from 1997 until 2000. He studied Psychology at thc Technical University of Berlin and the Business School of Dublin City University and Information Sciences at the Freie Universität Berlin. During the past years he has conducted several national and international projects in the fields of knowledge management, benchmarking and business process engineering. His special fields of interest are the issues of motivation, corporate culture and leadership in the context of knowledge management. Furthermore, he is involved in the development of the benchmarking method, business process modeling and the development and application of the balanced scorecard approach.

Since autumn 2000 Jens Vorbeck has continued his work as a consultant with the IBM consulting group *IBM Unternehmensberatung GmbH (UBG)*.

Contributors

Kay Alwert
Fraunhofer Institute IPK
Pascalstr. 8-9
D-10587 Berlin
Germany

Cornelia Baumbach
Key Project Manager
Hilti Aktiengesellschaft
Corporate Human Resources
Feldkircherstrasse 100
Postfach 333
FL-9494 Schaan
Liechtenstein

Christian Berg
Director of R&D
Phonak AG
Laubisrütistrasse 28
CH-8712 Stäfa
Switzerland

Agnes Bradier
European Commission –
DG Information Society
Rue de la Loi 200
B-1049 Bruxelles
Belgium

Robert C. Camp, PhD, PE
Chairman Global Benchmarking
Network (GBN)
Best Practice Institute™,
Suite 1-200
625 Panorama Trail
Rochester, NY
USA

Geoff Carter
Chargé de Mission
European Foundation for Quality
Management (EFQM)
Avenue des Pleiades 15
B-1200 Brussels
Belgium

Peter Drtina
Coordinator of Knowledge
Management & IPR
Phonak AG
Laubisrütistrasse 28
CH-8712 Stäfa
Switzerland

Dr. Jakob Edler
Fraunhofer Institute ISI
Breslauer Straße 48
D-76139 Karlsruhe
Germany

Prof. Leif Edvinsson
Professor of Intellectual Capital
University of Lund
Sveavägen 44
S-10350 Stockholm
Sweden

Ina Finke
Fraunhofer Institute IPK
Pascalstr. 8-9
D-10587 Berlin
Germany

Dr. Christoph Haxel
Head of Henkel InfoCenter
Henkel KGaA, Düsseldorf
Henkelstraße 67
D-40191 Düsseldorf
Germany

Paul Hearn
Project Officer for Knowledge Management
European Commission –
DG Information Society
Rue de la Loi 200
B-1049 Bruxelles
Belgium

Peter Heisig
Head of Competence Center Knowledge Management
Vice-Chairman Global Benchmarking Network (GBN)
Fraunhofer Institute IPK
Pascalstr. 8-9
D-10587 Berlin
Germany

Ingo Hoffmann
„Knowledge Raven"
Management GmbH
Lehmbruckstr. 13
D-10245 Berlin
Germany

Dr. Arne Jaitner
Fraunhofer Institute IPK
Pascalstr. 8-9
D-10587 Berlin
Germany

Anne Jubert
European Commission –
DG Information Society
Rue de la Loi 200
B-1049 Bruxelles
Belgium

Andrea Martin
IBM Deutschland
Informationssysteme GmbH
Anziger Straße 29
D-81671 München
Germany

Prof. Dr. Ing. Kai Mertins
Head of Division
Corporate Management
Fraunhofer Institute IPK
Pascalstr. 8-9
D-10587 Berlin
Germany

Manuela Müller
Jutastr. 17
D-80636 Munich
Germany

Dr. Jürgen Oldigs-Kerber
Aventis Pharma
Deutschland GmbH
DI&A Knowledge Management
Industriepark Höchst, Gebäude
H831, Raum C219
D-65926 Frankfurt am Main
Germany

Laurence Prusack
Formerly Executive Director
IBM Institute for
Knowledge Management
One Main Street, 6th Floor
Cambridge, MA 02142
USA

Robert Schindler
Fraunhofer Institute IPK
Pascalstr. 8-9
D-10587 Berlin
Germany

Prof. Dr. Wolfgang Scholl
Institut für Psychologie
Humboldt-Universität zu Berlin
Oranienburger Str. 18
D-10178 Berlin
Germany

Dr. Peter Schütt
IBM Global Services
Pascalstr. 100
D-70548 Stuttgart
Germany

Anja Schulze
Research Associate
Institute for
Technology Management
University of St. Gallen
Unterstrasse 16/22
CH-9000 St. Gallen
Switzerland

Alla Shpilsky
Aventis Pharma
Deutschland GmbH
DI&A Knowledge Management
Industriepark Höchst
D-65926 Frankfurt am Main
Germany

Stephen Sorensen
Aventis Pharma
Deutschland GmbH
DI&A Knowledge Management
Industriepark Höchst
D-65926 Frankfurt am Main
Germany

Patricia Spallek
Knowledge Research &
Information Services
Arthur D. Little International, Inc.
Gustav-Stesemann-Ring 1
D-65189 Wiesbaden
Germany

Rob van der Spek
CIBIT Adviseurs
Arthur van Schendelstraat 570
Postbus 19210
NL-3501 De Utrecht
Netherlands

Frank Spellerberg
Manager Europe, Knowledge
Research & Information Services
Arthur D. Little International, Inc.
Gustav-Stesemann-Ring 1
D-65189 Wiesbaden
Germany

Dr. Andreas Spielvogel
Director
Development Processes & Tools
Business Unit Original Equipment
Continental AG
Vahrenwalder Str. 9
D-30165 Hannover
Germany

Jens Vorbeck
Consultant
IBM Unternehmensberatung
GmbH
Kaiserswerther Str. 117
40474 Düsseldorf
Germany

Christina Ulbrich
Fraunhofer Institute IPK
Pascalstr. 8-9
D-10587 Berlin
Germany

Markus Will
Fraunhofer Institute IPK
Pascalstr. 8-9
D-10587 Berlin
Germany

Index